Methods in Molecular Biology

Series Editor
John M. Walker
School of Life Sciences
University of Hertfordshire
Hatfield, Hertfordshire, AL10 9AB, UK

For further volumes:
http://www.springer.com/series/7651

Stem Cell Protocols

Edited by

Ivan N. Rich

HemoGenix, Inc., Colorado Springs, CO, USA

Editor
Ivan N. Rich
HemoGenix, Inc.
Colorado Springs, CO, USA

ISSN 1064-3745 ISSN 1940-6029 (electronic)
ISBN 978-1-4939-1784-6 ISBN 978-1-4939-1785-3 (eBook)
DOI 10.1007/978-1-4939-1785-3
Springer New York Heidelberg Dordrecht London

Library of Congress Control Number: 2014954236

Printed on acid-free paper

Humana Press is a brand of Springer
Springer is part of Springer Science+Business Media (www.springer.com)

Preface

Primary stem cells represent a very small and unique population of cells. Indeed, the proportion of stem cells compared to other cells in an organ or tissue is so small that they are usually morphologically unidentifiable. Yet, their presence can be detected by functional properties and characteristics that make stem cells unique. The present volume in this successful series of Methods in Molecular Biology on Stem Cell Protocols focuses on methods that allow primary stem cells from a variety of sources to be isolated, cultured in vitro, detected, and measured for specific applications. These applications range from those in basic, stem cell and veterinary research to toxicology and cellular therapy and regenerative medicine.

If the volume demonstrates a slight bias towards the blood-forming system, it is no coincidence. More is known about the blood-forming or hematopoietic system than probably any other primary stem cell system. The hematopoietic system has been used as a model for many stem cell systems. However, although every stem cell system is unique in its own way, there are many commonalities that are demonstrated in the methods and techniques described in this volume.

All of these unique properties and characteristics are discussed and examined, mostly at the cellular level and in detail in this volume. Unlike non-definitive stem cell systems such as embryonic stem (ES) and induced pluripotent stem (iPS) cells, primary definitive stem cells still represent a black box of unknown biology and physiology. Although ES and iPS cells can be manipulated to produce large numbers of stem cells and functionally mature cells, such as hepatocytes, cardiomyocytes, and neurons, investigators still have to demonstrate that these artificially produced parent and progeny cells are similar in virtually every aspect to their primary counterparts.

It is therefore imperative to understand and characterize primary stem cell populations and how they are regulated. Primary, definitive stem cells emit and receive a multitude of signals, both internal and external, to maintain homeostasis. For applications such as cellular therapy and regenerative medicine, it is important to understand the complexity of the stem cell compartment in order to develop assays that provide informative and predictive information prior to use in the patient. Assays are already available for this purpose. But, new future technologies, such as nanotechnology, may provide the capability, accuracy, and sensitivity needed to probe far deeper into the intricacies of the stem cell compartment. The question is whether these new technologies will be accepted and trusted over other technologies that may have been in place for decades.

Those new to the field of stem cell biology will find a wealth of "go to" and "how to" information that will hopefully provide some of the fundamentals to spark further interest and incentive to develop new technologies that will be required for basic and clinical research in the future. Those who have worked and are established in the field will hopefully find some new methodologies that will provide a wider viewpoint and an even greater scope for their own research.

The editor wishes to thank all the authors who contributed their time and effort to this volume and stem cell biology field.

Colorado Springs, CO ***Ivan N. Rich***

Contents

Contributors

TILL ACKER • *Institute of Neuropathology, University of Giessen, Giessen, Germany*
ALBERTO ALBERTINI • *Instituto di Technologie Biomediche, Segrate Milan, Italy*
ECKHARD U. ALT • *Department of Surgery, Tulane University, New Orleans, LA, USA*
GIOVANNI BERTALOT • *Azienda Ospedaliera Desenzano d/G, Leno, Italy*
IVAN BERTONCELLO • *Lung Health Research Center, University of Melbourne, Melbourne, VIC, Australia*
DEEPA BHARTIYA • *Stem Cell Department, National Institute for Research and Reproductive Health, Mumbai, India*
SALLY BOXALL • *Leeds Institute of Rheumatic and Musculoskeletal Medicine, Leeds University, Leeds, UK*
MATTHEW J. BRANCH • *Ophthalmology DCN, University of Nottingham, Nottingham, UK*
IVONE G. BRUNO • *InGeneron, Inc., Houston, TX, USA*
VINCENT M. CASSONE • *Department of Biology, University of Kentucky, Lexington, KY, USA*
BENJAMIN CAO • *Department of Medical Biology, University of Tromsø, Tromsø, Norway*
CINZIA COCOLA • *Instituto di Technologie Biomediche, Segrate Milan, Italy*
MICHAEL E. COLEMAN • *InGeneron, Inc., Houston, TX, USA*
DAVID CRAEYE • *ReGenesys BVBA, Heverlee, Belgium*
JULE T. DANIELS • *Department of Ocular Biology and Therapeutics, University College London, London, UK*
ROBERT DEANS • *ReGenesys BVBA, Heverlee, Belgium*
Athersys, Inc, Cleveland, OH, USA
YANNE S. DOUCET • *Department of Dermatology, College of Physicians and Surgeons, Columbia University, New York, NY, USA; Ecole Doctorale des Sciences de la Vie et de la Santé, Université Aix-Marseille, Marseille, France*
MARC DZIASKO • *Department of Ocular Biology and Therapeutics, University College London, London, UK*
JAMES L. FUNDERBURGH • *Department of Ocular Biology and Therapeutics, University College London, London, UK*
BOYAN K. GARVALOV • *Institute of Neuropathology, University of Giessen, Giessen, Germany*
KRISTEL GIJBELS • *ReGenesys BVBA, Heverlee, Belgium*
BRIAN GRAY • *Molecular Targeting Technologies, Inc., West Chester, PA, USA*
HOLLI HARPER • *HemoGenix, Inc, Colorado Springs, CO, USA*
WOUTER VAN'T HOF • *ReGenesys BVBA, Heverlee, Belgium; Athersys, Inc, Cleveland, OH, USA; Cleveland Cord Blood Bank, Cleveland, OH, USA*
ANDREW HOPKINSON • *Ophthalmology DCN, University of Nottingham, Nottingham, UK*

WILLIAM HRUSHESKY • *Oncology Analytics Corporation, Plantation, FL, USA*
ELENA JONES • *Leeds Institute of Rheumatic and Musculoskeletal Medicine, Leeds University, Leeds, UK*
ALVENA KURESHI • *Department of Ocular Biology and Therapeutics, University College London, London, UK*
HANNAH J. LEVIS • *Department of Ocular Biology and Therapeutics, University College London, London, UK*
YING LIANG • *Department Internal Medicine, University of Kentucky, Lexington, KY, USA*
YI LIU • *Department Internal Medicine, University of Kentucky, Lexington, KY, USA*
RUDY F. MARTINEZ • *InGeneron, Inc., Houston, TX, USA*
ISOBEL MASSIE • *Department of Ocular Biology and Therapeutics, University College London, London, UK*
SCOTT R. MCCLURE • *Department of Veterinary Clinical Sciences, Iowa State University, Ames, IA, USA*
PETER A.G. MCCOURT • *Department Medical Biology, University of Tromsø, Tromsø, Norway*
JONATHAN L. MCQUARTER • *Lung Health Research Center, University of Melbourne, Melbourne, VIC, Australia*
TAKAMICHI MIYAZAKI • *Institute of Frontier Medical Sciences, University of Kyoto, Kyoto, Japan*
LOUISE MORGAN • *Department of Ocular Biology and Therapeutics, University College London, London, UK*
TIMO Z. NAZARI-SHAFTI • *Department of Surgery, Tulane University, New Orleans, LA, USA*
MICHAEL NEALE • *Department of Ocular Biology and Therapeutics, University College London, London, UK*
SUSAN K. NILSSON • *CSIRO, Clayton, VIC, Australia; Australian Regenerative Medicine Institute, Clayton, VIC, Australia*
ANA OTEIZA • *Department of Medical Biology, University of Tromsø, Tromsø, Norway*
DAVID M. OWENS • *Department of Pathology and Cell Biology, Columbia University, New York, NY, USA*
SEEMA PARTE • *Stem Cell Department, National Institute for Research and Reproductive Health, Mumbai, India*
HIREN PATEL • *Stem Cell Department, National Institute for Research and Reproductive Health, Mumbai, India*
JIFFIN K. PAULOSE • *Department of Biology, University of Kentucky, Lexington, KY, USA*
PARIDE PELUCCHI • *Instituto di Technologie Biomediche, Segrate Milan, Italy*
JEF PINXTEREN • *ReGenesys BVBA, Heverlee, Belgium*
ELEONORA PISCITELLI • *Instituto di Technologie Biomediche, Segrate Milan, Italy*
ROLLAND REINBOLD • *Instituto di Technologie Biomediche, Segrate Milan, Italy*
IVAN N. RICH • *HemoGenix, Inc, Colorado Springs, CO, USA*
EDMUND B. RUCKER • *Department of Biology, University of Kentucky, Lexington, KY, USA*
RICHARD SCHÄFER • *Institute of Transfusion Medicine and Immunohaematology, Johann-Wolfgang-Goethe University Hospital, Frankfurt/Mail, Germany*
SASCHA SEIDEL • *Institute of Neuropathology, University of Giessen, Giessen, Germany*

CARL SHERIDAN • *Institute of Ageing and Chronic Disease, University of Liverpool, Liverpool, UK*
RADHIKA SHETH • *Department of Ocular Biology and Therapeutics, University College London, London, UK*
KALPANA SRIRAMAN • *Stem Cell Department, National Institute for Research and Reproductive Health, Mumbai, India*
HIROFUMI SUEMORI • *Institute of Frontier Medical Sciences, University of Kyoto, Kyoto, Japan*
FRANK RÜDIGER THADEN • *Instituto di Technologie Biomediche, Segrate Milan, Italy*
VICTORIA E. TOVELL • *Department of Ocular Biology and Therapeutics, University College London, London, UK*
BART VAES • *ReGenesys BVBA, Heverlee, Belgium*
AMANDA J. VERNON • *Department of Ocular Biology and Therapeutics, University College London, London, UK*
SARA WALBERS • *ReGenesys BVBA, Heverlee, Belgium*
WING-YAN YU • *Ophthalmology DCN, University of Nottingham, Nottingham, UK*
GARY VAN ZANT • *Department Internal Medicine, University of Kentucky, Lexington, KY, USA*
ILEANA ZUCCHI • *Instituto di Technologie Biomediche, Segrate Milan, Italy*

Chapter 1

Short Primer in Stem Cell Biology

Ivan N. Rich

Key words Stem cell systems, Continuously proliferating systems, Partially proliferating systems, Stem cell organization, Stem cell hierarchy

Stem cells can be divided into two types, namely non-definitive and definitive stem cells. Non-definitive stem cells have the capacity of developing into any organ or tissue of the body. The best example is the fertilized egg, but the most commonly known are embryonic stem cells (ESC) and induced pluripotent stem (iPS) cells. Definitive stem cells are derived from non-definitive stem cells and are organ or tissue specific. They are responsible for maintaining the organ or tissue. Definitive stem cells can be divided into two types; those that maintain continuously proliferating cell systems and those that maintain partially proliferating cell systems. Continuously proliferating cell systems include the blood-forming or hematopoietic system, the gastrointestinal system, the reproductive system, the skin, and specific cells of the eye. Partially proliferating stem cell systems include the liver, kidney, lung, and neural systems to name but a few (Fig. 1). In fact, virtually every organ and tissue is associated with or has its own definitive stem cell system [1–3].

It should be emphasized that our knowledge and understanding of stem cell systems is based upon arbitrarily dividing these biological systems into compartments and sub-compartments that allow us to better visualize these systems (Fig. 2). In reality, all of these systems represent a continuum, in which one cell stage of development changes imperceptibly into the next [4]. Although some stem cell systems may appear to be static, they are continuously in flux. All stem cell systems are highly regulated by positive and negative internal and external feedback signals that maintain the system in homeostasis and in harmony with every other system of the body.

All definitive stem cell systems exhibit a common organization and hierarchy [3] consisting of four primary compartments (Fig. 2). The first is the stem cell compartment. Next comes the

Ivan N. Rich (ed.), *Stem Cell Protocols*, Methods in Molecular Biology, vol. 1235,
DOI 10.1007/978-1-4939-1785-3_1, © Springer Science+Business Media New York 2015

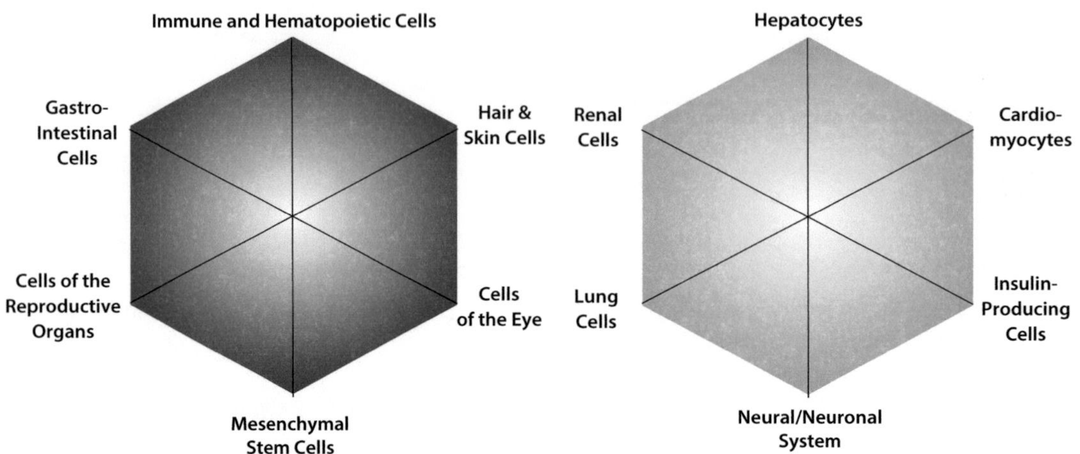

Fig. 1 Different types of proliferation biological systems

progenitor cell compartment. These are the direct progeny of the stem cells. Progenitor cells are the first cells to enter a differentiation lineage. For continuously proliferating stem cells systems, the progenitor cells are responsible for the amplification required to produce the requisite number of mature cells. The progenitor cells give rise to the precursor cells, which in many systems represent the first morphologically identifiable cells. The precursor cells differentiate and eventually produce mature, functional cells that make up most of the organ or tissue. Within each compartment there is a hierarchy comprising primitive to mature cells. These may be divided into sub-compartments that will contain cell populations that are identified and detected using many of the methods described in this volume.

Of particular importance are the properties and characteristics of the definitive stem cells [5] since these have implications not only for understanding the regulation of stem cells in basic research, but applications that involve toxicology, cellular therapy, regenerative medicine, and clinical treatments. Indeed, regulatory agencies are requiring greater focus on the properties and characteristics of stem cells, since these provide the backbone for designing and developing assays to determine the quality and potency of stem cells used for therapies (Fig. 2).

Regardless of whether stem cells are non-definitive or definitive and derived from primary tissues and organs, they all exhibit the same properties and characteristics.

- Stem cells are defined by their capacity for self-renewal.
- Stem cells are undifferentiated.

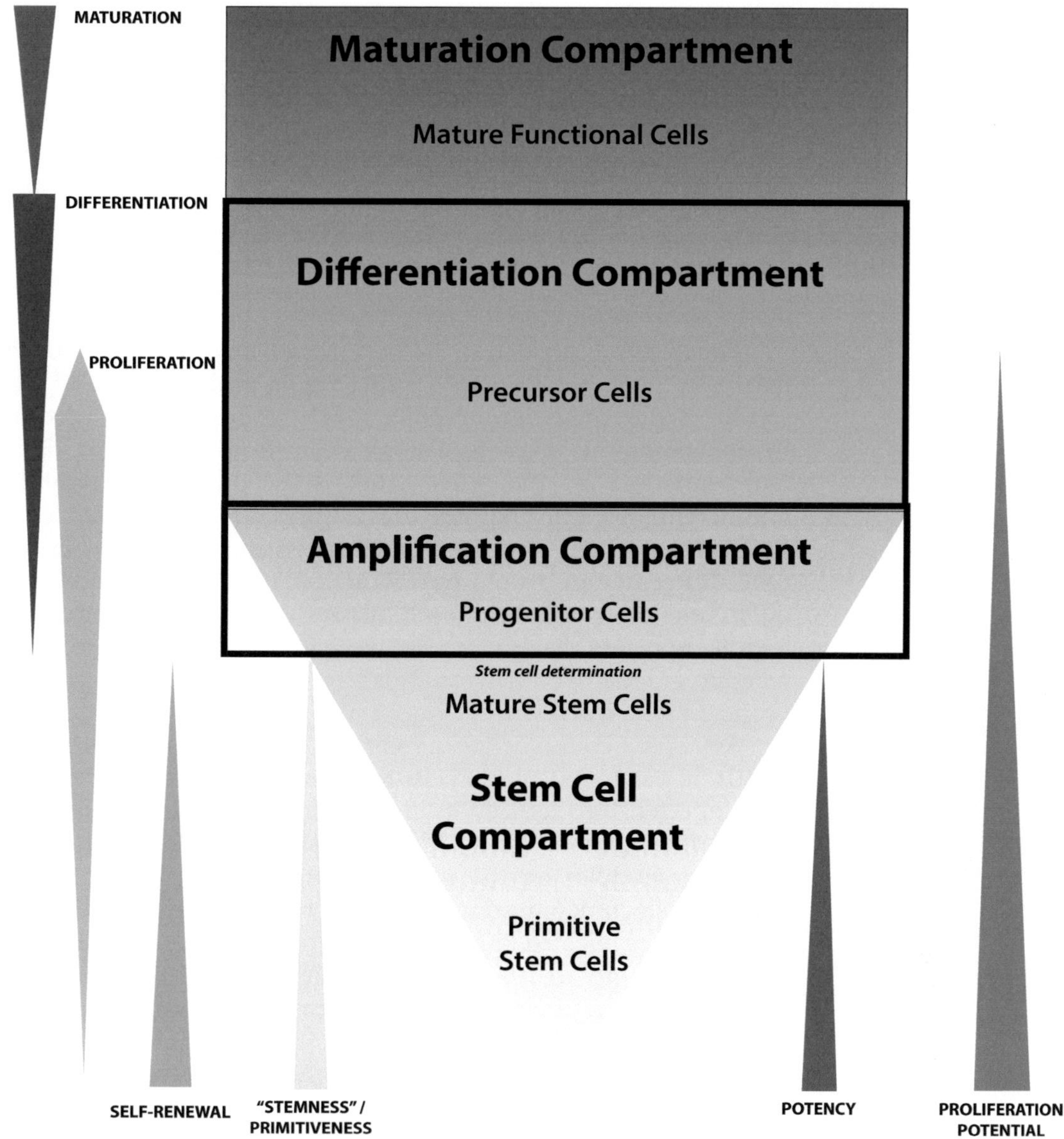

Fig. 2 The general organization of primary stem cell systems

- Stem cells only proliferate. Stem cells exhibit the greatest proliferation potential of all other cells in the body.
- Stem cells have the capability of becoming "determined," at which point they cease to be stem cells and enter the path of differentiation down a lineage that terminates as a mature functional cell.

Unlike the non-definitive ESC and iPSC that can be produced artificially in relatively large numbers, definitive stem cells are present in very small (<0.01 %) numbers. Yet, they are capable of maintaining a cell system throughout life. Probably the most notable example is the hematopoietic system. Under steady-state conditions a human adult produces approximately 2,000,000 red blood cells and 200,000 white blood cells every second of their life. Although the mechanism and regulation of the stem cell compartment is still a matter of active research, it is worth noting that most of our present knowledge and understanding of how a minute number of morphologically unidentifiable stem cells can accomplish this spectacular fete is due largely to the ingenuity of investigators to develop functional assays.

Functional assays detect the presence of stem cells by utilizing two basic biological processes. The first is proliferation and the second is differentiation. Proliferation is the expansion of cells by the continuous division of single cells into two daughter cells. Differentiation, on the other hand, is the process whereby an undifferentiated cell acquires the features of a specialized cell. Proliferation occurs prior to differentiation and without proliferation, differentiation would not occur. Although proliferation and differentiation overlap with each other, they are two separate biological processes that cannot be measured using the same assay readout; at least two assay readouts have to be used to detect each of these biological processes.

As a direct consequence of radiation research in the 1940s and 1950s, it was found that if cells from an untreated donor mouse were injected into lethally irradiated animals the latter survived. When these animals were later necropsied, investigators noticed bumps or nodules on the surface of the spleens. In 1961, Till and McCulloch demonstrated that this response could be quantified [6]. They injected known numbers of bone marrow cells from donor mice into lethally irradiated animals. They demonstrated that there was a direct correlation between the number of cells injected and the number of spleen nodules counted. However, the most important finding was that when the linear regression of this correlation was extrapolated backwards, the line intersected with the origin (Fig. 3). This observation demonstrated that each of the spleen nodules counted was derived from the clonal proliferation of a single cell; a stem cell. Although the stem cell within the spleen nodule could not be identified, the cells within each nodule were a colony of cells that had produced identifiable blood cell types. In other words, the stem cell was demonstrating its functional ability to both proliferate to produce daughter cells that could eventually differentiate and mature into functional mature cells. This original type of functional assay is still used today and has morphed into other types of in vivo functional assays [7]. Later it was found that cells with proliferation capability could produce colonies of cells in vitro in a semisolid medium such as agar or methylcellulose or even as adherent cells (Fig. 3) [8, 9]. However, it should be emphasized that it is the functional ability and capacity

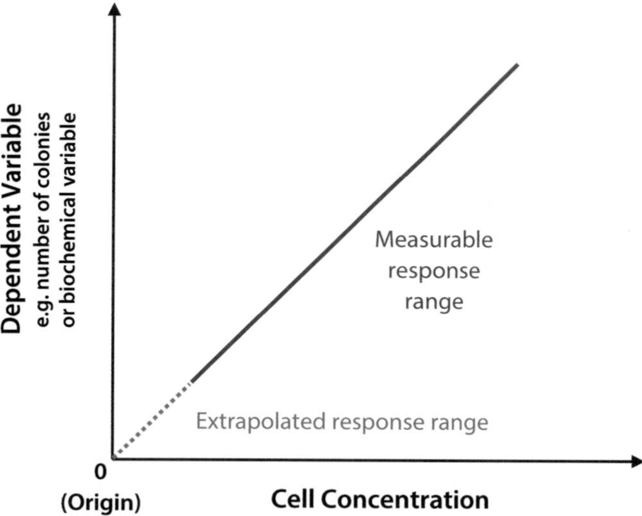

Fig. 3 Primary, definitive stem cell systems

of the original cell to proliferate in a clonal fashion that eventually forms a colony of cells that can be identified as originating from a specific cell type. Such colony-forming assays detect and measure differentiation ability and potential, but do not actually measure proliferation ability and potential, even though proliferation is required for colony formation [10]. Proliferation within a colony can only be measured using a proliferation assay readout.

One of the most significant achievements of identifying the presence of stem cells and their functional ability and potential has been to incorporate that knowledge into stem cell transplantation treatments for numerous malignancies and other diseases. Within the United States, the Stem Cell Therapeutic and Research Act of 2005 [11] recognized the power of stem cell treatment. Regulatory agencies and standards organizations around the world are rapidly producing guidelines and updating standards so that stem cell therapy can be performed with increased efficacy and reduced risk to the patient. There are about 4,000 stem cell clinical trials that have or are occurring worldwide. Yet, much is still trial and error because stem cell-specific assays are used either minimally or not at all.

In 1962, Thomas Kuhn used the term "paradigm shift" in his influential book, The Structure of Scientific Revolutions [12]. The commitment by all involved, law makers, regulatory agencies, standards organizations and the investigator, might just provide the "paradigm shift" required to harness and safely use the unique power of the stem cell, by incorporating standardized and validated methodologies to treat patients.

References

1. Blanpain C, Simons BD (2013) Unraveling stem cell dynamics by lineage tracing. Nat Rev Mol Cell Biol 14:489–502
2. Gancz D, Gilboa L (2013) Hormonal control of stem cell systems. Annu Rev Cell Dev Biol 29:137–162
3. Slukvin LL (2013) Deciphering the hierarchy of angiohematopoietic progenitors from human pluripotent stem cells. Cell Cycle 12: 720–727
4. Quesenberry PJ, Dooner MS, Aliotta JM (2010) Stem cell plasticity revisited: the continuum marrow model and phenotypic changes mediated by microvesicles. Exp Hematol 38:581–592
5. National Institutes of Health. Stem Cell Information. http://stemcells.nih.gov/info/basics/pages/basics2.aspx
6. Till JE, McCulloch EA (2011) A direct measurement of the radiation sensitivity of normal mouse bone marrow cells. 1961. Radiat Res 175:145–149
7. Lapidot T, Fajerman Y, Kollet O (1997) Immune-deficient SCID and NOD/SCID mice models as functional assays for studying normal and malignant human hematopoiesis. J Mol Med 75:664–673
8. Bradley TR, Metcalf D (1966) The growth of mouse bone marrow cells in vitro. Aust J Exp Biol Med 44:287–299
9. Luria EA, Panasyuk AF, Friedenstein AY (1971) Fibroblast colony formation from monolayer cultures of blood cells. Transfusion 11:345–349
10. Rich IN (2013) Potency, proliferation and engraftment potential of stem cell therapeutics: the relationship between potency and clinical outcome for hematopoietic stem cell products. J Cell Sci Ther. http://dx.doi.org/10.4172/2157-7013.S13-001
11. https://www.govtrack.us/congress/bills/109/hr2520
12. Kuhn TS (1962) The structure of scientific revolutions. University of Chicago Press, Chicago

Chapter 2

Measurement of Hematopoietic Stem Cell Proliferation, Self-Renewal, and Expansion Potential

Ivan N. Rich

Abstract

All stem cells exhibit the capacity and ability to proliferate. This is a fundamental property of stem cells, since it is required not only for self-renewal, but also for expansion and the ability for stem cells to engraft in a patient. The capacity or potential for proliferation by stem cells defines their degree of primitives or stemness. This, in turn, is directly related to stem cell self-renewal. Using a highly sensitive, accurate, and reliable ATP bioluminescence signal detection system that can be multiplexed with other assay readouts, it has been possible to determine three stem cell parameters using a single assay. Using primitive hematopoietic bone marrow stem cells as an example, an in vitro protocol is described that incorporates initially culturing primitive stem cells to induce them into cell cycle, followed by a secondary re-plating step that demonstrates both self-renewal capability and expansion potential.

Key words Stem cells, Stem cell self-renewal capacity, Proliferation potential, Expansion potential, ATP bioluminescence assays, Flow cytometry

1 Introduction

All stem cell systems consist of four primary compartments; the stem cell compartment, the amplification compartment, the differentiation compartment, and the maturation compartment. These systems are continuums in that one stage of cell development moves imperceptibly into the next. To try and understand these systems, assays have been used that divide the system into arbitrary and manageable compartments, sub-compartments, and different cell populations so that the organization and hierarchy can be better investigated and understood. These divisions are usually based on properties and characteristics specific to the cell populations detected and may include, for example, physical properties, response to drugs and other agents, as well as their response to growth factor and/or cytokines that may affect the proliferation or differentiation processes.

Ivan N. Rich (ed.), *Stem Cell Protocols*, Methods in Molecular Biology, vol. 1235,
DOI 10.1007/978-1-4939-1785-3_2, © Springer Science+Business Media New York 2015

Stem cells in primary, definitive biological systems represent a minute proportion of the total cellularity of the tissue or organ. Since these primary stem cells are usually morphologically unidentifiable in vivo, functional in vitro assays have been used to indicate their presence and their characteristics [1–3]. The hematopoietic system is often used as a model system for other stem cell systems. Indeed, many of the assays developed to study the hematopoietic system are now being applied to other stem cell systems [4–6]. It has been possible, for instance, to differentiate between primitive, quiescent, but high proliferative potential stem cells from more mature stem cells that are in cell cycle and exhibit a lower proliferation potential [7, 8]. Until a stem cell becomes determined and begins its development to a functional, mature end cell, it demonstrates two primary characteristics. The first is proliferation ability, which increases as the stem cells become less quiescent and more are recruited to enter the cell cycle. The proliferation ability of a stem cell is determined at a specific point in time of development, or rather at a specific stage of primitiveness or maturity. Thus, the proliferation ability of quiescent stem cells would be essentially zero, while the proliferation ability of stem cells near the point of entering a differentiation cell lineage would be much greater. The other parameter is proliferation potential or the capacity for proliferation. Stem cell proliferation potential is the exact opposite of stem cell proliferation ability; that is, the more primitive a stem cell, the greater its proliferation potential.

The ability to measure these two parameters has numerous applications, from in vitro toxicology to cellular therapy and regenerative medicine [9–11]. This is because proliferation ability and potential also define stem cell quality and potency, respectively.

The general protocol described below is divided into five primary steps.

1. Cell preparation.
2. Stem cell "priming" to induce primitive stem cells into cell cycle.
3. Analysis of "primed" stem cells.
4. Secondary re-plating of "primed" stem cells to demonstrate self-renewal and measure expansion potential.
5. Analysis of the secondary re-plated cells.

2 Materials

2.1 Equipment

1. Biohazard safety hood, also called a laminar air flow hood, for manipulating cells under sterile conditions.
2. Tissue culture incubator set at 37 °C, fully humidified and containing an atmosphere of 5 % carbon dioxide (CO_2) and 5 % oxygen (O_2) (*see* **Note 1**).

3. Cell counter for counting mononuclear cells (e.g., hemocytometer, electronic particle or cell counter, flow cytometer. *See* **Note 2**).
4. Dye exclusion cell viability counter (e.g., hemocytometer, fluorescence microscope, or flow cytometer. *See* **Note 3**).
5. Luminescence plate reader (Molecular Devices, SpectraMax L or Berthold, CentroLia) (*See* **Note 4**).
6. A full set of calibrated (preferably electronic) pipettes to accurately dispense liquids from 1 μL to 1 mL. Manual pipettes for mixing reagents and resuspending cells (*see* **Note 5**).
7. Multichannel (8-channel, preferably electronic) pipettes for dispensing into multiwell plates.

2.2 Supplies

1. Reagent reservoirs for an 8- or 12-channel pipette.
2. An assortment of sterile tips for different pipettes.
3. Sterile plastic tubes (e.g., 5, 10, 15, 50 mL).
4. Nonsterile 5 mL tubes with caps.
5. Stem cell proliferation potential, self-renewal, and expansion potential ATP bioluminescence assay (HALO-96 PREP, HemoGenix) (*see* **Note 6**).
6. Antibodies conjugated to different fluorophores to determine different stem cell, progenitor, and mature cell populations.

2.3 Reagents

1. Phosphate buffered saline (PBS).
2. Iscove's Modified Dulbecco's Medium (IMDM).
3. Serum (e.g., fetal bovine serum, FBS).
4. Cell fractionation medium, e.g., for density gradient centrifugation (NycoPrep, Axis-Shield; Ficoll (GE Healthcare Life Sciences)).

3 Methods

3.1 Cell Preparation

This protocol can be performed using bone marrow cells derived from multiple species. The protocol works best with fresh, primary cells. Cryopreserved bone marrow cells can also be used, but lower proliferation activity of the stem cells compared to fresh cells should be expected.

1. Remove the femora and tibia under aseptic conditions from mice and rats.
2. Cut the proximal and distal ends from the bones and flush out the bone marrow with 1–2 mL of IMDM using a syringe and needle directly into a 5 or 10 mL sterile tube containing

1–2 mL IMDM. Flush the bones at least 3–5 times to ensure that most of the marrow has been removed. For rat bone marrow cells, it is usually necessary to remove red blood cells, granulocytes, and other cells to produce a mononuclear cell (MNC) fraction by density gradient centrifugation according to the manufacturer's protocol.

3. Canine, primate, and human bone marrow can be obtained from vendors either as whole bone marrow or as a MNC fraction. If whole bone marrow aspirates are obtained, an MNC fraction must be prepared using density gradient centrifugation.
4. When a MNC fraction has been obtained, ascertain the nucleated cell concentration and viability by dye exclusion. Do not use cells that exhibit a viability of less than 85 %.
5. Adjust the cell concentration to a working concentration of 1×10^6 cells/mL using IMDM.
6. Prepare serial dilutions from 1×10^6 cells/mL to working cell dilutions in IMDM of 5×10^5 and 2.5×10^5 cells/mL. Ensure that for all three cell concentrations a volume of at least 2 mL is prepared.

3.2 In Vitro Stem Cell "Priming"

The MNC cell suspensions contain stem and progenitor cells at different stages of primitiveness. The PREP assay contains two culture reagents, one to "prime" the primitive quiescent stem cells into cell cycle and the other to expand the "primed" cells. Stem cell "priming" is performed in the 96-well plate provided with the kit. When setting up the "priming" cultures, the total number of replicate wells will be 16, of which 6–8 will be used to measure the proliferation potential and the cells in the remaining replicate wells will be used for secondary re-plating and assay multiplexing.

1. Label and prepare 3×5 mL sterile tubes (one for each cell dose) and accurately dispense 1.8 mL of the "priming" stem cell culture reagent.
2. 0.2 mL from each cell dose is dispensed into the tubes producing a final volume of 2 mL. At this point, the cell concentration from each dose has been reduced tenfold.
3. Gently vortex the contents of the tubes.
4. Remove a sterile 96-well plate from its plastic cover and, if available, using the repeater function on the pipette, dispense 0.1 mL into 16 replicate wells in columns across the plate. Each dose will then take up 2 columns of the 96-well plate. The working cell concentrations will be reduced 100-fold so that the final cell concentrations are 2,500, 5,000, and 10,000 cells/mL.

5. Replace the lid onto the 96-well plate and transfer the plate to the incubator. For animal cells, with the exception of primate, culture the cells for 4–5 days. For primate and human cells, the incubation time is extended between 5 and 7 days.

3.3 Determining Proliferation Potential of "Primed" Stem Cells

In Subheading 3.2, 16 wells from each cell dose were cultured. In this part of the procedure, the proliferation potential of the cells in 6–8 replicates is measured. The remaining wells are used in Subheading 3.4 for secondary re-plating.

1. Prior to determining the proliferation of the "primed," cultured stem cells, the assay is calibrated and standardized using the ATP standard and controls provided with the PREP kit. The ATP-Enumeration Reagent (ATP-ER) that contains the components to produce bioluminescence is thawed at room temperature and transferred to a reservoir for an 8-channel pipette. Each well receives 0.1 mL of the ATP-ER and after mixing, the plate is transferred to the luminometer and incubated in the dark for 2 min prior to measurement. The slope of the ATP standard curve must lie within the specified range provided with the kit to proceed to the sample measurement. The values of the controls must also lie on the standard curve and exhibit values within a specific range (*see* **Note 7**).
2. Transfer the plate containing the "primed" stem cells from the incubator to a sterile hood. Remove the lid from the plate. Remove a sterile adhesive foil included with the kit and peel away the backing paper. Place the foil over the whole plate and press it down so that it adheres. Using a sharp knife or scalpel, cut away the foil covering the wells that will be used for measuring proliferation and peel the foil away from the plate. Be careful not to remove the foil away from those wells containing cells that will be used for secondary re-plating. All unused wells will also remain sterile.
3. Using an 8-channel pipette, dispense 0.1 mL of the ATP-ER into the replicate wells to be used to determine proliferation. Mixing of the well contents is important (*see* **Note 8**).
4. Once all the wells have received ATP-ER, transfer the plate to the luminometer and incubate in the dark for 10 min prior to measuring the luminescence.
5. Using the ATP standard curve, interpolate the results from the instrument (as Relative Luminescence Units, RLU) into standardized ATP concentrations (*see* **Note 9**).
6. Plot the ATP concentration (μM) on the *Y*-axis against the cell dose on the *X*-axis. Use linear regression analysis to produce a straight line curve and note the slope of the fitted curve. The slope of the cell dose response linear regression is a measure of

the proliferation potential and therefore the primitiveness of the stem cells being measured. The steeper the slope, the more primitive the stem cells (*see* **Note 10**).

3.4 Preparing Cells for Secondary Re-plating

The cells in the remaining wells from each cell dose are removed and re-plated.

1. Transfer the 96-well plate to a sterile hood.
2. Label and prepare 3 × 10 mL sterile plastic tubes with caps, one for each cell dose.
3. Carefully remove the foil covering the wells that were not used to measure proliferation.
4. Using a manual pipette add 0.1 mL of IMDM (at room temperature) to each well. Each well will now contain a total of 0.2 mL.
5. Using a manual pipette and sterile tips, mix the contents of each well several times so that the cells are in a cell suspension. Take care not to cause too many bubbles.
6. Transfer 0.1 mL of the contents of each well from a specific cell dose to the respective 10 mL tube.
7. Add another 0.1 mL of IMDM to each of the wells, mix, and transfer all of the contents from each well to the 10 mL tube.
8. Repeat this procedure for each cell dose transferring the cells to the respective 10 mL tubes.
9. Centrifuge the cells at 200 ×*g* for 10 min at room temperature.
10. Discard the supernatant and resuspend the cells in each tube with 0.5–1.0 mL IMDM.
11. Perform a nucleated cell count and prepare 0.3 mL from each original cell suspension so that a new cell dose response of 2.5×10^5, 5×10^5, and 1×10^6 is prepared in 3 × 5 mL tubes.
12. Any remaining cells from each of the original cell doses can be used for assay multiplexing. For example, phenotypic analysis by flow cytometry can be performed on the remaining cells to determine the presence of different stem and progenitor cell populations.

3.5 Secondary Re-plating to Investigate Self-Renewal Capacity and Expansion Potential of "Primed" Stem Cells

The "primed" stem cells that have been collected from each cell dose are now re-cultured in an Expansion Culture Medium (ECM) (*see* **Note 11**).

1. Remove the ECM (included with the assay kit) and allow it to thaw at room temperature.
2. While the ECM is thawing, remove the foil from the unused wells of the 96-well plate. Using a new sterile adhesive foil cover the already used wells so that the contents of these wells

do not contaminate the unused wells that will be used for secondary re-plating.

3. Label and prepare 3 × 5 mL sterile tubes and dispense 1.8 mL of the ECM into each of the three tubes.
4. Dispense 0.2 mL from the tube containing 2.5×10^5 cells/mL into one of the tubes containing 1.8 mL ECM. Repeat this for each of the other cell doses.
5. Mix the contents of each tube gently on a vortex mixer.
6. For each cell dose, dispense 0.1 mL into 16 replicate wells of the 96-well plate.
7. Replace the lid and transfer the plate to the incubator. Culture the cells for another 5–7 days.

3.6 Determination of Stem Cell Self-Renewal and Expansion Potential

This procedure is essentially the same as described in Subheading 3.3 and part of Subheading 3.4. If primitive, quiescent stem cells were present in the "priming" culture step, stem cells will also be present in the secondary re-plating step. However, in this case, the "primed" cells should also demonstrate extensive expansion.

1. Transfer sufficient ATP-ER to a reservoir for the number of replicate wells to be assessed for proliferation.
2. Prior to measuring proliferation for each cell dose, perform the ATP standard curve and measure the controls. Without this procedure it will not be possible to ensure that the assay is working correctly, convert RLU values obtained from the instrument to standardized ATP concentrations, and compare the results from the "priming" step to those of the secondary re-plating step.
3. Dispense 0.1 mL of the ATP-ER into each of the replicate wells. Mix according to the proper protocol (*see* **Note 8**) and transfer the plate to the plate luminometer.
4. Incubate the plate in the dark for 10 min prior to reading the luminescence.
5. Analyze the resulting cell dose response in the same manner as described in Subheading 3.3, **step 6**. If stem cell expansion has occurred, the ATP concentrations for all 3 cell doses should be greater than that for the "primed" cells. However, the slope of the linear regression cell dose response may be lower than that for the "primed" stem cells. This is an indication that the "primed" stem cells were more primitive (greater proliferation potential) than after secondary culture. As the stem cells expand and mature, the slope of the linear regression cell dose response will decline. If little or no proliferation is observed after secondary culture, this is an indication that few if any primitive stem cells were originally present, thereby implying that no self-renewal occurred.

6. The cells in the replicate wells that were not used to measure luminescence can be removed as described in Subheading 3.4 and the compliment of stem and progenitor cells ascertained using flow cytometry to determine the expansion and differentiation provided by the "primed" stem cells. Instead of preparing the cells for flow cytometry, gene expression analysis could be performed (*see* **Note 12**).
7. Alternatively, detection of other stem and progenitor cell proliferation can be investigated using other proliferation assays.

4 Notes

1. Culturing cells under low oxygen tension is not a necessity, but for many cell types (e.g., hematopoietic cells and mesenchymal stem cells), it helps significantly. This is because most cells exist under oxygen tensions that are below atmospheric conditions. The partial oxygen tension in the venous system is between 40 and 45 mmHg. This is approximately equivalent to 5 % oxygen. Culturing cells between 3 and 7 % oxygen is advantageous because it reduces oxygen toxicity due to the production of free oxygen radicals. Cell growth media often contains agents that keep molecules in a reduced state. Examples are reduced glutathione, vitamin E, β-mercaptoethanol, and α-thioglycerol. Culturing cells under low oxygen tension produces an additive effective with these agents resulting in a higher culture plating efficiency.
2. It is important to distinguish between the total nucleated cell (TNC) count and a mononuclear cell count (MNC). The TNC may contain not only stem cells, but other cell types that can dilute the stem cell fraction. An MNC fraction usually contains the stem cells and other primitive proliferating cells. It is produced by separating these cells from the remaining cells using, for example, density gradient centrifugation or other physical separation medium. When working with stem cells from different organs and tissues, it is therefore important to decide the type of cell count upon which the results will be based. It is not recommended to use a cell count based on TNC since this will mean that far fewer stem cells will be used for the assay readout. The consequence will be that an extremely low signal will be produced that will influence the result, interpretation, and conclusions.
3. Dye exclusion viability detects membrane integrity. Cells demonstrating a viability of less than 85 % should not be used since the chances of these cells being capable of sustaining cell proliferation are low. Dyes such as trypan blue, propidium iodide

(PI), acridine orange, and 7-aminoactinomycin D (7-AAD) can be used. The latter agent is usually used in combination with flow cytometry. It should be noted that dye exclusion viability does not detect cellular and metabolic integrity. In other words, it does not determine the "health" of the cell. To do this, a metabolic viability assay should be used. However, if an absorbance, fluorescence, or bioluminescence viability assay is performed, these assays combine metabolic viability with measuring cell proliferation.

4. Plate readers are available as multimode readers or as individual readers that can measure absorbance, fluorescence, or luminescence. It is important to determine which plate reader is available in the laboratory. In some cases a multimode reader (capable of measuring all three readouts) may need an extra adaptor in order to measure luminescence. It is also necessary to determine if the luminescence plate reader can measure "flash," "glow," or both types of luminescence output. A "flash" readout is produced by chemiluminescence reagents, while a "glow" luminescence is usually produced by bioluminescence reagents. For this protocol, an instrument that measures "glow" bioluminescence is required.
5. Accurate dispensing, especially for microcultures such as those prepared in 96-well plates, is extremely important. Small pipetting errors can produce large variations that can significantly affect the resulting statistics and conclusions.
6. Stem cell Proliferation, Renewal and Expansion Potential (PREP) assays are part of a family of ATP bioluminescence proliferation assays for hematopoietic stem and progenitor cells. Cells produce intracellular ATP for their chemical energy requirements. Without ATP, the cells would die. When stem cells are induced to proliferate, there is a concomitant increase in the intracellular ATP concentration from basal levels of ATP. After culture, the ATP is released by cell lysis and acts as a limiting substrate for a luciferin/luciferase reaction. Both the lysis and luciferin/luciferase reagents are combined in the ATP-Enumeration Reagent (ATP-ER) used in these assays. The reaction produces bioluminescence in the form of light that can be measured in a plate luminometer.
7. Performing the ATP standard curve and measuring the controls is an integral part of the assay. A specific range is allowed for the slope of the ATP standard curve and the values for the controls. If the values are outside the specified ranges, this part of the assay should be repeated. If the values continue to be outside of the specified parameters, new reagents will have to be used. Only when the ATP standard curve and controls

comply with the specified parameters should samples be processed. In this way, the user knows whether the assay is working correctly.

8. A specific procedure is used to mix the ATP-ER with the cells. This involves mixing the contents in the center and at each "corner" of the well twice. In this way, the lysis buffer reacts with the cells to release intracellular ATP. If mixing is insufficient, not all the cells will be lyzed and a low ATP value will be obtained. If the cells are mixed too well, the ATP can be degraded, again producing low ATP values.
9. The output of a luminescence plate reader is Relative Luminescence Units (RLU). This is a non-standardized output that does not allow results from one experiment to be compared to that of another. Performing the ATP standard curve allows the RLU values to be interpolated into standardized ATP concentration, which can then be compared over time and between stem cell "priming" and secondary re-plating.
10. The ATP standard curve is plotted as a log-log linear regression. This does not mean that the *X*- and *Y*-axes can be simply converted from a linear scale to a log scale. All the values (RLU and ATP concentrations) must be transformed into log values and then plotted. Some instruments include software that can be programmed to perform this function automatically. If this is not available, third-party software such as Microsoft Excel, GraphPad Prism, or Systat SigmaPlot or TableCurve 2D software can be used to perform these functions and convert the RLU values from the samples into ATP concentrations.
11. The "priming" stem cell culture reagent contains growth factors and cytokines that are used to induce the stem cells out of quiescence and into cell cycle. The Expansion Culture Medium (ECM) contains similar growth factors and cytokines to that of the "priming" culture reagent, but in addition also includes factors such as erythropoietin (EPO), granulocyte-macrophage colony-stimulating factor (GM-CSF), thrombopoietin (TPO), interleuin-2 (IL-2), and interleukin-7 (IL-7). These factors act on progenitor cells from different lympho-hematopoietic lineages thereby resulting in cell expansion. The presence of these progenitors as well as the stem cells can be detected using flow cytometry.
12. Assay multiplexing is an important tool used to obtain as much information from a single sample as possible. Multiplexing involves combining different assay readouts to measure different parameters on the same cells.

References

1. Bradley TR, Metcalf D (1966) The growth of mouse bone marrow cells in vitro. Aust J Exp Biol Med Sci 44:287–299
2. Pluznik DH, Sachs L (1966) The induction of clones of normal mast cells by a substance from conditioned medium. Exp Cell Res 43:553–563
3. Iscove NN, Sieber F, Winterhalter KH (1974) Erythroid colony formation in cultures of mouse and human bone marrow: analysis of the requirement for erythropoietin by gel filtration and affinity chromatography on agarose-concanavalin A. J Cell Physiol 83: 309–320
4. McQualter JL, Yuen K, Williams B et al (2010) Evidence of an epithelial stem/progenitor cell hierarchy in the adult mouse lung. Proc Natl Acad Sci U S A 107:1414–1419
5. Povsic TJ, Zavodni KL, Vainorius E et al (2009) Common endothelial progenitor cell assays identify discrete endothelial progenitor cell populations. Am Heart J 157:335–344
6. Zhang L, Valdez JM, Zhang B et al (2011) ROCK inhibitor Y-27632 suppresses dissociation-induced apoptosis of murine prostate stem/progenitor cells and increases their cloning efficiency. PLoS One 6:e18271. doi:10.1371/journal.pone.0018271
7. Rosendaal M, Hodgson GS, Bradley TR (1979) Organization of haemopoietic stem cells: the generation-age hypothesis. Cell Tissue Kinet 12:7–29
8. Hodgson GS, Bradley TR (1979) Properties of haematopoietic stem cells surviving 5-fluorouracil treatment: evidence for a pre-CFU-S cell? Nature 381–2
9. Rich IN, Hall KM (2005) Validation and development of a predictive paradigm for hemotoxicology using a multifunctional bioluminescence colony-forming proliferation assay. Toxicol Sci 87:427–441
10. Hall KM, Harper H, Rich IN (2012) Hematopoietic stem cell potency for cellular therapeutic transplantation. In: Pelayo R (Ed.) Advances in hematopoietic stem cell research. InTech. ISBN: 978-953-307-930-1
11. Rich IN (2013) Potency, proliferation and engraftment potential of stem cell therapeutics: The relationship between potency and clinical outcome for hematopoietic stem cell products. J Cell Sci Ther. S13-001. doi:10.4172/2157-7013

Chapter 3

Measuring the Aging Process in Stem Cells

Yi Liu, Gary Van Zant, and Ying Liang

Abstract

Stem cells persist in replenishing functional mature cells throughout life by self-renewal and multilineage differentiation. Hematopoietic stem cells (HSCs) are among the best-characterized and understood stem cells, and they are responsible for the life-long production of all lineages of blood cells. HSCs are a heterogeneous population containing lymphoid-biased, myeloid-biased, and balanced subsets. HSCs undergo age-associated phenotypic and functional changes, and the composition of the HSC pool alters with aging. HSCs and their lineage-biased subfractions can be identified and analyzed by flow cytometry based on cell surface makers. Fluorescence-activated cell sorting (FACS) enables the isolation and purification of HSCs that greatly facilitates the mechanistic study of HSCs and their aging process at both cellular and molecular levels. The mouse model has been extensively used in HSC aging study. Bone marrow cells are isolated from young and old mice and stained with fluorescence-conjugated antibodies specific for differentiated and stem cells. HSCs are selected based on the negative expression of lineage markers and positive selection for several sets of stem cell markers. Lineage-biased HSCs can be further distinguished by the level of SLAM/CD150 expression and the extent of Hoechst efflux.

Key words Stem cells, Hematopoietic stem cells, Aging, Flow cytometry, Fluorescence-activated cell sorting

1 Introduction

Stem cells are rare and self-renewing cells that give rise to all types of mature cells. In any tissue or organ with high cell turnover, stem cells should be long lived in order to constantly replenish cells lost throughout the lifetime and to maintain optimal tissue function. Therefore, stem cells are exposed to the noxious effects of both intrinsic and extrinsic effectors of damage during organismal aging [1]. As a result, stem cells may undergo functional decline, and their repair and renewal capacity may be impaired, which in turn contributes to overall organismal aging [2, 3]. Because of the unprecedented experimental model systems that are available for the exploration of hematopoietic stem cells (HSCs), stem cell aging research in the field of hematology has been the subject of extensive studies and has advanced dramatically in the past several years [4].

Ivan N. Rich (ed.), *Stem Cell Protocols*, Methods in Molecular Biology, vol. 1235,
DOI 10.1007/978-1-4939-1785-3_3, © Springer Science+Business Media New York 2015

It is likely that the same broad concepts that define and characterize blood-forming stem cells will apply to stem cell populations found elsewhere.

HSCs reside in the bone marrow and provide life-long production of hematopoietic progenitors (HPCs) and peripheral blood lymphoid and myeloid cells. At the same time, HSCs undergo self-renewal divisions in order to sustain the stem cell pool. Precisely regulated blood cell production is vital for organismal survival; therefore functional failure of HSCs can potentially threaten the longevity of an organism. Accumulating evidence in the study of mouse models has suggested that HSCs undergo age-related changes in phenotype, function, and clonal composition. The changes of aged HSCs include increased HSC number [5–9]; reduced self-renewal capacity [10, 11]; skewed differentiation towards myeloid lineage at the replacement of lymphoid cells [5, 7, 12]; enhanced mobilization from bone marrow to peripheral blood [13]; reduced homing back to bone marrow [14]; decreased proliferative response to cytokines [9]; and loss of cell polarity [15]. The HSC population is heterogeneous and is composed of three subfractions with distinct differentiation potentials [16–18]. These subfractions are (1) myeloid-biased HSCs with a high myeloid differentiation potential, (2) lymphoid-biased HSCs with a preferred lymphoid differentiation, and (3) balanced HSCs with equal lineage outputs. With aging, myeloid-biased HSCs become dominant in the old bone marrow, resulting in a skewed myeloid output in the circulation. These phenotypic and functional alterations in old HSCs have been ascribed to the age-associated accumulation of a variety of damages that are intrinsic to HSCs as well as extrinsic to their microenvironment [19–21]. DNA mutations [22–24], telomere shortening [25], and oxidative stress [26, 27] are among the most significant cellular changes in old HSCs; these changes trigger signaling cascades that lead to cell cycle checkpoint activation [28, 29], apoptosis [30], senescence [31, 32], or differentiation [33]. At the molecular level, young and old HSCs demonstrate distinct profiles in both transcriptome and epigenome, resulting in the identification of genes and pathways that correlate with HSC aging [34–37].

Characterization of HSCs and their aging process requires the isolation and purification of HSCs. The advent of flow cytometry has allowed this task to be successfully implemented and enables researchers to isolate HSCs and other types of blood cells from young and old subjects (mice or humans) for further functional analysis. In this procedure, bone marrow cells are stained with fluorescence-conjugated monoclonal antibodies that bind specific cell surface proteins. HSCs are analyzed and sorted by fluorescence-activated cell sorting (FACS) based on the expression level of these markers. In the mouse model, HSCs and HPCs are enriched in the population negative for the markers of all differentiated lineages cells (*L*ineage-) and positive for stem cell markers *S*ca-1 and c-*K*it (LSK cells) [38]. HSCs are further purified from LSK population by several sets of cell surface proteins, including (1) Flk2$^-$ CD34$^-$ LSK [39]; (2) CD150$^+$ CD48$^-$ CD41$^-$ LSK [40, 41]; (3) SP^{LSK} (*S*ide *P*opulation with high Hoechst efflux) [42]; and (4) EPCR$^+$ CD150$^+$ CD48$^-$ CD34$^-$ LSK [10]. In these phenotypically defined HSCs, lineage-biased HSCs can be distinguished by the expression of CD150 protein or the extent of Hoechst efflux [17, 43, 44]. Myeloid-biased HSCs are Flk2$^-$ CD34$^-$ LSK CD150^{high} or SP^{LSK} with lower Hoechst efflux (Lower-SP^{LSK}), whereas Flk2$^-$ CD34$^-$ LSK CD150$^{negative/low}$ or SP^{LSK} with higher Hoechst efflux (Upper-SP^{LSK}) population contains a majority of lymphoid-biased HSCs. Balanced HSCs have not been clearly defined, but cells with Flk2$^-$ CD34$^-$ LSK CD150^{low} immunophenotype demonstrate a balanced output at both lymphoid and myeloid lineages. Another SLAM maker, CD229, has recently been found to distinguish lymphoid-biased HSCs from myeloid-biased cells [45]. Although the functionality of these phenotypically defined HSCs needs to be confirmed by the gold-standard transplantation assay, flow cytometry-mediated HSC purification greatly facilitates analysis of the cellular and molecular mechanisms of HSCs and their aging process. HSC aging in humans has not been well studied partially due to lack of markers to define highly purified HSCs [46]. Despite this limitation, studies have shown that HSCs from older people have some characters similar to those in old mice: increased numbers, myeloid differentiation skewing [47, 48], and accumulation with damages [26, 32, 47, 49], for example. Human HSCs and HPCs are identified by a different set of markers, which are lineage$^-$ CD34$^+$ CD38$^-$ CD45RA$^-$ CD90$^+$. The size of this population has been shown to increase with aging [48]. CD49 has been recently discovered to present in a group of highly primitive HSCs, leading to a further purification of human HSCs [50]. Since CD150 cannot be used for labeling human HSCs, the clonal composition of HSC population in human and its shift during the aging process is not clear [51].

Flow cytometry-mediated HSC sorting is a multi-step process (Fig. 1). HSCs are very rare in the adult bone marrow, comprising less than 0.005 % of total bone marrow cells. This low frequency requires a large number of cells to be processed in order to obtain sufficient HSCs for experimental uses. For this reason, pre-enrichment steps are typically used to remove differentiated cells and reduce the sample size and subsequent sorting time. Ficoll-mediated density separation and/or red blood cell lysis are two procedures commonly used to remove granulocytes and red blood cells. The most prominent step for pre-enrichment of stem and progenitor cells from the bulk of bone marrow cells is the immunomagnetic depletion of all types of differentiated cells. HSCs are identified based on the antibody-mediated negative and positive

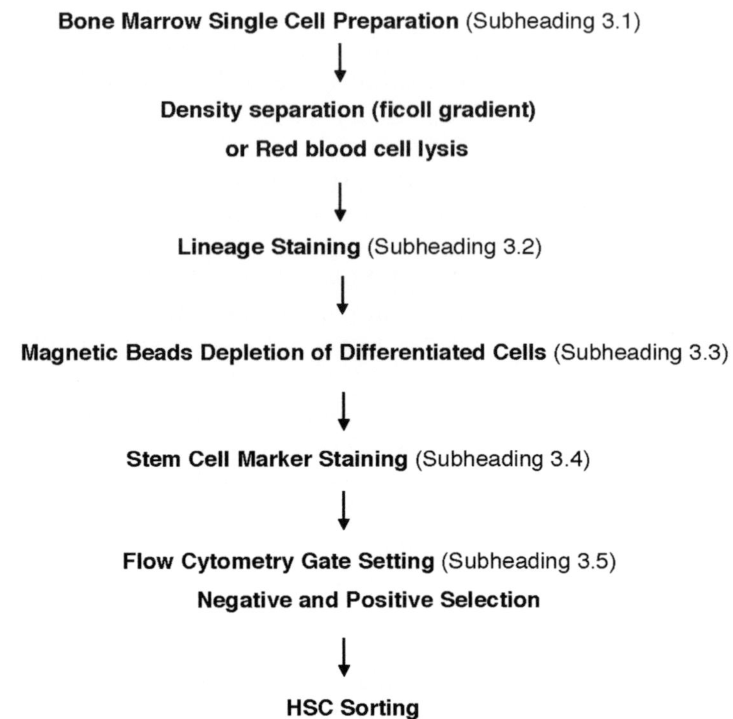

Fig. 1 Flowchart of mouse HSC isolation and purification by flow cytometry

selections, in which differentiated cells are antibody-labeled and removed, whereas HSCs are stained with specific makers and recovered. This is a multiple-step cell separation process requiring multiple fluorochromes to be used. Therefore, the precise compensation among different fluorescent signals will ensure that these signals are correctly detected by the flow cytometer and HSCs will be selected by the defined immunophenotype and sorted with a high purity. In this chapter, we will focus on FACS and describe the procedures involved in the HSC staining and sorting from the mouse bone marrow.

2 Materials

2.1 Animals

C57BL/6 mice are the strain commonly used for aging studies. This strain has a mean lifespan of 800 days in males and 750 days in females [52, 53]. Typically, young C57BL/6 mice are 6–8 weeks old, and old ones used in aging studies are usually more than 24 months old. If other inbred strains are used, the age of old mice needs to be practically determined based on their actual mean lifespan.

2.2 Reagents and Supplies

1. Medium: Hank's Balanced Salt Solution (HBSS) with 2 % heat-inactivated fetal bovine serum, phosphate buffered saline (PBS) without calcium and magnesium.
2. Syringes with 23_G1-gauge needles to flush marrow out of femurs and tibias of young mice. Marrow cavity in old mice is larger; therefore the use of larger needles, such as 20_G1 1/2-gauge, is recommended.
3. Cell strainer with 70 μm nylon screen to filter bone marrow single cell suspensions.
4. Density-separation medium: Ficoll-Paque PREMIUM 1,084; store and use at room temperature.
5. Red blood cell lysis buffer: 8.26 g of NH_4Cl, 1.19 g of $NaHCO_3$, and 200 μl of EDTA (0.5 M, pH 8) in 1 litre (L) distilled H_2O. Adjust pH to 7.3 and filter-sterilize through 0.2 μm filter. Store stock solution at 4 °C.
6. 15 or 50 mL conical tubes for holding bone marrow cells during antibody staining.
7. 5 mL polystyrene round-bottom tube (FACS tube) for holding cells during fluorescence-activated cell sorting.
8. 5 mL polypropylene round-bottom tubes (collection tube) or 1.5 mL Eppendorf tubes for colleting sorted cells.
9. Magnetic stand for 15 and 50 mL centrifuge tubes.

2.3 Fluorescence-Conjugated Antibodies

1. Lineage marker antibodies: 53-7.3 (anti-CD5), 53-6.7 (anti-CD8a), RA3-6B2 (anti-B220; CD45R), M1/70 (anti-CD11b; Mac-1), RB6-8C5 (anti-Gr-1; Ly-6G and Ly-6C), and Ter119 (anti-erythroid antigen; Ly76). Note that all these lineage antibodies (used in the author's lab) are biotinylated and anti-mouse antibodies.
2. Streptavidin-conjugated allophycocyanin-Cy7 (APC-Cy7) antibody for the secondary immunofluorescent staining of lineage cells, labeled with biotinylated primary antibodies.
3. Alexa Fluor® 700-conjugated anti-mouse c-Kit (CD117) antibody.
4. Pacific blue (PB)-conjugated anti-mouse Sca-1 (Ly6A/E) antibody.
5. Phycoerythrin (PE)-conjugated anti-mouse Flk2/Flt3 (CD135) antibody.
6. Fluorescein-5-isothiocyanate (FITC)-conjugated anti-mouse CD34 antibody.
7. Alexa Fluor® 647 anti-mouse CD150 (SLAM) antibody.
8. Hoechst 33342 is a DNA dye that can be pumped out by stem and progenitor cells. Stem/progenitor cells thus have a

Hoechst low fluorescence in both the blue and red regions of the spectrum.

9. A viability dye, such as propidium iodide (PI) or 7-aminoactinomycin D (7-AAD).
10. 2.4G2 antibody against FcgII/III receptors for blocking non-specific antibody binding.

2.4 Magnetic Beads for Depletion of Differentiated Cells

Magnet bead-mediated depletion of differentiated cells is recommended for HSC sorting, because it will pre-enrich HSCs and HPCs and reduce the sorting time. Two types of magnetic beads are commercially available for this purpose.

1. Dynabeads® Sheep Anti-Rat IgG: 4.5 μm superparamagnetic beads covalently bound with affinity purified polyclonal sheep anti-rat IgG.
2. Streptavidin-conjugated paramagnetic beads used for MACS columns and cell separation unit from Miltenyi Biotec (Auburn, CA).

2.5 FACS Instrument

Flow cytometers/cell sorters are equipped with lasers, filters, and detectors and can simultaneously select for the presence or absence of several cell surface markers. Several instruments are available from different manufacturers. For the technique described in this chapter, sorting is performed using the BD FACSAria™ II.

3 Methods

3.1 Preparation of Bone Marrow Cells

3.1.1 Sample

1. Euthanize young or old mice by either cervical dislocation or isoflurane inhalation.
2. Dissect the femurs and tibias, and cut the ends off the bones.
3. Put 10 mL medium into a 50 mL conical tube. Put 23_G1-gauge (young marrow) or 20_G1 1/2-gauge (old marrow) needles onto a 3 mL syringe and pre-fill the syringe with 1 mL medium. Note that the amount of medium used is for collecting bone marrow cells from ten mice.
4. Insert the needle into one end of the bone and flush the marrow out of the bone cavity with pre-filled medium several times. Repeat the same procedure from the other end of the bone to ensure the maximal collection of marrow cells.
5. Prepare a single-cell suspension by drawing and expelling the marrow and medium through the needle several times; the marrow will tend to dissociate as it passes through the needle.
6. Filter the single cell suspension through a nylon screen (cell strainer, 70 μm).

3.1.2 Density Separation Procedure (Ficoll)

1. Dilute samples 1:2 in medium, and make the final volume 30 mL.
2. Pour 20 mL Ficoll into a 50 mL conical tube, and then slowly layer (tilting tube and running the cells down the side of the tube) 30 mL of diluted marrow cells on top (*see* **Note 1**).
3. Centrifuge at 600 × *g* for 30 min at room temperature with the "Brake Off" setting.
4. Remove half of the top layer and discard.
5. Carefully pipet off "cloudy" interface layer (approximate 10 mL) and transfer into a clean 50 mL tube. Wash these cells with 50 mL medium twice; a pellet should be seen at the bottom of the tube (*see* **Note 2**).

3.1.3 Red Cell Lysis Procedure (Optional)

1. Resuspend cells in red blood cell lysis buffer at 3–4 times the original sample volume.
2. Incubate on ice for 10 min.
3. Centrifuge cells at 400 × *g*, 4 °C for 5 min, wash twice, and resuspend cells in medium with a density of 10^8 cells per mL. Meanwhile, aliquot 0.5×10^6 cells into a FACS tube and use them as a nonstaining control (Control 1) (*see* **Note 3**). Note that all control cells are resuspended in 50 μL of medium.

3.2 Staining Bone Marrow Cells with Lineage Antibodies

1. Cells are first stained with 2.4G4 antibody directed against FcgII/III receptors on ice for 15 min, washed twice, and resuspended into 3 mL medium.
2. Cells are stained with biotinylated antibodies against lineage markers, including CD5, CD8a, B220, Mac-1, Gr-1, and Ter119. All antibodies should be titrated before use and dilutions are selected that can brightly stain antigen-positive cells without nonspecifically staining antigen-negative cells. The concentrations of antibodies used in the authors' laboratory are CD5 (1:200), CD8a (1:200), B220 (1:300), Mac-1 (1:320), Gr-1 (1:350), and Ter119 (1:320). All the antibodies are mixed to make lineage antibody cocktail according to these titrations.
3. Add lineage antibody cocktail to bone marrow cells prepared in Subheading 3.1.3 with a concentration of 112.4 μL per 1×10^8 cells.
4. Incubate cells with antibody for 30 min at 4 °C on a rocker.
5. Wash cells twice and resuspend cells in medium, with a density of 10^8 cells per mL in a 15 mL conical tube. Also aliquot 0.5×10^6 cells into a FACS tube for a lineage staining control (Control 2).

3.3 Magnetic Beads Depletion of Differentiated Lineage Cells

3.3.1 Dynabeads® Sheep Anti-Rat IgG

1. Wash Dynabeads® Sheep Anti-Rat IgG before use according to the manufacturer's instruction. Resuspend Dynabeads with a density of 4×10^8 beads per 1 mL PBS and store at 4 °C.
2. Mix Dynabeads well and add them to cells (stained with lineage antibodies) with a concentration of 1×10^8 cells per 1 mL beads solution (4×10^8 beads, i.e., 4 beads per single cell).
3. Incubate cells with Dynabeads for 20 min at 4 °C on a rocker.
4. Bring the total volume to 6 mL. Put cells into the magnet stand. The Dynabeads, along with differentiated cells, are attracted and attached to the magnet during a period of 2–3 min.
5. Carefully remove the cell suspension without disturbing the attached beads, and transfer it to a new 15 mL conical tube for the second round of magnet separation. Repeat the same procedure one more time for the maximal removal of remaining beads from the cell suspension.
6. In order to enhance the cell yield, the attached beads are washed again with 6 mL medium and subjected to three rounds of magnet separations.
7. After two rounds of washes and three rounds of magnet separations, all cells are collected into a 50 mL conical tube, spun down, and resuspended into the medium at a concentration of 10×10^6/mL.

3.3.2 Miltenyi Streptavidin-Conjugated Paramagnetic Beads

1. Miltenyi beads are already in the ready-to-use solution and can be directly added into the cells prepared from Subheading 3.2. For 10^8 cells, use 0.4 mL medium plus 0.1 mL magnetic beads. Incubate for 15 min at 4 °C on a rocker.
2. During this incubation period, place the column in the magnet and prepare a miniMACS column (capacity 10^7 cells in the magnetic fraction) by running medium through it. This column size is appropriate for enriching progenitors from up to 2×10^8 bone marrow cells. If larger numbers of bone marrow cells are being processed, midiMACS columns with a capacity of 10^8 cells in the magnetic fraction can be used.
3. Load the cells onto a MACS column and allow the cells to pass through the column. Return the cell suspension to the column twice, allowing the cells to pass through the column a total of three times for maximal removal of lineage$^+$ cells. Unbound cells are lineage-depleted (or stem/progenitor-enriched) cells (*see* **Note 4**). Dispense six aliquots of 0.5×10^6 cells into six FACS tubes that will be used for stem cell marker single-color controls (Controls 3–7) and for PI viability marker staining (Control 8).

3.4 Stem Cell Marker Staining

1. Add CD34-FITC to lineage-depleted cells (main sample) at a concentration of 1:25. Incubate for 90 min at 4 °C on a rocker. Add 2 μL of CD34-FITC into Control 3 tube.
2. Incubate for 1 h, and add other markers for stem cell staining. The panel of markers and the corresponding concentrations used in the authors' laboratory are Sca-1 (1:100), c-Kit (1:100), Flk2/Flt3 (1:100), and CD150 (1:100). Streptavidin is also added to the cell suspension to bind biotinylated lineage antibodies (1:150). Meanwhile, add Sca1 (0.5 μL), c-Kit (0.5 μL), Flk2/Flt3 (0.5 μL), and CD150 (0.5 μL) to Control tubes 4–7, respectively, and add Streptavidin (0.33 μL) to Control tube 2. Incubate cells with these antibodies for 30 min.
3. Wash the sample and control cells twice.
4. Resuspend sample cells into the medium at a density of 10×10^6/mL. Add 2 μL PI (1 mg/mL stock concentration) per mL cells. Add 0.1 μL PI into control tube 8 for PI single-color control.
5. Cells must be transferred to 5 mL polystyrene round-bottom tubes before FACS.

3.5 Flow Cytometry and FACS

1. FACS instrument alignment procedure and instrument settings

   The FACS instrument must be aligned using procedures recommended by the manufacturer. The use of calibration beads that mimic fluorescently labeled cells facilitates selection of instrument settings in the range of the type of cells to be sorted. For the techniques described in the chapter, access to a two-laser (Octagon 488 nm and Trigon 633 nm) instrument capable of sorting cells on the basis of ten independent variables is recommended.
2. FACS windows

   Figure 2 shows typical flow cytometry profiles about serial gating for the selection of HSCs. Cells are gated on the forward and side scatter and PI staining (Fig. 2a). PI negative viable cells (PI^-) cells are then gated for the negative expression of lineage antibodies-APC Cy7 (Fig. 2b). In lineage- cells, Sc-1 and c-Kit positive cells are selected (Fig. 2c). $Lineage^-$ $Sca\text{-}1^+$ $c\text{-}Kit^+$ cells are further gated for the negative expression of Flk2/flt3 and CD34 (Fig. 2d). Therefore, PI^- $lineage^-$ $Sca1^+$ $c\text{-}Kit^+$ $Flk2/flt3^-$ $CD34^-$ cells represent a population of highly purified HSCs, and they comprise 0.0032 % of whole bone marrow cells in young mice. In this population, three subfractions are divided based on the level of CD150 expression in which myeloid-biased, balanced, and lymphoid-biased HSCs are enriched in the populations with high, low, and negative expression of CD150, respectively (Fig. 2e). With aging, numbers of HPCs and HSCs, like LSK (Fig. 2f) or more greatly purified LSK $Flk2/flt3^-$

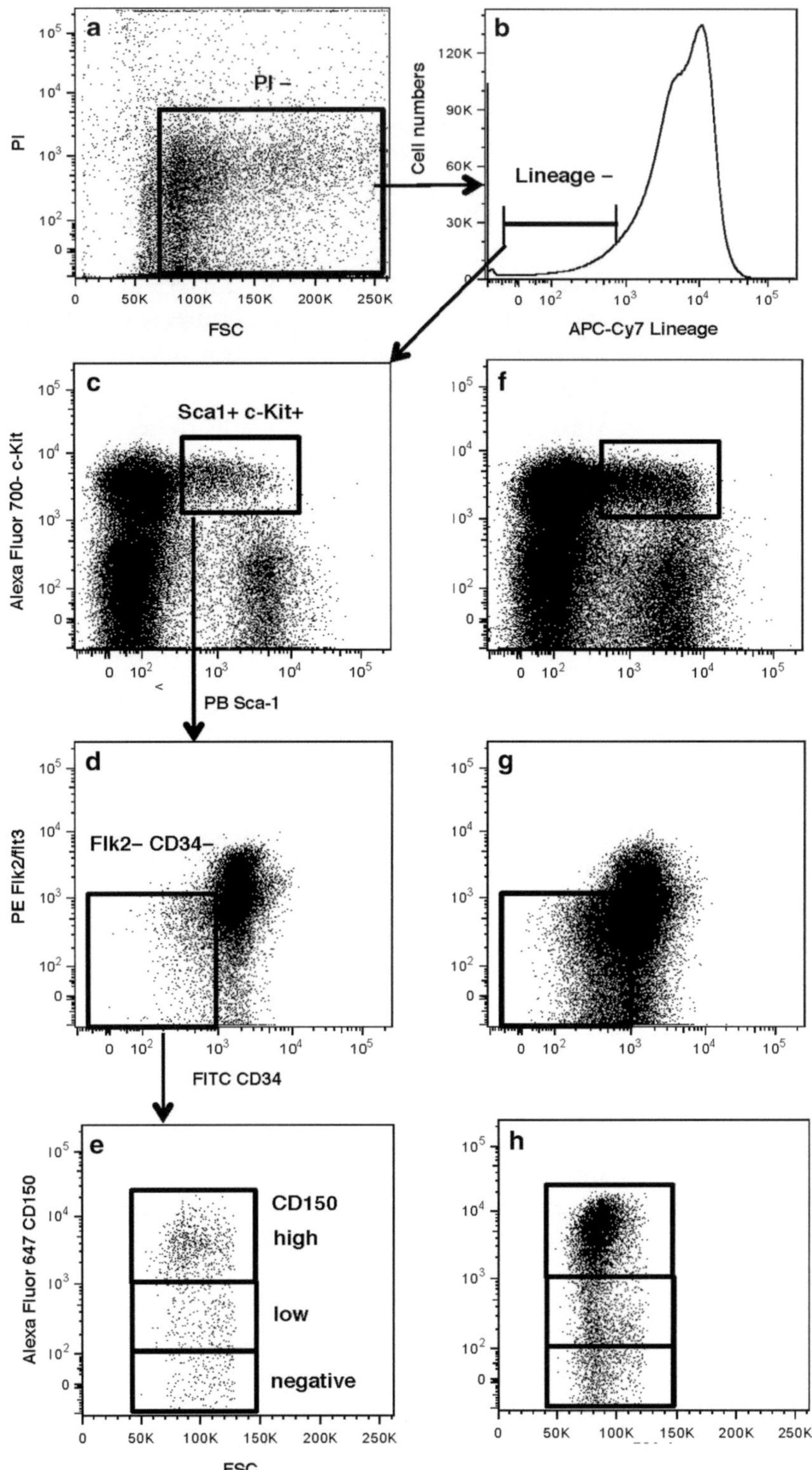

Fig. 2 Flow cytometry profiles and gate settings for selection of young and old lineage- Sca1$^+$ c-Kit$^+$ Flk2$^-$ CD34$^-$ HSCs and lineage-biased subpopulations. (**a**) Low-density mononucleated cells without PI expression (PI-). (**b**) Selection of lineage negative cells (lineage-). (**c**) Positive selection of cells expressing Sca1 and

CD34⁻ (Fig. 2g) cells, are significantly increased, and the proportion of LSK Flk2/flt3⁻ CD34⁻ cells in the aged bone marrow (0.0331 %) is nearly tenfold higher than in young marrow. With aging, myeloid-biased HSCs progressively dominate the old bone marrow, whereas the proportions of balanced and lymphoid-biased HSCs gradually diminish (Fig. 2h).

3. Collection medium and vessels

   The choice of collection tubes and media will depend on the number of cells needed and the intended use of the sorted cells. Note that medium needs to be added to the collection tubes. If less than 500,000 cells are collected, 1.5 mL Eppendorf tube pre-filled with 0.5 mL medium will be suitable. 5 mL polypropylene round-bottom tubes containing about 1.5 mL medium will be required for collecting 0.5 to 1.5×10^6 cells. Cells can also be directly sorted into multi-well plates using the program Automated Cell Deposition Unit (ACDU). Suitable collection media includes HBSS plus 2 % FBS, long-term culture (LTC) media, or serum-free media.

4. Verify the purity of sorted cells

   Even though flow cytometry is expected to reliably sort an HSC product of high purity with an acceptable yield, post-sort analysis is necessary to determine the purity of sorted cells. An aliquot of the sorted product should be reanalyzed on the same instrument on which it was sorted. In general, if greater than 90 % of cells fall into the original gates, the sorting is considered to have high fidelity. In addition to purity, cell viability and recovery should also be measured by independent quantitative analysis in sorted cells. For example, a hemocytometer with Trypan blue staining can be used to determine the number and viability of cells recovered after sorting.

4 Notes

1. Add the Ficoll-Paque into the tube first, and then put the bone marrow cells on the top. It is extremely important to load the cells slowly—especially the first several millimeters of cell solution. After a correct loading, a clear interface between cells and Ficoll-Paque should be seen, and this will ensure a successful gradient separation.

Fig. 2 (continued) c-Kit antigens ($Sca1^+$ $c\text{-}Kit^+$). (**d**) Selection of cells with undetectable levels of Flk2 and CD34 ($Flk2^-$ $CD34^-$) from lineage- $Sca1^+$ $c\text{-}Kit^+$ (LSK) cells. LSK $Flk2^-$ $CD34^-$ cells are highly purified HSCs. (**e**) Subdivision of LSK $Flk2^-$ $CD34^-$ cells into myeloid-biased, balanced, and lymphoid-biased populations based on the expression level of CD150. (**f**) LSK cell selection in the bone marrow of old mice (>24 month old). (**g**) Dramatic increase in the number of LSK $Flk2^-$ $CD34^-$ cells in old bone marrow. (**h**) Dominance of myeloid-biased $CD150^{high}$ LSK $Flk2^-$ $CD34^-$ cells at the replace of lymphoid-biased and balanced HSCs in old bone marrow

2. All procedures are performed on ice from this step. After washing and spinning, a white pellet containing mononucleated cells should be seen at the bottom of the tube. The presence of a red pellet indicates the incomplete separation of red blood cells from mononucleated cells; red blood cell lysis is recommended. After the gradient separation, around 10 % of total bone marrow cells will be recovered.
3. Single-color controls are used for the background staining and compensation that are required for any type of flow cytometry. The excitation spectrum of a fluorochrome is a range of light wavelengths. When multiple fluorochromes are used in flow cytometry, the emission spectra from different fluorochromes could be overlapped. As a result, fluorescence could leak to other filters, causing false positive signals. To correct this spectral overlap, a procedure called fluorescence compensation is needed. Cells are stained with a single fluorochrome and run through the flow cytometer. The compensation settings are adjusted to ensure that the signal detected in a particular detector derives solely from the fluorochrome that is being measured. In the described HSC sorting protocol, eight tubes of control cells are set up, and each tube contains 0.5×10^6 cells in 50 μL medium. Tube 1 is unstained (background) control. Tube 2 is APC-Cy7 lineage control. Tube 3 is FITC CD34 control. Tube 4 is Pacific Blue Sca1 control. Tube 5 is Alexa Fluor 700 c-Kit control. Tube 6 is the PE Flk2 control. Tube 7 is the Alexa Fluor 647 CD150 control. Tube 8 is the PI control.
4. About 10 % of mononucleated cells will be recovered from the Ficoll-gradient separation. Another 10 % cells will be recovered from the magnetic lineage depletion. Therefore, only about 1 % of whole bone marrow cells will be obtained after Ficoll-gradient separation and beads depletion, and bone marrow stem and progenitor cells are significantly enriched.

Acknowledgements

This work was supported by grants from the Edward P. Evans Foundation (to G.V.Z.) and The National Center for Advancing Translational Sciences, National Institutes of Health, through grant number KL2TR000116 (to Y. L.). The content is solely the responsibility of the authors and does not necessarily represent the official views of the NIH.

We acknowledge Jennifer F. Rogers for editing the manuscript.

References

1. Liu L, Rando TA (2011) Manifestations and mechanisms of stem cell aging. J Cell Biol 193:257–266
2. Beerman I, Maloney WJ, Weissmann IL et al (2010) Stem cells and the aging hematopoietic system. Curr Opin Immunol 22:500–506
3. Rando TA (2006) Stem cells, ageing and the quest for immortality. Nature 441: 1080–1086
4. Geiger H, de Haan G, Florian MC (2013) The ageing haematopoietic stem cell compartment. Nat Rev Immunol 13:376–389
5. Morrison SJ, Wandycz AM, Akashi K et al (1996) The aging of hematopoietic stem cells. Nat Med 2:1011–1016
6. de Haan G, Van Zant G (1999) Dynamic changes in mouse hematopoietic stem cell numbers during aging. Blood 93:3294–3301
7. Sudo K, Ema H, Morita Y et al (2000) Age-associated characteristics of murine hematopoietic stem cells. J Exp Med 192:1273–1280
8. Geiger H, True JM, de Haan G et al (2001) Age- and stage-specific regulation patterns in the hematopoietic stem cell hierarchy. Blood 98:2966–2972
9. Henckaerts E, Geiger H, Langer JC et al (2002) Genetically determined variation in the number of phenotypically defined hematopoietic progenitor and stem cells and in their response to early-acting cytokines. Blood 99:3947–3954
10. Dykstra B, Olthof S, Schreuder J et al (2011) Clonal analysis reveals multiple functional defects of aged murine hematopoietic stem cells. J Exp Med 208:2691–2703
11. Levi BP, Morrison SJ (2008) Stem cells use distinct self-renewal programs at different ages. Cold Spring Harbor Symp Quant Biol 73:539–553
12. Kim M, Moon HB, Spangrude GJ (2003) Major age-related changes of mouse hematopoietic stem/progenitor cells. Ann N Y Acad Sci 996:195–208
13. Xing Z, Ryan MA, Daria D et al (2006) Increased hematopoietic stem cell mobilization in aged mice. Blood 108:2190–2197
14. Liang Y, Van Zant G, Szilvassy SJ (2005) Effects of aging on the homing and engraftment of murine hematopoietic stem and progenitor cells. Blood 106:1479–1487
15. Florian MC, Dorr K, Niebel A et al (2012) Cdc42 activity regulates hematopoietic stem cell aging and rejuvenation. Cell Stem Cell 10:520–530
16. Muller-Sieburg CE, Cho RH, Karlsson L et al (2004) Myeloid-biased hematopoietic stem cells have extensive self-renewal capacity but generate diminished lymphoid progeny with impaired IL-7 responsiveness. Blood 103: 4111–4118
17. Beerman I, Bhattacharya D, Zandi S et al (2010) Functionally distinct hematopoietic stem cells modulate hematopoietic lineage potential during aging by a mechanism of clonal expansion. Proc Natl Acad Sci U S A 107:5465–5470
18. Muller-Sieburg CE, Sieburg HB, Bernitz JM et al (2012) Stem cell heterogeneity: implications for aging and regenerative medicine. Blood 119:3900–3907
19. Wagner W, Horn P, Bork S et al (2008) Aging of hematopoietic stem cells is regulated by the stem cell niche. Exp Gerontol 43:974–980
20. Mayack SR, Shadrach JL, Kim FS et al (2010) Systemic signals regulate ageing and rejuvenation of blood stem cell niches. Nature 463: 495–500
21. Oakley EJ, Van Zant G (2010) Age-related changes in niche cells influence hematopoietic stem cell function. Cell Stem Cell 6:93–94
22. Rossi DJ, Bryder D, Seita J et al (2007) Deficiencies in DNA damage repair limit the function of haematopoietic stem cells with age. Nature 447:725–729
23. Garinis GA, van der Horst GT, Vijg J et al (2008) DNA damage and ageing: new-age ideas for an age-old problem. Nat Cell Biol 10:1241–1247
24. Rossi DJ, Seita J, Czechowicz A et al (2007) Hematopoietic stem cell quiescence attenuates DNA damage response and permits DNA damage accumulation during aging. Cell Cycle 6:2371–2376
25. Sahin E, Depinho RA (2010) Linking functional decline of telomeres, mitochondria and stem cells during ageing. Nature 464:520–528
26. Yahata T, Takanashi T, Muguruma Y et al (2011) Accumulation of oxidative DNA damage restricts the self-renewal capacity of human hematopoietic stem cells. Blood 118: 2941–2950
27. Garrison BS, Rossi DJ (2012) Reactive oxygen species resulting from mitochondrial mutation diminishes stem and progenitor cell function. Cell Metab 15:2–3
28. Janzen V, Forkert R, Fleming HE et al (2006) Stem-cell ageing modified by the

cyclin-dependent kinase inhibitor p16INK4a. Nature 443:421–426

29. Xiao N, Jani K, Morgan K et al (2012) Hematopoietic stem cells lacking Ott1 display aspects associated with aging and are unable to maintain quiescence during proliferative stress. Blood 119:4898–4907
30. Rossi DJ, Bryder D, Weissman IL (2007) Hematopoietic stem cell aging: mechanism and consequence. Exp Gerontol 42:385–390
31. Chen J, Astle CM, Harrison DE (2000) Genetic regulation of primitive hematopoietic stem cell senescence. Exp Hematol 28: 442–450
32. Wagner W, Bork S, Horn P et al (2009) Aging and replicative senescence have related effects on human stem and progenitor cells. PLoS One 4:e5846
33. Wang J, Sun Q, Morita Y et al (2012) A differentiation checkpoint limits hematopoietic stem cell self-renewal in response to DNA damage. Cell 148:1001–1014
34. Beerman I, Bock C, Garrison BS et al (2013) Proliferation-dependent alterations of the DNA methylation landscape underlie hematopoietic stem cell aging. Cell Stem Cell 12: 413–425
35. Rossi DJ, Bryder D, Zahn JM et al (2005) Cell intrinsic alterations underlie hematopoietic stem cell aging. Proc Natl Acad Sci U S A 102:9194–9199
36. Chambers SM, Shaw CA, Gatza C et al (2007) Aging hematopoietic stem cells decline in function and exhibit epigenetic dysregulation. PLoS Biol 5:e201
37. Tadokoro Y, Ema H, Okano M et al (2007) De novo DNA methyltransferase is essential for self-renewal, but not for differentiation, in hematopoietic stem cells. J Exp Med 204: 715–722
38. Okada S, Nakauchi H, Nagayoshi K et al (1992) In vivo and in vitro stem cell function of c-kit- and Sca-1-positive murine hematopoietic cells. Blood 80:3044–3050
39. Adolfsson J, Borge OJ, Bryder D et al (2001) Upregulation of Flt3 expression within the bone marrow Lin(-)Sca1(+)c-kit(+) stem cell compartment is accompanied by loss of self-renewal capacity. Immunity 15:659–669
40. Kiel MJ, Yilmaz OH, Iwashita T et al (2005) SLAM family receptors distinguish hematopoietic stem and progenitor cells and reveal endothelial niches for stem cells. Cell 121: 1109–1121
41. Yilmaz OH, Kiel MJ, Morrison SJ (2006) SLAM family markers are conserved among hematopoietic stem cells from old and reconstituted mice and markedly increase their purity. Blood 107:924–930
42. Goodell MA, Brose K, Paradis G et al (1996) Isolation and functional properties of murine hematopoietic stem cells that are replicating in vivo. J Exp Med 183:1797–1806
43. Challen GA, Boles NC, Chambers SM et al (2010) Distinct hematopoietic stem cell subtypes are differentially regulated by TGF-beta1. Cell Stem Cell 6:265–278
44. Morita Y, Ema H, Nakauchi H (2010) Heterogeneity and hierarchy within the most primitive hematopoietic stem cell compartment. J Exp Med 207:1173–1182
45. Oguro H, Ding L, Morrison SJ (2013) SLAM family markers resolve functionally distinct subpopulations of hematopoietic stem cells and multipotent progenitors. Cell Stem Cell 13:102–116
46. Doulatov S, Notta F, Laurenti E et al (2012) Hematopoiesis: a human perspective. Cell Stem Cell 10:120–136
47. Kuranda K, Vargaftig J, de la Rochere P et al (2011) Age-related changes in human hematopoietic stem/progenitor cells. Aging Cell 10:542–546
48. Pang WW, Price EA, Sahoo D et al (2011) Human bone marrow hematopoietic stem cells are increased in frequency and myeloid-biased with age. Proc Natl Acad Sci U S A 108: 20012–20017
49. Vaziri H, Dragowska W, Allsopp RC et al (1994) Evidence for a mitotic clock in human hematopoietic stem cells: loss of telomeric DNA with age. Proc Natl Acad Sci U S A 91: 9857–9860
50. Notta F, Doulatov S, Laurenti E et al (2011) Isolation of single human hematopoietic stem cells capable of long-term multilineage engraftment. Science 333:218–221
51. Larochelle A, Savona M, Wiggins M et al (2011) Human and rhesus macaque hematopoietic stem cells cannot be purified based only on SLAM family markers. Blood 117:1550–1554
52. Goodrick CL (1975) Life-span and the inheritance of longevity of inbred mice. J Gerontol 30:257–263
53. Rowlatt C, Chesterman FC, Sheriff MU (1976) Lifespan, age changes and tumour incidence in an ageing C57BL mouse colony. Lab Anim 10:419–442

Chapter 4

Measuring the Potency of a Stem Cell Therapeutic

Holli Harper and Ivan N. Rich

Abstract

The potency of a drug is one of the most important parameters of a therapeutic. Besides providing the basis for manufacturing consistency and product stability, the potency can predict product failure or toxicity due to incorrect potency, provide release criteria, and the dose that will ensure that it can be used as intended. Recently, cellular therapeutics, in particular, stem cell therapy products, have being designated as "drugs" by regulatory agencies if they produce a systemic effect in the patient. Regulatory agencies are becoming increasingly stringent with respect to the manufacture, production, and testing of these products prior to being used in a patient. A clear understanding of what potency is and how it can be measured should help erase the misunderstandings and misconceptions that have accrued within the cellular therapy field. This protocol describes how the potency of hematopoietic stem cell therapy products is determined. The same principles apply to any proliferating stem cell therapeutic product.

Key words Potency, Stem cell potency, Cellular therapy, Drug potency, Stem cell proliferation, Stem cells, Umbilical cord blood, Bone marrow, Mobilized peripheral blood, Mesenchymal stem cells

1 Introduction

Potency may be defined as the quantitative and validated measurement of biological activity of the "active" ingredients or components of a product, which, when administered to a patient, produces the intended effect or response. This definition is part and parcel of the United States Food and Drug Administration (U.S. FDA) requirements and regulations for a drug potency assay. It also includes many of the aspects used by the European Medicines Agency (EMA) [1, 2].

Potency determination of traditional drugs has been a routine exercise for pharmaceutical companies for many years. With the designation of umbilical cord blood (UCB) as a "drug" by the FDA in 2009 [3], because it results in systemic effects after transplantation into patients with blood malignancies, there has been an increased effort to designate other cellular therapies as "drugs." For academic institutions that have viewed hematopoietic stem cell transplantation as a routine procedure since the 1970s, the

Ivan N. Rich (ed.), *Stem Cell Protocols*, Methods in Molecular Biology, vol. 1235,
DOI 10.1007/978-1-4939-1785-3_4,

introduction of regulations and the requirement for potency testing has resulted in considerable misunderstanding of how potency should be measured and the misconception that potency must correlate with clinical outcome [4, 5] (*see* **Note 1**). Potency should correlate with a biological response, but this response need not correlate with clinical outcome.

There are three primary requirements for a potency assay. The first is knowledge of the biological properties and characteristics of the "active" ingredients or components that are responsible for the response that has to be measured as a basis for the assay. The second is a reference standard (RS) of the same material or source as the sample being measured and against which the sample can be compared to determine the potency ratio. The final requirement is a quantitative and validated assay that can measure the "active" components in both the reference standard and the sample.

For a cell therapeutic that requires the administration of stem cells, it is these that represent the "active" ingredients or components, since it is only the stem cells that are responsible for engraftment and later reconstitution. Therefore, sufficient information must be available that allows the assay to identify and measure a stem cell response that can be used for potency measurement. When developing any new cellular therapeutic, the earlier a potency assay can be developed, the better. Furthermore, as more information is accrued, the potency assay may pass through several iterations that will better define how the response should be measured.

For traditional drugs, establishing an RS is not a difficult task since sufficient amounts of the drug are usually available for several RS batches to be produced at different times. Cellular RSs are more difficult to establish. Large numbers of cells are not available to establish "global" RSs that would allow, for example, all cord blood banks worldwide to use the same RS. Nevertheless, providing a standardized assay readout is available that allows direct comparison of samples, both within and between laboratories over time, in-house RSs for different tissues can be established and used to calculate the potency ratio.

The following protocol was designed and developed for hematopoietic stem cell therapy products derived from human umbilical cord blood, normal or mobilized peripheral blood, bone marrow, and purified stem cell fractions (e.g., $CD34^{+}$ cells) from these tissues. However, by changing the method by which cells are cultured in vitro, the same basic protocol can be used for mesenchymal stem/stromal cells (MSC) as well as other stem cell sources, including other primary stem cell preparations, embryonic stem (ES) cells and induced pluripotential stem (iPS) cells that might be used as a cellular therapeutic (Fig. 1). All protocols use the same fully calibrated, standardized, and validated ATP bioluminescence

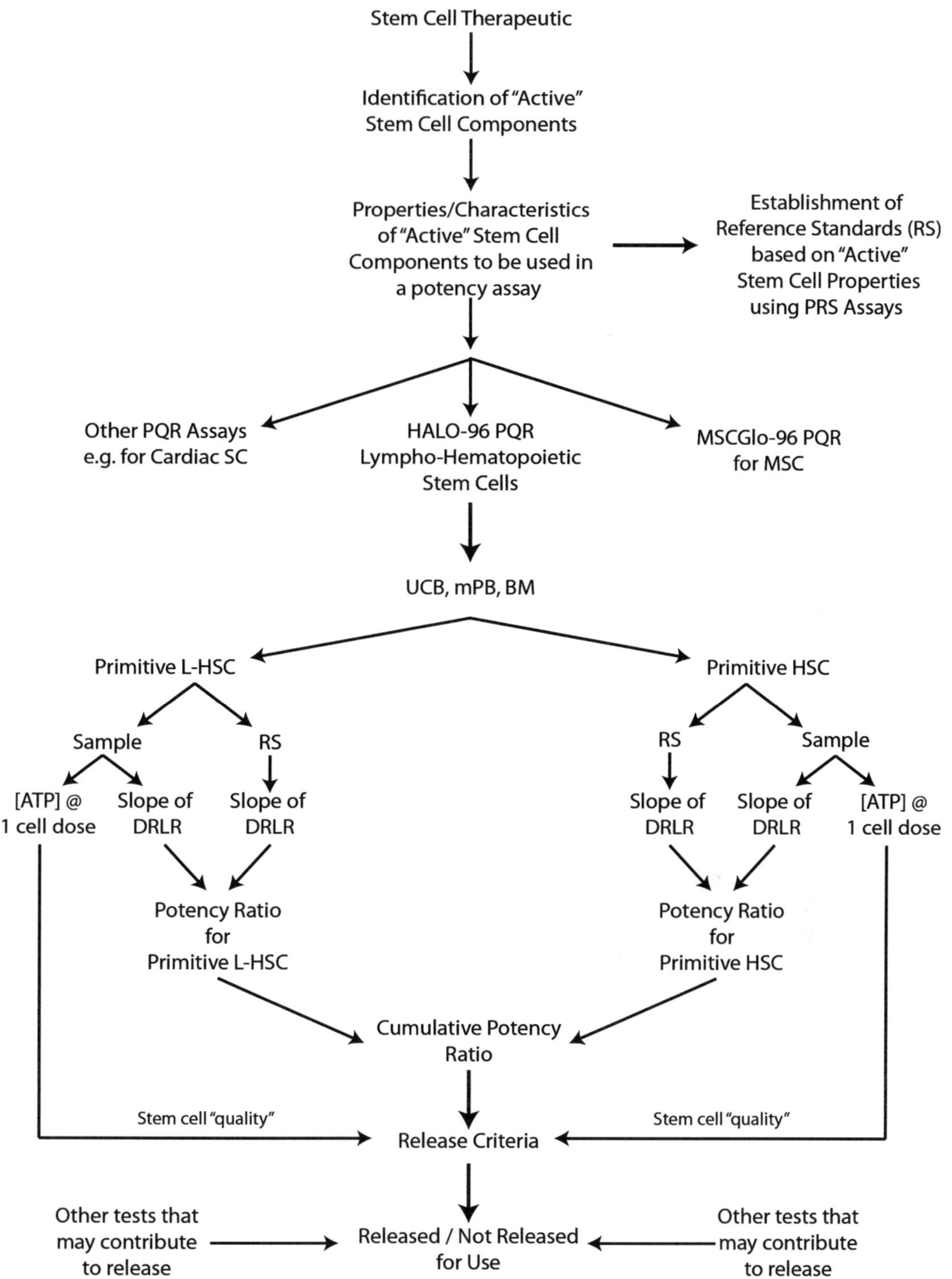

Fig. 1 Potency assay paradigm. This is a flow diagram demonstrating how stem cell potency is measured. The paradigm demonstrates the steps for measuring potency of hematopoietic stem cell using two stem cell populations as the "active" components. It also shows how the release criteria for a cell therapeutic are determined from both the potency ratio and "quality" of the stem cell populations. Abbreviations used: *RS* reference standard, *PQR* potency, quality, release, *UCB* umbilical cord blood, *mPB* mobilized peripheral blood, *BM* bone marrow, *Primitive L-HSC* primitive lympho-hematopoietic stem cell, *Primitive HSC* primitive hematopoietic stem cell, *DRLR* dose response linear regression

signal detection system for measuring stem cell proliferation potential and ability. The present protocol focuses on detecting a minimum of two "active" stem cell populations derived from human, lympho-hematopoietic tissues (*see* **Note 2**).

2 Materials

2.1 Equipment

1. To measure potency of lympho-hematopoietic stem cells for cellular therapy, two assays are required. The first, HALO-96 PRS [4, 6], is used to establish the first primary, in-house RS. Once the potency of this RS has been established, the potency of all other RS batches or lots and samples can be measured using HALO-96 PQR. For MSCs, MSCGlo-96 PRS and MSCGlo-96 PQR are available. All assays incorporate a fully calibrated, standardized, and validated ATP bioluminescence readout (*see* **Note 3**). These assay kits contain Master Mixes (*see* **Note 4**) or growth medium needed to culture the cells in vitro and measure their proliferation response (*see* **Note 5**). Only an appropriate cell sample is required as well as ancillary equipment and supplies. All assay kits are available from HemoGenix.
2. Luminescence plate reader (plate luminometer, Molecular Devices, SpectraMax L; Promega Corporation; Berthold, CentroLia) or a multimode plate reader that not only may have the capability of measuring absorbance and/or fluorescence, but also luminescence.
3. Fully humidified incubator at 37 °C with an atmosphere containing 5 % carbon dioxide (CO_2) and, if possible, nitrogen to reduce the oxygen tension from atmospheric concentrations to 5 % O_2. Both the CO_2 and O_2 concentrations should be monitored using a Fyrite analyzer.
4. Biohazard/laminar flow hood for sterile cell/tissue culture.
5. Laboratory centrifuge.
6. Single channel, electronic pipettes for volumes from 0.1 to 1 mL include sterile pipette tips.
7. An 8-channel electronic pipette to dispense volumes of 0.1 mL using non-sterile tips.
8. Electronic cell counter or flow cytometer that can determine nucleated cell counts.
9. Method for determining dye exclusion viability (e.g., hemocytometer, flow cytometer, viability counter).
10. Sterile, conical 15 and/or 50 mL tubes with screw lid.
11. Sterile, 5 mL tubes with snap on lid.
12. 1.5 mL Eppendorf conical vials.

2.2 Reagents

1. Dulbecco's phosphate buffered saline (dPBS).
2. Iscove's Modified Dulbecco's Medium (IMDM).
3. Density gradient centrifugation (DGC) medium (NycoPrep 1.077, Axis-Shield; or Ficoll, GE Healthcare).
4. Cryopreservation medium: IMDM + 10 % fetal bovine serum (FBS) + dimethylsulfoxide (DMSO). Alternatively, a commercially available cryopreservation medium can be used, e.g., CryoStor (BioLife Solutions).
5. DNase (Sigma-Aldrich).
6. Thaw medium: IMDM + 10 % FBS.

3 Methods

3.1 Establishing a Reference Standard

3.1.1 Preparation of the Mononuclear Cell (MNC) Fraction (See **Note 6**)

1. Collect or procure fresh, whole tissue that will be used for the primary RS.
2. Split the tissue between two or more sterile 50 mL tubes (maximum 25 mL/tube) and dilute with the same amount of dPBS. Mix the contents by inversion several times.
3. In a new sterile, 50 mL tube, add 15 mL of the DGC medium, and tilting the tube at about 45°, carefully and very slowly dispense the diluted cells on top of the DGC layer making sure that the cells do not mix with the DGC medium. For each 15 mL of DGC medium, add between 15 and 20 mL of diluted cells to be fractionated.
4. Centrifuge the tubes using the conditions specified by the manufacturer of the DGC medium. After centrifugation, allow the centrifuge to slow down without the brake. After removal of the tube(s), there should be a distinct layer of cells at the interface between the plasma and the DGC medium. Granulocytes and platelets are usually contained in the DGC medium layer. The red blood cells will be a pelleted at the base of the tube.
5. Remove and discard the top layer of plasma until about 5 mm is left above the interface.
6. Using a 1 mL manual pipette with sterile tip, gently remove the cells at the interface and transfer them to a new 50 mL tube. Interface cells from at least 2–3 tubes can be pooled.
7. Fill the tube with dPBS to a total of 50 mL, replace the cap, mix by inversion, and centrifuge the cells for 10 min at $200 \times g$ at room temperature.
8. Remove and discard the supernatant above the cell pellet and resuspend the cells in about 2–5 mL IMDM.

9. Perform a nucleated cell count and viability assessment. The viability should be at least 90–95 % or greater. This MNC fraction contains stem cells and other primitive nucleated cells, and will represent the first or primary (1°) RS.

*3.1.2 Cryopreservation of the Primary Reference Standard (See **Note 7**)*

1. Prepare the cryopreservation medium or use a commercially available medium and keep on ice or at 4 °C.
2. After determining the nucleated cell count, calculate the volume of the cell concentration that will produce 2×10^6 cells/mL. Keep the cells on ice or at 4 °C.
3. Calculate the number of cryopreservation vials required to freeze a certain number of cell suspension aliquots and label each vial.
4. Calculate the amount of cryopreservation reagent needed to make up 1 mL in each vial and dispense this amount of reagent into the vials.
5. Dispense the volume of cell suspension into each vial.
6. Cap each vial securely and mix by inversion.
7. Place the vials in a CoolCell, transfer the CoolCell to a −80 °C freezer, and leave overnight. Alternatively, cryopreserve the cells using an automated rate-controlled freezer.
8. After freezing, transfer the vials to an LN2 container for storage.

*3.1.3 Cell Thawing to Establishing the Primary, In-house Reference Standard (See **Note 8**)*

In order to establish the primary, in-house RS, it is necessary to thaw the vial of cells that came with the HALO-96 PRS assay kit and a vial of cells that had been cryopreserved for use as the future RS.

1. Thaw the vials in a 37 °C water bath, by swirling the vial for approx. 1 min.
2. When a small ball of ice remains (1–2 min), remove the vial from the water bath, sterilize the outside of the vial by spraying with 70 % ethanol, and carefully unscrew the vial lid.
3. It is possible that clumping can occur at this stage. To avoid this, add DNase to the total volume in the vial to achieve a concentration of 6 μg/mL before proceeding to the next step and mix.
4. Use a 1 mL pipette to gently mix the contents of the vial and transfer to a 50 mL tube containing 20 mL of thaw medium. Up to three vials of the same cells can be added to this 20 mL of thaw medium. However, clumping can still occur. Add DNase at a final concentration of 6 μg/mL before proceeding to the next step.
5. Gently mix the cells by swirling the contents of the tube. Do not use repeat pipetting to mix the cells. This could cause

further rupture of cells and the release of DNA resulting in increased clumping.

6. Centrifuge the cells at 200 × *g* for 10 min at room temperature and discard the supernatant after centrifugation.
7. Resuspend the cells in 1 mL of IMDM. If necessary, add 6 μg/mL DNase.
8. Perform a nucleated cell count and viability assessment.
9. Prepare working serial cell dilutions containing 750,000, 500,000, and 250,000 cells/mL in IMDM for both the kit RS and the future 1° RS. A total volume of 1 mL is sufficient.
10. The method of determining the potency of the primary, in-house RS is the same as measuring the potency of an unknown sample. Follow the instructions in Subheading 3.2.

3.2 Measuring the Potency and Quality of Lympho-Hematopoietic Stem Cells

Two stem cell populations for both the RS and sample (or the RS from the PRS kit and the primary RS sample from Subheading 3.1) will each be cultured at three cell doses. The two stem cell populations are designated HPP-SP and CFC-GEMM (*see* **Note 9**). Master Mixes containing all components, including the growth factor cocktails needed to stimulate the two stem cell populations, are included in the kits.

3.2.1 Stem Cell Culture

1. Thaw a vial of RS cells and the unknown sample as described in Subheading 3.1.3 and for both the RS and sample, prepare working serial cell dilutions of 750,000, 500,000, and 250,000 cells/mL in IMDM. If possible, use electronic pipettes. A total volume of 1 mL for each cell concentration is sufficient.
2. Remove the two Master Mixes for the two stem cell populations to be determined from the freezer and thaw in a beaker of water or at room temperature. When thawed, mix gently by inversion.
3. For both the RS and sample, 3 × 5 mL tubes are labeled for HPP-SP and 3 for CFC-GEMM; i.e., a total of six tubes are necessary to test the RS and six for the sample.
4. Using an electronic pipette (if possible), dispense 0.9 mL of the HPP-SP Master Mix into three tubes for the RS and three for the sample. Similarly, dispense 0.9 mL of the CFC-GEMM Master Mix into the other three tubes for the RS and sample.
5. Transfer 0.1 mL of the lowest cell dose (250,000 cells/mL) of the RS cells into the HPP-SP tube and the first CFC-GEMM tube. Follow this by transferring 0.1 mL from the 500,000 and 750,000 cells/mL into the individual tubes. This procedure reduces the cell concentration in each tube by tenfold.
6. Repeat this process for the sample cells.
7. All 12 tubes will now contain 1 mL of Culture Master Mix.

8. Using an electronic pipette and repeat pipetting function, transfer 0.1 mL from the lowest HPP-SP stem cell dose for RS cells to all eight replicate wells (*see* **Note 10**) in the first column of the sterile, 96-well plate provided. Similarly, transfer 0.1 mL from the middle cell dose for the RS, HPP-SP stem cells into eight replicate wells in column two of the plate. Finally, transfer the highest dose of RS, HPP-SP stem cells to column 3.
9. Repeat this procedure for the RS, CFC-GEMM stem cells filling columns 4–6.
10. Repeat **steps 8** and **9** for the sample cells filling columns 7–9 for the HPP-SP stem cells and columns 10–12 for the CFC-GEMM stem cells. The final cell concentrations are now 2,500, 5,000, and 7,500 cells/well.
11. Replace the lid and transfer the plate to the incubator. Culture the cells for 5 days. This incubation period can be extended to 7 days if required for increased assay sensitivity and/or to accommodate work schedules (*see* **Note 11**).

*3.2.2 Assay Calibration and Standardization (See **Note 12**)*

Prior to measuring proliferation of the cultured cells, the assay must be calibrated and standardized. It is essential that laboratory gloves are worn during this and the following steps, since ATP is present on the skin and can cause erroneous results.

1. Remove a vial of ATP standard, one vial each of high and low controls and a bottle of ATP-Enumeration Reagent (ATP-ER) from the freezer and thaw the contents at room temperature.
2. Label five Eppendorf 1.5 mL vials and perform a serial dilution of the ATP Standard (10 μM) in IMDM to produce ATP dilutions of 1, 0.5, 0.1, 0.05, and 0.01 μM.
3. Using an electronic pipette with repeat dispensing function, dispense 0.1 mL of IMDM alone into the first four wells of the first column into the non-sterile, 96-well plate provided with the assay kit. This will be the background.
4. Dispense 0.1 mL of the lowest ATP concentration (0.01 μM) into the next four replicate wells in the first column of the plate.
5. The remaining ATP dilutions are then transferred into each of four replicate wells so that the ATP standard curve occupies the first three columns of the 96-well plate.
6. Dispense 0.1 mL of the low ATP control into the first four wells of column 4 and 0.1 mL of the high ATP control into the last four wells of column 4. In this way, three ATP standard curves and controls can be performed on a single 96-well plate.
7. Pour the ATP-ER into a reservoir and, using an 8-channel pipette, dispense 0.1 mL of the ATP-ER into all eight wells of

the first column. Mix the contents making sure that no bubbles are produced. Discard the tips after using.

8. Using new pipette tips for each column, continue adding 0.1 mL of ATP-ER into the other wells and mix. Discard the tips after each ATP-ER addition.
9. Transfer the plate to the luminescence plate reader and incubate in the dark at room temperature for 2 min.
10. Measure the luminescence using an integration time of 2 ms. If a multimode instrument is used, it will probably be necessary to establish the "gain" required to achieve optimum sensitivity using the method provided by the instrument manufacturer.
11. If the instrument software cannot be programed to calculate and graph the ATP standard curve, it will be necessary to export the data to third-party software (e.g., Microsoft Excel) and plot the curve as a log-log linear regression (*see* **Note 13**).
12. The goodness of fit of the regression line should be 0.995 or greater. The slope of the log-log linear regression should be within the range specified in the assay manual. The two control values should lie directly on the curve and also be within specific limits specified in the manual (Fig. 2). If these specifications are met, the instrument has been calibrated and the assay standardized so that the remaining part of the assay (measurement of the samples) can be completed.

3.2.3 Processing and Measurement of Proliferation of the Reference Standard and Sample

1. The calibration of the instrument and standardization of the assay must conform to the parameters given in the assay manual in order to measure the response of the RS and sample. If the ATP standard curve and controls do not comply with these parameters, it will be necessary to either repeat the previous step or troubleshoot why these parameters are not obtained.
2. Remove the sample plate from the incubator and let it come to room temperature.
3. Using the 8-channel pipette, dispense 0.1 mL of the ATP-ER into all wells of the first column and mix without causing bubbles. Mixing is extremely important at this stage since the ATP-ER contains a lysis buffer that releases the iATP so that it can react with luciferin and luciferase. If the cells are not lysed properly, incorrect ATP values will result.
4. Change the tips on the pipette and add 0.1 mL of the ATP-ER to the next column of wells. Repeat this for the whole plate.
5. Transfer the plate to the luminometer and incubate in the dark for 10 min, prior to reading the plate.

3.2.4 Analysis of Results

1. The output of a plate luminometer is Relative Luminescence Units or RLU. This is a nonstandardized value because different instruments will produce different RLU values for the

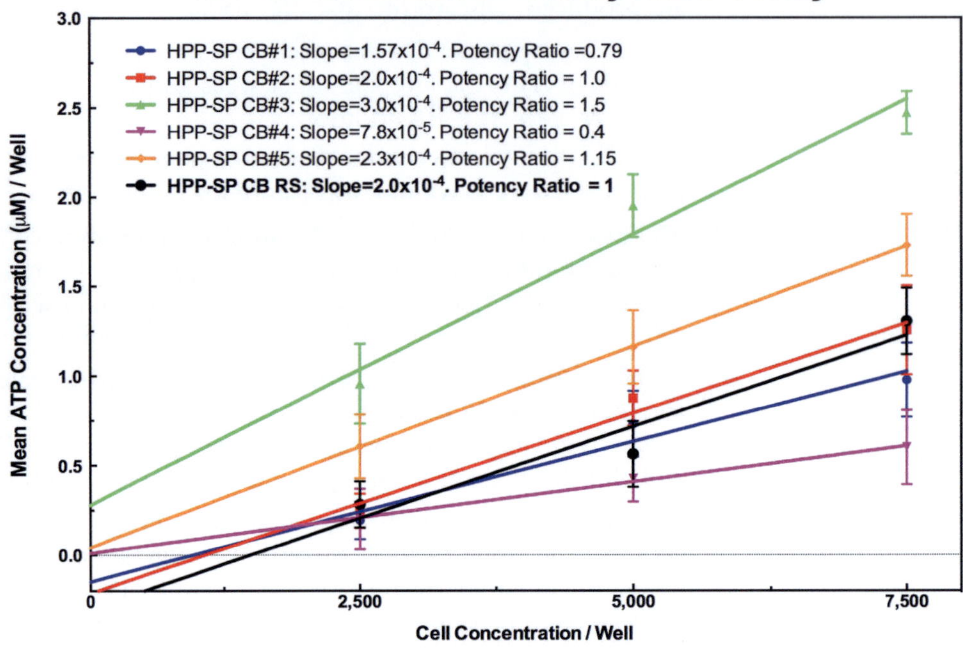

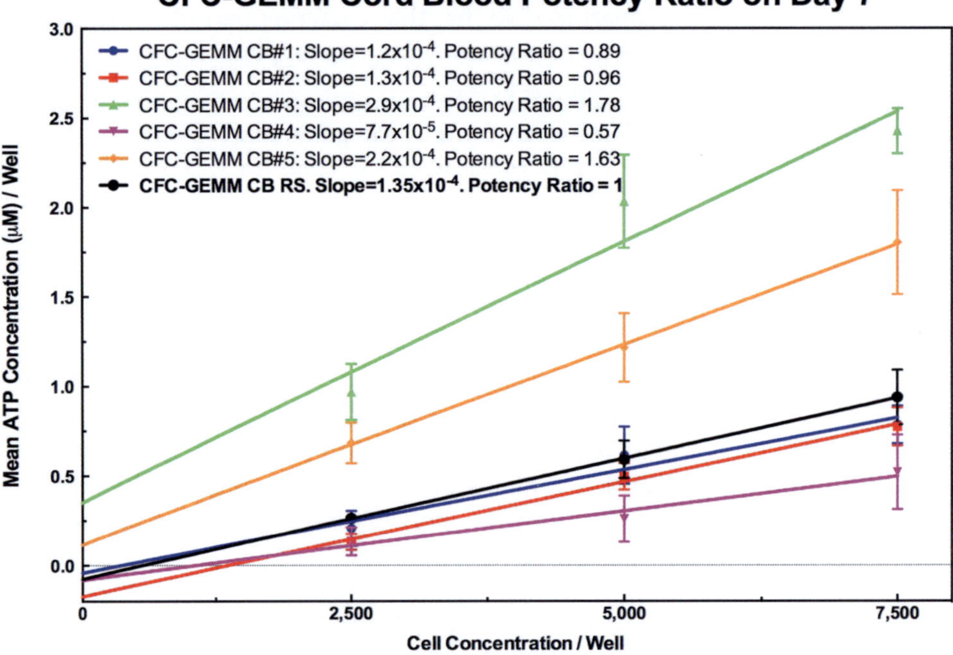

Fig. 2 Measurement of cell proliferation using ATP bioluminescence. When cells are stimulated to proliferate, the concentration of iATP increases in proportion to the amount of stimulation. When the iATP is released from the cells it acts as a limiting substrate for a luciferin/luciferase reaction that produces bioluminescence. The "glow" luminescence produced is measured as light by a plate luminometer. The concentration of ATP produced is calculated from an ATP standard curve that is performed prior to measuring any samples. Inclusion of controls allows the instrument to be calibrated, which together with the standards provides a the means to validate the assay

same samples. To compare results, it is necessary to transform RLU values into standardized ATP concentrations (μM). This is performed using the ATP standard curve. If the software allows, it can be programed to interpolate all RLU values into ATP concentrations for each individual well and then calculate the mean, standard deviation, and %CV values for each group of replicates. If the software cannot perform these functions, the results will have to be exported into a third-party software (e.g., Microsoft Excel) and the mathematical and statistical functions used to perform the interpolations. Other software packages such as GraphPad Prism, Systat SigmaPlot, or OriginLab's Origin can also be used.

2. Once all the results are in ATP concentrations, perform a linear regression analysis for the RS dose response for HPP-SP and CFC-GEMM and the sample HPP-SP and CFC-GEMM cell populations and calculate the slope of each dose response curve.

3. (*See* **Note 14**) The potency ratio for each unknown stem cell population is given by:

$$\text{Potency Ratio} = \frac{\text{Slope of the sample stem cell population}}{\text{Slope of the RF stem cell population}}.$$

The potency ratio will either be (a) similar or =1, (b) >1 or (c) <1. The potency of the RS is always 1. Therefore a sample demonstrating a potency ratio >1 indicates that less of the sample can be used to obtain the same response as the RS for that stem cell population, and vice versa for a sample potency ratio <1. However, since two stem cell populations are being measured, a cumulative potency ratio is used. It therefore follows that the cumulative potency ratio of both stem cell populations should be greater than that of the RS. It is important that the potency ratio for the CFC-GEMM alone is similar to or greater than that for the RS, since it is this stem cell population that will be primarily responsible for short-term engraftment and reconstitution.

4. (*See* **Note 14**) From the cell dose response curves for each stem cell population, determine the ATP concentration at a specific cell dose, e.g., 5,000 cells/well. This ATP concentration gives the proliferation ability or "quality" of the stem cell population. If the ATP concentration value is >0.04 μM, this is an indication that the stem cells will sustain proliferation. If the ATP concentration value is <0.04 μM, but >0.01 μM, the cells might proliferate, but they will not be able to sustain proliferation to result in engraftment if the cells were transplanted into a patient. If the ATP concentration <0.1 μM, the cells will be metabolically dead. Thus, the proliferation ability value is also a metabolic viability measurement.

4 Notes

1. Measuring stem cell potency and the potency of other cellular therapeutics is a new area with a lot of uncertainties, not only for the institutions and companies that work in the field, but also the agencies that regulate it. It has been assumed by the regulatory agencies that the same regulations that apply to traditional drugs will also apply to cellular therapeutics. In most cases this is correct. One of the exceptions is the requirement to measure all "active" ingredients or components (*see* **Note 2**). However, it is also apparent that many false assumptions have or are being made with respect to how cell potency can be measured without taking present regulations and guidelines into account. For example, any potency assay must include a dose response for both the sample and the reference standard. This allows determination of the potency ratio (*see* Subheading 3.2.4). The potency ratio cannot be determined from a single point. A non-validated assay cannot be used to measure potency, since the purpose of assay validation is to obtain scientific data about the assay that allows the results to be trusted. An assay, such as the colony-forming unit (CFU) assay, may have been used for decades and "grandfathered" in [5], but that does not mean that the results can be trusted or, more importantly, used to release a product for patient use [4, 6, 7]. Potency requires that a biological response has to be measured. A basic laboratory test such as cell number or viability [8] is required to perform a cell potency assay, but these, by themselves, are not potency assays [4, 5]. It follows that considerable thought has to be given when developing a potency assay and should be one of the first things to be considered when developing a new cellular therapeutic.
2. Regulatory agencies require that the potency of all "active" ingredients or components shall be determined. This is certainly possible for traditional drugs but is impossible for stem cells. This is because the stem cell compartment of biological systems contains an unspecified number of stem cells at different degrees of primitiveness that constitute a continuum. That is, one degree of stem cell primitiveness passes imperceptibly into the next. In vitro culture technology is incapable of separating and detecting this continuum at its lowest possible level. The technology described in this protocol is, at present, only capable of detecting up to eight different stem cell populations within the lympho-hematopoietic stem cell continuum. For other stem cell systems, it may not even be possible to differentiate two stem cell populations. Although testing the potency of more stem cell populations might provide greater accuracy to predict a response, the requirement to measure all

"active" stem cell components would not only be impractical but prohibitively expensive and time consuming.

3. Bioluminescence is the most sensitive, nonradioactive signal detection system available. The principle of using bioluminescence to measure stem cell proliferation or the proliferation of other cells types is shown in Fig. 2. All cells produce chemical energy in the form of adenosine triphosphate or ATP. This is produced in the mitochondria of the cells. Some cells (e.g., stem cells) have very low levels of intracellular ATP while others (e.g., hepatocytes) have high basal levels of ATP because they are actively metabolizing. When stem cells are induced into proliferation, their iATP levels increase significantly. This increase correlates directly with their proliferation. Intracellular ATP can therefore be used as a marker for proliferation. After stimulation and incubation of the cells, the iATP is released and acts as a limiting substrate for a luciferin/luciferase reaction to produce bioluminescence. This is in the form of light that is then measured in the plate luminometer.

4. The HALO assay platform was originally derived from the conventional colony-forming unit (CFU) or cell (CFC) assay, first described in 1966 [9, 10] and later modified to use a methylcellulose, semisolid medium [11]. Due to the many problems that have been encountered over the years with the CFU/CFC assay that include subjectivity, high coefficients of variation, lack or standards and controls and therefore the inability to validate the assay, and the fact that it detects cell differentiation ability rather than cell proliferation, a Suspension Expansion Culture (SEC) Technology was developed specifically for the HALO Platform. Removing the viscous methylcellulose medium from the assay allows reagents to be dispensed rapidly and accurately using normal pipettes or even a liquid handler. By allowing the cells to interact, not only is the onset of cell proliferation reduced by at least 24 h, but the sensitivity of assay is also increased.

5. Measuring potency of stem cells uses a basic characteristic common to all stem cells; their capacity and ability to proliferate. The degree of primitiveness or "stemness," their capacity to self-renew (a property that defines stem cells), and even their ability to engraft in a patient after transplantation all rely on the proliferation process. The response that is determined using this protocol is made up of two components, namely proliferation potential and proliferation ability. Proliferation potential measures the primitiveness of the stem cell or its capacity to proliferate. This is equivalent to stem cell potency, which, in turn, predicts engraftment potential. Proliferation ability is a measure of stem cell "quality"; that is, the proliferation status of the stem cells at a specific point in time.

Both stem cell potency and "quality" are used to release the product for use in a patient.

6. It is very important that when adding the cells to the DGC medium, the cells form a distinct and sharp layer on the top of the medium. The sharper this layer, the better the quality of the cell fractionation. In addition, when removing the cells from the interface after centrifugation, care should be taken not to withdraw cells underneath the interface since these contain granulocytes and platelets. The granulocytes, in particular, can cause clumping problems when the cells are eventually thawed after cryopreservation.
7. There are many ways to cryopreserve cells. There is no standardized method for cryopreserving stem cells. Many centers use an automated rate-freezing system and a preformulated cryopreservation medium (e.g., CryoStore, BioLife Solutions). Alternatively, ampoules of cells can be frozen in CoolCells, which are placed in a −80 °C freezer overnight. In both cases the cells are stored in liquid nitrogen (LN2). Regardless of the method used to cryopreserve the cells, $2.5–5 \times 10^6$ MNCs/mL should be cryopreserved/vial. Approximately 50 % of the cells will be lost upon thawing.
8. To determine whether the cryopreserved batch or lot of cells can be used as an RS, the stem cells should be tested and compared with an external RS of the same material or source. For hematopoietic cells and MSCs, HALO-96 PRS or MSCGlo-96 PRS assay kits are used, respectively. These kits contain a vial of cryopreserved cells against which the in-house primary RS can be tested. The quality and potency of the future, primary RS should be similar or greater than that of the assay kit reference standard. The method by which this is performed is exactly the same as the procedure for measuring the potency of an unknown sample. If the primary reference standard exhibits similar or greater potency and quality as the external RS, this 1° RS is now used to establish a secondary (2°) and even a tertiary (3°) RS batch. The 2° RS is tested against the 1° RS and the 3° RS is tested against the 2° RS. To test unknown samples, the most recent RS established is used. Prior to using the last RS vials of cells, a new RS batch is established and tested against the previous batch of cells. In this way the 1° and even the 2° RS will not be depleted and will be available for many years.
9. Two stem cell populations are detected and measured for hematopoietic stem cell potency. The first is a primitive lymphohematopoietic stem cell designated the High Proliferation Potential—Stem and Progenitor cells or HPP-SP. This stem cell is capable of producing both lymphopoietic and hematopoietic cell lineages. The second is the Colony-Forming

Cell—Granulocyte, Erythroid, Macrophage, Megakaryocyte, or CFC-GEMM. As its name implies, this primitive hematopoietic stem cell produces cells of all three hematopoietic lineages, but not cells of the lymphopoietic system.

10. The reason for performing eight replicate wells is for statistical purposes. When measuring the response of rare, primary cell populations, such as stem cells, outliers can occur. This is due to different numbers of primitive cells being dispensed into neighboring, replicate wells despite attempts to produce a homogenous cell suspension. By performing eight replicates of the same sample, it is possible to remove or mask the results of up to three outliers, if necessary, to reduce variation without significantly affecting the overall results. If the number of replicates is reduced to six, a maximum of two outliers can be removed from the results. The fewer the replicate number, the greater the chance of high variations in the results. From an economic standpoint, it is cheaper to increase the number of replicates than to repeat a complete experiment. However, the number of cells available may be limited. If this is the case, then it is best to reduce the number of replicates to a minimum of four and ensure that a cell dose response for both stem cell populations is performed.

11. By increasing the culture time from 5 to 7 days, an approximate threefold increase in stem cell proliferation can be expected. This increases the assay sensitivity threefold. However, an increase in percent coefficients of variation (%CV) may also occur.

12. Performing the ATP standard curve and controls is an integral part of the assay and should always be completed. If the ATP standard curve and control specifications are not met, it will be necessary to repeat this step. The calibration and standardization ensures that the assay is working correctly prior to measuring the samples. It also allows the assay to be validated. A validated assay is a prerequisite to performing a potency assay, since it is a regulatory requirement. When an assay is validated according to regulatory guidelines, the results can be trusted. Validation is not the same assay verification. The latter occurs when a new, perhaps more advanced assay, is tested against an established assay. If the results from the two assays correlate, the newer assay is said to be "verified" against the established assay [12, 13]. Assay validation is a more complicated procedure that involves many experiments to document the parameters of accuracy, sensitivity, selectivity or specificity, precision (reliability and reproducibility), and robustness. Assay validation can only be performed if standards and controls are available. An assay without standards and controls cannot be considered standardized and therefore cannot be validated.

13. A log-log linear regression is calculated from the equation: $\log Y = A + B \times \log X$, where A is the intercept and B is the slope of the linear regression and uses a least-squares method to fit the points.
14. There is a direct correlation between the stem cell potency ratio and "quality." This means that both parameters are needed to allow release of a product for use in the patient. Neither potency ratio alone nor "quality" alone is sufficient. Together, these parameters provide the information that will predict whether the cells have both the capacity and ability to engraft in the patient. This is the engraftment potential and is the response that will correlate with potency.

References

1. U.S. Food and Drug Administration (FDA) (2011) Guidance for industry. Potency tests for cellular and gene therapy products. http://www.fda.gov/downloads/BiologicsBloodVaccines/GuidanceComplianceRegulatoryInformation/Guidances/CellularandGeneTherapy/UCM243392.pdf
2. European Medicines Agency (EMA) (2008) Guideline on potency testing of cell based immunotherapy medicinal products for the treatment of cancer. http://www.tga.gov.au/pdf/euguide/bwp27147506en.pdf
3. U.S. Food and Drug Administration (FDA) (2009) Guidance for industry. Minimally manipulated, unrelated allogeneic placental/umbilical cord blood intended for hematopoietic reconstitution for specified indications. http://www.fda.gov/downloads/BiologicsBloodVaccines/GuidanceComplianceRegulatoryInformation/Guidances/Blood/UCM187144.pdf
4. Rich IN (2013) Potency, proliferation and engraftment potential of stem cell therapeutics: the relationship between potency and clinical outcome for hematopoietic stem cell products. J Cell Sci Ther S13:001. doi:10.4172/2157-7013.S13-001
5. Page KM, Zhang L, Mendizabal A et al (2011) Total colony-forming units are a strong, independent predictor or neutrophil and platelet engraftment after unrelated cord blood transplantation: a single center analysis of 435 cord blood transplants. Biol Blood Marrow Transplant 17:1362–1374
6. Hall KM, Rich IN (2011) Hematopoietic stem cell potency for cellular therapeutic transplantation. In: Pelayo R (ed) Advances in hematopoietic stem cell research. InTech, Croatia, pp 384–406
7. Rich IN (2013) The difference between stem cell viability and potency: a short guide for parents and patients. Parent's guide to cord blood foundation. http://parentsguidecordblood.org/newsletter_archive/newsletters_2013-10.php#potency
8. Page KM, Zhang L, Mendizabal A (2012) The cord blood Apgar: a novel scoring system to optimize selection of banked cord blood grafts for transplantation (CME). Transfusion 52:272–283
9. Bradley TR, Metcalf D (1966) The growth of mouse bone marrow cells in vitro. Aust J Exp Biol Med Sci 44:287–299
10. Pluznik DH, Sachs J (1966) The induction of clones of normal mast cells by a substance from conditioned medium. Exp Cell Res 43:553–563
11. Iscove NN, Sieber F, Winterhalter KH (1974) Erythroid colony formation in cultures of mouse and human bone marrow: analysis of the requirement for erythropoietin by gel filtration and affinity chromatography on agarose-concanavalin A. J Cell Physiol 83:309–320
12. Rich IN (2003) In vitro hematotoxicity testing in drug development: a review of past and future applications. Curr Opin Drug Discov Dev 6:100–109
13. Rich IN (2007) High-throughput in vitro hemotoxicity testing and in vitro cross-platform comparative toxicity. Expert Opin Drug Metab Toxicol 3:295–307

Chapter 5

Culturing Protocols for Human Multipotent Adult Stem Cells

Bart Vaes, Sara Walbers, Kristel Gijbels, David Craeye, Robert Deans, Jef Pinxteren, and Wouter van't Hof

Abstract

Culture procedures are presented that support the initiation and controlled expansion of the multipotent adult progenitor cell (MAPC) population within the human bone marrow derived multipotent mesenchymal stromal cell compartment. Culture procedures or conditions and characterization assays that maintain and survey the distinctive primitive MAPC properties are discussed in the context of cell culturing platforms that facilitate controlled expansion of clinical grade human MAPC product to levels required for mid to late stage clinical testing.

Key words Mesenchymal stromal cells, Multipotent adult progenitor cells, Immune modulation, Cell therapy, Manufacture, Clinical trials

1 Introduction

Over the past years, various types of bone marrow (BM)-derived multipotent mesenchymal stromal cells (MSC) have been isolated and characterized. This includes the multipotent adult progenitor cell (MAPC) population which appears to be more primitive than classically cultured MSC [1]. When expanded under the culture conditions specified in this chapter, human MAPC display enhanced proliferation and regenerative properties, while displaying strong immunomodulatory properties characteristic of cells in the MSC compartment [2, 3]. The unique MAPC properties have triggered great interest in their utility as a cell-based therapy platform for modulation of inflammatory conditions, immune dysregulation and tissue repair [4, 5]. Protocols for culturing rodent MAPC have been provided previously [6, 7]. Here we discuss culture procedures for human MAPC, amenable to the manufacture of human cGMP MAPC technology-based cell therapy products for use in clinical trials, with emphasis on culture platforms that accommodate controlled scale-up, standardization, and automation.

Ivan N. Rich (ed.), *Stem Cell Protocols*, Methods in Molecular Biology, vol. 1235,
DOI 10.1007/978-1-4939-1785-3_5, © Springer Science+Business Media New York 2015

2 Materials

2.1 Culture Vessels

1. 0.20 μm Filtropur filter (Sarstedt).
2. Sterile syringe (BD Biosciences).
3. 15 mL polypropylene centrifuge tube (Corning).
4. 50 mL polypropylene centrifuge tube (Corning).
5. 70 μm cell strainer (BD Biosciences).
6. 500 mL Vacuum Filter/Storage Bottle System, 0.22 μm Pore 33.2 cm^2 PES Membrane (Corning).
7. 150 mL Tube Top Vacuum Filter System, 0.22 μm Pore 13.6 cm^2 CA Membrane (Corning).
8. 75 cm^2 Rectangular Canted Neck Cell Culture Flask (T75, Corning).
9. Cryovials (1.5 mL system 100™; Fisher Scientific).

2.2 Reagents

1. Dexamethasone (Sigma-Aldrich).
2. Dulbecco's Modified Eagle medium (DMEM; 1 g/L glucose; Life Technologies).
3. Dimethyl sulfoxide (DMSO; Sigma-Aldrich).
4. Fetal Bovine Serum (FBS; Lonza).
5. Fibronectin (FN; 0.1 %; Sigma).
6. Acetic Acid (HAc; VWR Prolabo).
7. Hydrochloric Acid (HCl; VWR Prolabo).
8. Histopaque-1077® (Sigma-Aldrich).
9. Human epidermal growth factor (hEGF), 1 mg (R&D Systems).
10. Human platelet derived growth factor-BB (hPDGF-BB), 2 mg (R&D Systems).
11. Insulin-Transferrin-Selenium (ITS; 500×; Sigma-Aldrich).
12. Linoleic acid-bovine serum albumin (LA-BSA; 0.5×; Sigma-Aldrich).
13. L-Ascorbic acid 2-phosphate sesquimagnesium salt hydrate (Sigma-Aldrich).
14. MCDB-201 (Sigma-Aldrich).
15. NaOH pellets (VWR Prolabo).
16. Phosphate Buffered Saline (PBS; Lonza).
17. Penicillin-Streptomycin (10,000 U/mL; Life Technologies).
18. Trypsin-Ethylenediaminetetraacetic acid (Trypsin-EDTA; Lonza).

3 Methods

3.1 Preparation of Growth Factor Solutions

1. Human PDGF-BB stocks
   (a) A 100 mM HCl solution is prepared by adding 1 mL HCl to 119 mL Milli-Q water. Next a 4 mM HCl solution is made by adding 2 mL of 100 mM HCl to 48 mL Milli-Q water. Filter the solution through a 150 mL Tube Top Vacuum Filter System.
   (b) 2 mg hPDGF-BB is delivered in a solution of 30 % v/v acetonitrile and 0.1 % v/v TFA. This is diluted to 40 mL with 4 mM HCl to achieve a final concentration of 50 μg/mL.
   (c) Prepare 100 μL aliquots and store at −80 °C. Each aliquot should only be thawed once to avoid repeated freeze–thaw cycles.
   (d) Frozen hPDGF-BB aliquots remain stable for up to 12 months at −80 °C.
2. Human EGF stocks
   (a) A 1 M HAc solution is prepared by adding 1 mL HAc to 16.5 mL Milli-Q water. Next a 10 mM HAc solution is made by adding 0.5 mL 1 M HAc to 49.5 mL Milli-Q water. Filter the solution through a 150 mL Tube Top Vacuum filter system.
   (b) Reconstitute 1 mg hEGF with 40 mL 10 mM HAc to obtain a final concentration of 25 μg/mL.
   (c) Prepare 200 μL aliquots and store at −80 °C. Each aliquot should only be thawed once to avoid repeated freeze–thaw cycles.
   (d) Frozen hEGF aliquots remain stable for up to 12 months at −80 °C.

3.2 Preparation of Ascorbic Acid

(a) To make a stock solution of 10^{-4} M L-ascorbic acid, dissolve 1.45 g L-Ascorbic acid in 500 mL PBS. Filter the solution through a 500 mL Vacuum Filter/Storage Bottle System.

(b) Prepare 5 mL aliquots and store at −20 °C. Each aliquot should only be thawed once to avoid repeated freeze–thaw cycles.

(c) Frozen L-ascorbic acid aliquots remain stable for up to 3 months in the dark at −80 °C.

3.3 Preparation of Dexamethasone

(a) Prepare a 0.25 mM stock solution by dissolving 0.981 mg dexamethasone per 10 mL Milli-Q water.

(b) Calculate the exact amount of dexamethasone needed. The supplied powder contains on average 65 mg dexamethasone per gram powder. The accurate amount is provided for each batch.

(c) Example for 6.8 mg per 100 mg powder: dissolve 70 mg total powder (4.76 mg dexamethasone) in 48.52 mL Milli-Q water.

(d) Filter through a 150 mL Tube Top Vacuum Filter System.

(e) Prepare 100 μL aliquots and store at −80 °C. Each aliquot should only be thawed once to avoid repeated freeze–thaw cycles.

(f) Frozen dexamethasone aliquots remain stable for up to 12 months at −80 °C.

3.4 Preparation of MAPC Expansion Medium

1. MCDB-201 is prepared by dissolving 17.7 g MCDB-201 in 1 L Milli-Q water.
2. Adjust the solution to pH 7.2 using NaOH pellets, and sterile filter using a 500 mL Vacuum Filter/Storage Bottle System. The solution can be stored for up to 4 weeks at 4 °C.
3. Dulbecco's Modified Eagle Medium is mixed with MCDB-201 solution at a 60:40 vol/vol ratio.
4. 500 mL base medium is supplemented with the following reagents;
   (a) 1 mL of 500× Insulin-Transferrin-Selenium.
   (b) 2.5 mL of 100× Linoleic Acid-Bovine Serum Albumin.
   (c) 5 mL of 10,000 U/mL Penicillin-Streptomycin.
   (d) 5 mL of 10^{-4} M L-Ascorbic Acid.
   (e) 100 μL of 50 μg/mL hPDGF.
   (f) 200 μL of 25 μg/mL hEGF.
   (g) 100 μL 0.25 mM dexamethasone.
   (h) 10 mL (2 %) Fetal Bovine Serum (*see* **Note 1**).
5. Complete MAPC expansion medium is then filtered through a 500 mL Vacuum Filter/Storage Bottle System and can be used for up to 1 month when stored at 4 °C.

3.5 Preparation of Fibronectin (FN) Coating Solution

1. Prepare 1× FN (100 ng/mL) coating solution by diluting 50 μL of 0.1 % FN into 500 mL of PBS. Store solution at 4 °C.
2. Coat T75 culture flasks for at least 30 min in a 37 °C, 5.5 % CO_2 incubator with 5 mL coating solution.

3.6 Preparation of Freezing Medium

A 2× stock of freezing medium (20 % DMSO, 20 % FBS in expansion medium) is prepared as described below. Required volumes of freezing solution are routinely prepared freshly for each freezing procedure. The 2× stock may be used for up to a week when stored at 4 °C.

1. Determine the required volume of 2× stock needed for the cells.
2. Mix 20 % FBS, 20 % DMSO and 60 % MAPC expansion medium (Subheading 3.4).

3. Filter the 2× stock of freezing medium through a 0.20 μm Filtropur filter and precool on ice.
4. Concentrated DMSO solutions will solidify on ice and freezing medium should only be chilled after DMSO dilution.
5. For freezing cells, mix equal volumes of cell suspension with 2× freezing medium to achieve the required cell concentrations in a final freezing solution containing 10 % DMSO and 10 % FBS.

3.7 Thawing of Cryopreserved Human MAPC

1. Remove a cell vial from a cryogenic storage tank (e.g., liquid nitrogen), slightly open the cap to release the pressure.
2. Thaw the cell suspension in a 37 °C water bath until all ice crystals have disappeared. This should take 2–3 min and no more than 5 min for a 1.5 mL cryovial.
3. Transfer the thawed cell suspension to a 15 mL conical tube.
4. Dropwise add 10 mL MAPC expansion medium.
5. Wash the cryovial with 1 mL of medium to ensure that all cells are transferred.
6. Add fresh MAPC expansion medium to fill up the 15 mL conical tube.
7. Centrifuge for 5 min, 350 ×*g* at RT.
8. Remove the supernatant, resuspend the cell pellet in medium by gently pipetting.
9. Count cells using a hemacytometer or nucleocounter.

3.8 MAPC Isolation from Bone Marrow Aspirate

1. Fresh marrow will provide optimal results but successful isolations have also been performed using aspirates that were shipped overnight.
2. Transfer the bone marrow aspirate into a 50 mL conical centrifuge tube and add an equal volume of PBS.
3. Carefully layer 20 mL of the diluted bone marrow on top of 20 mL Histopaque-1077®, use multiple tubes if needed.
4. Centrifuge at 1,000 ×*g* for 20 min at RT with the centrifuge break set to 0.
5. Collect the mononuclear layer and dilute to 50 mL with PBS in a fresh 50 mL conical tube.
6. Centrifuge at 350 ×*g* for 5 min at RT.
7. Remove supernatant and resuspend cells in 50 mL PBS.
8. If cell clumps are observed, the cell suspension may be filtered over a 70 μm cell strainer.
9. Centrifuge at 350 ×*g* for 5 min at RT.

10. Remove the supernatant and resuspend cells in a small volume of PBS, e.g., 1 or 2 mL and count cells.
11. For MAPC isolations, the cells are seeded at a density of $0.5–1.0 \times 10^6$ cells/cm^2 in 15 mL pre-warmed medium on 1× FN (100 ng/mL) coated T75 flasks.
12. Place the cells in an incubator under hypoxic conditions (5.5 % CO_2, 5 % O_2) at 37 °C (*see* **Note 2**).
13. Remove the medium 24 h after seeding and rinse each T75 flask three times with 5 mL PBS to remove non adherent cells.
14. For expansion, 10 mL fresh medium is added and cells are cultured for 5–8 days depending on growth. Replace the culture medium every 2–3 days.
15. Cell growth and expansion
    (a) Cells undergo clonal expansion and will become visible as distinct cell clusters.
    (b) When clonal expansion clusters reach a confluency of 50–70 % (within the clusters) cells need to be passaged.

3.9 MAPC Culture from Thaw

1. Pre-warm complete MAPC culture medium in a 37 °C water bath.
2. Add 10 mL medium to a 1× FN pre-coated T75 flask.
3. Seed the cells at a density of 500 cells per cm^2 (*see* **Note 3**).
4. Incubate the cells in a gas-controlled incubator at 37 °C, 5.5 % CO_2 and 5 % O_2 (*see* **Note 2**).

3.10 MAPC Subculturing and Expansion (See Note 3, Cell Density)

1. Collect and save conditioned growth medium from 2–3-day-old MAPC cultures in a sterile recipient.
2. Rinse each T75 flask with 3 mL PBS and remove PBS.
3. Add 2 mL of 0.05 % Trypsin-EDTA (*see* **Note 4**). Incubate for 2–5 min at 37 °C until the cells have detached. If necessary, gently tap the flask to detach all cells.
4. Stop the reaction by adding 3 mL of the collected expansion medium from **step 1** to each T75 flask.
5. Collect and transfer cells to a conical tube and centrifuge for 5 min, at 350 × *g*, RT.
6. Remove supernatant, resuspend cells in MAPC expansion medium and count cells.
7. Seed at a density of 500 cells per cm^2 in 10 mL pre-warmed medium in a 1× FN coated flask and place in the incubator.
8. Passage MAPC every other day in order to maintain cultures at low density (*see* **Note 3**).
9. The type of culture vessel used will change depending on the clinical development stage of the cell product as it drives the number of cells required per expansion (*see* **Note 5**).

3.11 Cell Characterization

MAPC cultures are characterized using a variety of assays, discussed under **Note 6**.

3.12 MAPC Cryopreservation Procedure

1. Resuspend the appropriate amount of cells (typically 1×10^6) in 750 μL of expansion medium and transfer to a cryovial.
2. Place the cells on ice to cool.
3. Add 750 μL of prechilled freezing medium (Subheading 3.6) in one quick movement and gently flick the vial.
4. Place the cryovial in a Nalgene container filled with 100 % isopropanol.
5. Place at −80 °C to freeze cells at a rate of 1 °C per minute (*see* **Note 7**).
6. After a minimum of 6 h the vials can be stored in a cryopreservation tank.

4 Notes

1. *Serum supplementation.* MAPC culture protocols were developed using fetal bovine serum. It is well known that serum as a culture medium supplement can provide a significant source of variability. Optimal serum concentrations may vary depending on serum batch characteristics and different cell types may require different serum concentrations to accommodate preservation of their unique properties. Since MAPC cultures display a fast doubling time of less than 24 h, proliferation rate is a rapid method to screen different serum lots for their capacity to support MAPC expansion, followed by additional cell characterization (*see* **Note 6**). Once an appropriate lot has been identified a large quantity of serum from that batch should be reserved to provide as much consistency as possible. For manufacturing purposes, the amount of serum purchased or allocated should permit multiple series of manufacturing runs. A main practical reason to avoid serum for human cell therapy culturing is the fact that the world serum supply may become insufficient for clinical production of an off-the-shelf allogeneic product. In addition to practical incentives, there is significant need for development of serum-free culturing alternatives for scientific and safety reasons. Currently it is largely unknown which (growth) factors present in animal serum influence particular properties and/or behavior of human cells, especially when intended for clinical use. It is essential that a clinical human cell product displays high consistency and an important step towards consistent production would be the use of serum-free defined medium.

2. *Oxygen levels.* MAPC populations are cultured under hypoxic conditions (3–5 % O_2), whereas MSC are routinely expanded under normoxic conditions (21 % O_2).
3. *Cell density.* Cell seeding density and the time frame at which cultures reach confluency are critical factors that can drastically alter the outcome of culture protocols. At confluency, cell populations start to undergo changes based on cross talk mechanisms (e.g., WNT or TGF-signaling pathways) that will impact stemness and proliferation characteristics. Classically, MSC are seeded at higher density (e.g., 5,000 cells per cm^2) and passaged at confluency. Under these conditions, MSC culture can be routinely expanded for 2–5 passages. MAPC and MAPC platform-based products are seeded at much lower densities (200–2,000 cells/cm^2) and they are consistently passaged at sub-confluency (30–70 %). Under these conditions, in combination with the culture media and oxygenation levels described here, cell populations can be routinely expanded for up to 15–20 passages (50–70 population doublings).
4. *Trypsin.* Historically, trypsin used for cell expansions has been porcine derived, thus in effect providing a xenobiotic-containing expansion procedure. Xeno-free protocols are under development using a variety of trypsin alternatives, such as Accutase™ or TrypLE™ Select. The optimal choice of trypsin should be determined for each cell culture protocol, as the cell type and culture plastic surface used will impact the efficiency of cell detachment for expansion.
5. *Culture vessel system.* For production of research or preclinical grade cell therapy products, culture protocols based on T75 culture vessels suffice. Ultimately, for production of clinical grade materials to be used in clinical trials, larger scale expansion platforms are required [5]. Available 2-dimensional culture technologies include the use of cell factories, i.e., 10- or 40-tier culture vessels, which can be implemented for generation of consistent clinical product at levels sufficient for early and mid-stage clinical evaluations (10–100 patients). Critical late-stage, Phase 3, clinical studies will require cell numbers that are anticipated to depend on different types of cell expansion platforms providing controlled scale-up. Technologies under evaluation for this include various bioreactor systems and three-dimensional cell culture modules based on permeable nanotube configurations.
6. *MAPC product characterization.* A variety of assays can be carried out for physiochemical or biological characterization of cell populations resulting from specific culturing protocols. Physiochemical parameters refer to size, morphology, confluency, and gene and protein expression. Biological characterization

measures biological functions such as proliferation, differentiation, or migration [8]. The International Society for Cellular Therapy (ISCT) has defined minimal criteria to define mesenchymal stromal cell products [9]. These criteria include the adherence to plastic in standard culture conditions, the phenotypic profile consisting of CD105^{+}, CD73^{+}, CD90^{+}, CD45^{-}, CD34^{-}, CD14^{-} or CD11b^{-}, CD79α^{-} or CD19^{-}, HLA-DR^{-}, and multipotent functionality defined by in vitro differentiation protocols guiding osteoblast, adipocyte, and chondrocyte development. MAPC cultures display certain differences in phenotype and routinely show enhanced proliferation capacity compared to MSC. In addition, in comparison to MSC, MAPC cultures display distinct transcriptome profiles, which are deemed to underlie differences in in vivo activity. For example, MAPC have demonstrated the capacity to induce the formation of functional blood vessels when implanted in a Matrigel plug with VEGF and bFGF under the skin of nude mice, while vessels induced by MSC are not consistently functional in this manner [10]. Thus, specific cell characterization assays are important for development and application of therapeutic stem cells products. For instance, donor/batch variation and the effects of process improvement can be understood more readily using a routinely implemented cell comparability testing panel. Such a panel may include markers (protein, mRNA, miRNA) or profiling assays to establish cell identity, while biologic potency assays are indicative for in vivo activity. At present, immunosuppression and pro-angiogenic activity have received the most attention in terms of providing regenerative capacity of therapeutic stromal stem cells. Immunosuppressive capacity of stromal cells is most frequently evaluated in vitro in mixed lymphocyte reactions [3]. Measurement of pro-angiogenic activity in vitro has included the ability of stromal cells or conditioned media from cell cultures to induce tube formation of human umbilical vein endothelial cells (HUVEC). A main drawback of this type of potency assay for routine use is the need for accessory or target cell types (e.g., human PBMC or HUVEC), which adds assay variability. The development of simplified, surrogate potency assays are needed to overcome these limitations. For MAPC based culture platforms, the expression of the cytokines CXCL5, VEGF, and IL8 has been reported to specifically correlate with their pro-angiogenic capacity [11]. This supports the development of a matrix of ELISA assays as a simple and efficient potency screen for specific MAPC cell therapy activity.

7. *Cryopreservation*. Alternative systems are available, such as CoolCell® alcohol-free containers, or controlled rate freezer protocols per manufacturer instructions.

References

1. Verfaillie CM, Schwartz R, Reyes M et al (2003) Unexpected potential of adult stem cells. Ann N Y Acad Sci 996:231–234
2. Jacobs SA, Pinxteren J, Roobrouck VD et al (2013) Human multipotent adult progenitor cells are nonimmunogenic and exert potent immunomodulatory effects on alloreactive T-cell responses. Cell Transplant 22: 1915–1928
3. Jacobs SA, Roobrouck VD, Verfaillie CM et al (2013) Immunological characteristics of human mesenchymal stem cells and multipotent adult progenitor cells. Immunol Cell Biol 91:32–39
4. Auletta JJ, Cooke KR, Solchaga LA et al (2010) Regenerative stromal cell therapy in allogeneic hematopoietic stem cell transplantation: current impact and future directions. Biol Blood Marrow Transplant 16:891–906
5. Vaes B, Van't Hof W, Deans R (2012) Application of MultiStem((R)) allogeneic cells for immunomodulatory therapy: clinical progress and pre-clinical challenges in prophylaxis for graft versus host disease. Front Immunol 3:345
6. Jiang Y, Vaessen B, Lenvik T et al (2002) Multipotent progenitor cells can be isolated from postnatal murine bone marrow, muscle, and brain. Exp Hematol 30:896–904
7. Subramanian K, Geraerts M, Pauwelyn KW et al (2010) Isolation procedure and characterization of multipotent adult progenitor cells from rat bone marrow. Methods Mol Biol 636:55–78
8. Bravery CA, Carmen J, Fong T et al (2013) Potency assay development for cellular therapy products: an ISCT review of the requirements and experiences in the industry. Cytotherapy 15:9–19
9. Dominici M, Le Blanc K, Mueller I et al (2006) Minimal criteria for defining multipotent mesenchymal stromal cells. The International Society for Cellular Therapy position statement. Cytotherapy 8:315–317
10. Roobrouck VD, Clavel C, Jacobs SA et al (2011) Differentiation potential of human postnatal mesenchymal stem cells, mesoangioblasts, and multipotent adult progenitor cells reflected in their transcriptome and partially influenced by the culture conditions. Stem Cells 29:871–882
11. Lehman N, Cutrone R, Raber A et al (2012) Development of a surrogate angiogenic potency assay for clinical-grade stem cell production. Cytotherapy 14:994–1004

Chapter 6

Isolation of Murine Bone Marrow Scavenging Sinusoidal Endothelial Cells

Peter A.G. McCourt, Ana Oteiza, Benjamin Cao, and Susan K. Nilsson

Abstract

The bone marrow (BM) is permeated with sinusoidal vessels lined with sinusoidal endothelial cells (SEC), which are crucial for BM physiology and hemopoietic stem cell (HSC) renewal. However, little is known about the characteristics or functional capacity of bone marrow sinusoidal endothelial cells (BMSEC). One significant barrier to the study of BMSEC is the lack of specific cell surface markers that can be used to isolate these cells. Nevertheless, BMSEC possess one exceptional property, namely, the ability to scavenge large amounts of soluble waste molecules such as advanced glycation end-products (AGE) and we have utilized this to label BMSEC for cell sorting and isolation. We describe the means to produce and fluorescently label AGE, its use as a functional in vivo marker of BMSEC and the isolation of these cells from murine BM using fluorescent activated cell separation (FACS).

Key words Bone marrow niche, Sinusoidal endothelial cell, Scavenger, Advanced glycation end-products

1 Introduction

Bone marrow (BM) hemopoietic stem cells (HSC) are ultimately responsible for the production of all blood cells and reside within a specialized microenvironment within endosteal BM termed the stem cell "niche" [1–3]. Previously, we demonstrated that endosteal BM HSC (Lin$^-$Sca-1$^+$c-Kit$^+$CD150$^+$CD48$^-$ [4]) home more efficiently and have significantly higher hemopoietic reconstitution potential than their central counterparts [5, 6], suggesting the microenvironment within this region is different to other BM areas. Interfacing between the blood and the niche are BM vessels lined with sinusoidal endothelial cells (SEC) [4, 7], which are crucial for BM ultrastructure, for self-renewal and reconstitution of HSC and their transport from and to the BM extravascular space [7–10]. However, the prospective isolation of BMSEC has proven to be incredibly difficult due to the lack of specific cell surface markers. Recently, BMSEC were shown to scavenge advanced

Ivan N. Rich (ed.), *Stem Cell Protocols*, Methods in Molecular Biology, vol. 1235,
DOI 10.1007/978-1-4939-1785-3_6,

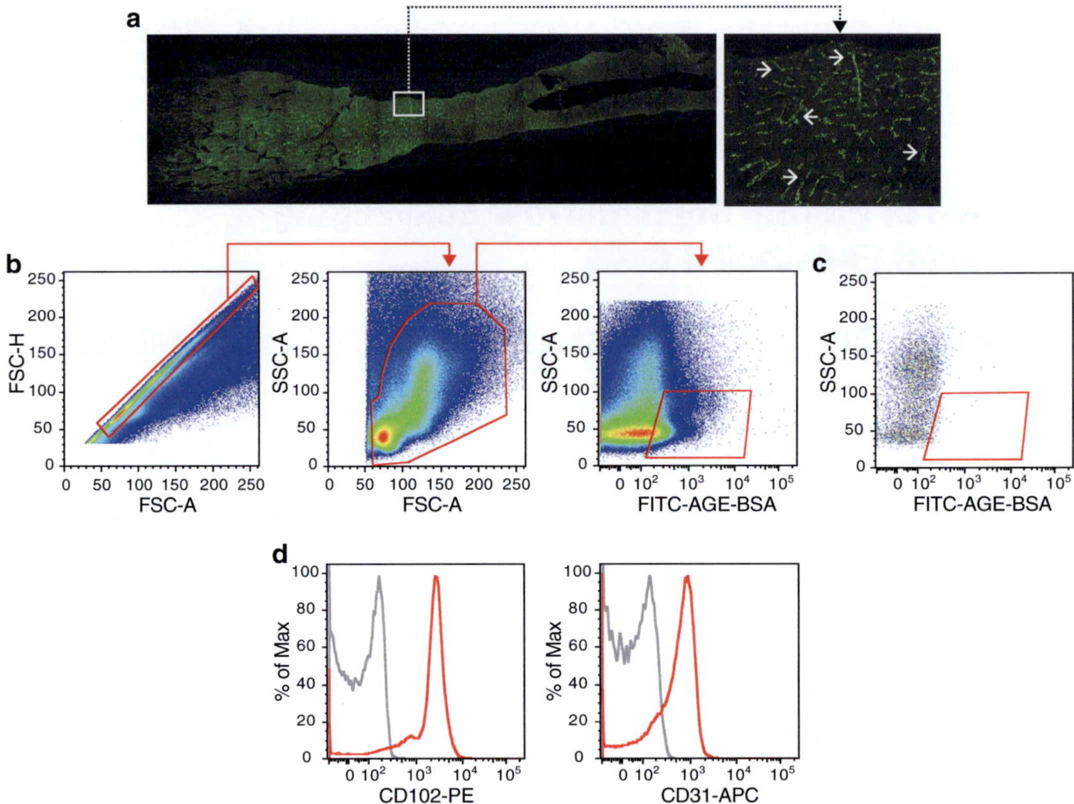

Fig. 1 (**a**) Mouse femoral BM 1 h post FITC-AGE-BSA injection. FITC-AGE-BSA uptake is restricted to BM sinusoidal endothelium (green fluorescence). *Arrows* indicate BM sinusoids. (**b**) Dot plots showing the sequential sorting strategy for FITC-AGE-BSA⁺ BM. From *right* to *left*: Single cells, BM nucleated cells, FITC-AGE-BSA⁺ cells for sorting. (**c**) Unstained BM nucleated cells from a non-injected mouse. (**d**) CD102 and CD31 expression on FITC-AGE-BSA⁺ BM (*Red line*). Isotype control (*Grey line*)

glycation end-products (AGE) with high avidity [11] in a similar manner to liver SEC (LSEC) [12, 13]. Both BMSEC and LSEC are classified as "scavenging endothelial cells," maintaining vertebrate circulatory homeostasis via the rapid removal of vast amounts of soluble (i.e., non-particulate) waste molecules, such as AGE and oxidized LDL, via clathrin mediated endocytosis [14]. By attaching fluorescent tags to these waste molecules, one can specifically label the sinusoidal endothelium in BM and liver via a specific function, without any uptake in other endothelial structures (e.g., arterial vessels in BM) or extravascular cells/structures [11, 14].

Herein we describe the production [13] and use of FITC-AGE-BSA to specifically label and prospectively isolate murine endosteal BMSEC. Injected FITC-AGE-BSA is stable in vivo and highly specific for BMSEC [11] (Fig. 1a). We demonstrate that this in vivo labelling provides an easy method for the isolation of murine BMSEC using fluorescent activated cell separation (FACS).

2 Materials

2.1 Production of AGE-BSA

1. 25 % ammonia solution to wash all glassware, stirrer bars, and lids/stoppers and allow drying in dust-free conditions.
2. Bovine serum albumin, essentially fatty acid free, 99 % pure (BSA).
3. 150 mM D-glucose made by adding 0.675 g D-glucose, ultrapure to reverse osmosis (RO) water to a final volume of 25 mL.
4. 2 M NaOH: 3.2 g sodium hydroxide, ultrapure (NaOH) added to (RO) water to a final volume of 40 mL.
5. 1 M Na_2HPO_4: 7.098 g disodium hydrogen phosphate, ACS grade (Na_2HPO_4) added to RO water to a final volume of 50 mL (*see* **Note 1**).
6. 100 mL Buchner (side-arm) flask with oversize rubber stopper (*see* **Note 2**).
7. Schott Duran borosilicate 100 mL glass bottle with screw cap.
8. Teflon coated stirrer bar (2 cm).
9. Magnetic stirrer plate.
10. Sterile filters (0.22/0.45 μm) (*see* **Note 3**).
11. Laminar flow hood for completing **step 1** (*see* **Note 4**).
12. Nitrogen gas.
13. Inflatable glove chamber (*see* **Note 5**).
14. Incubator capable of heating to 50 °C and housing a magnetic stirrer plate.
15. Regenerated cellulose dialysis membrane with MWCO 25,000 Da.
16. Phosphate buffered saline pH 7.4, 280 mOsm (PBS).
17. Latex condom.

2.2 FITC Labelling of AGE-BSA

1. Fluorescein isothiocyanate isomer 1 (FITC, >90 % pure).
2. 1.0 M Na_2CO_3: 21.2 g sodium carbonate (anhydrous) added to RO water to a final volume of 200 mL.
3. 1.0 M $NaHCO_3$: 16.8 g sodium bicarbonate added to RO water to a final volume of 200 mL.
4. 1.0 M Na_2CO_3/$NaHCO_3$ pH 9.5 buffer: 30 mL 1.0 M Na_2CO_3 added to 70 mL 1.0 M $NaHCO_3$. Filter through a 0.22 μm filter.
5. Dialysis cassette, MWCO 10,000 Da (*see* **Note 6**).

2.3 Characterisation of AGE-BSA and FITC-AGE-BSA

1. 1 mg/ml stock solutions of BSA, AGE-BSA and FITC-AGE-BSA.
2. PBS.

3. 10 mm Cuvette cell suitable for UV-VIS and fluorescence emission spectroscopy.
4. UV-VIS-NIR spectrophotometer (we use a Varian Cary 500).
5. Luminescence Spectrophotometer (we use a Perkin Elmer LS50B).

2.4 FITC-AGE-BSA Intravenous Injection

1. Adult C57BL/6J (Ly5.2) mice, 6–8 week old.
2. 1 mL syringe attached to 27.5-gauge needle.
3. Heat lamp.
4. 70 % Ethanol made in distilled water.
5. Facial tissue wipes.
6. Sodium chloride (NaCl) (0.9 %) for injection.
7. FITC-AGE-BSA diluted to 375 mg/ml in saline for injection.
8. Apparatus to immobilize mouse during injection.

2.5 Sampling of Endosteal Bone Marrow

1. Sterile #11 surgical blade and #3 handle.
2. 50 mL polypropylene conical centrifuge tubes.
3. Phosphate buffered saline (PBS): 310 mOsm, pH 7.2 (*see* **Note 7**).
4. PBS, 310 mOsm, pH 7.2, supplemented with 2 % bovine calf serum (PBS-2 %Se).
5. 1 mL syringes attached to 23-gauge and 21-gauge needles to flush marrow from bones.
6. Sterile porcelain mortar and pestle.
7. 40 μm nylon cell strainer.
8. Digestive enzymes: 3 mg/mL Collagenase I/4 mg/mL Dispase II in PBS. Enzymatic mixture needs to be made fresh every day of use. (*Clostridium Histolycum* Collagenase I, Worthington Biochemical and *Bacillus Polymyxa* neutral protease, grade II Dispase II, Roche Applied Science.)
9. 37 °C orbital shaker (Eppendorf Thermomixer comfort model).
10. Hemocytometer and microscope equipped with phase contrast or an automated cell counter (Sysmex model KX-21N).

2.6 BMSEC Pre-enrichment by Immunomagnetic Cell Separation

2.6.1 Immuno-labelling Cells with a Cocktail of Lineage Antibodies

1. Lineage depletion antibody cocktail: a mixture of purified rat anti-mouse antibodies recognizing the following cell surface antigens: GR-1, MAC-1, and TER119 diluted in PBS-2 %Se. Antibody concentrations are all ≤1 μg/mL.

2.6.2 Immunomagnetic Cell Separation

1. Magnetic beads for cell labelling (Dynal Dynabeads: 4.5 μm diameter sheep anti-rat IgG Dynabeads, 4×10^8 beads/mL).
2. Magnetic bead working buffer: PBS, 310 mOsm supplemented with 2 mM EDTA and 0.1 % (w/v) fraction V bovine serum albumen (BSA; pH 7.4).
3. Magnet (Dynal MPC-L for 1–8 mL samples).
4. Tube rotator or similar suspension mixer: allowing both tilting and rotation at 4 °C for Dynabeads incubation step (MACSmix Tube Rotator in a fridge).
5. 1.5 mL microtubes.
6. Polypropylene 5 and 14 mL round bottom tubes.

2.7 BMSEC Fluorescence Activated Cell Separation

1. An optimally titrated mixture of the following antibodies: Allophycocyanin (APC) conjugated rat anti-mouse CD31 (MEC13.3 clone) and Phycoerythrin (PE) conjugated rat anti-mouse CD102 (ICAM-2, 3C4 clone) in PBS-2 %Se (*see* **Note 8**).
2. 5 mL polystyrene round bottom tube with cell strainer (40 μm) caps. The cap is swapped to a 5 mL polypropylene tube for filtering cells.
3. Flow cytometer with sorting capability (Cytopeia Influx 516SH cell sorter equipped with five solid state lasers (355, 405, 488, 561, and 633 nm). Band pass filter settings for the detection of fluorescence for FITC, APC, and PE are 528 ± 19, 660 ± 10, and 605 ± 20 respectively). BMSEC are sorted using a 70 μm nozzle at 30 psi, drop delay frequency of 61 kHz. Sorting speed 25,000 cells/s.
4. Collection tubes post-sorting: 5 mL polypropylene tubes or 1.5 mL microtubes.

3 Methods

3.1 Production of AGE-BSA

1. Dispense 50 mL ultrapure RO water (water vacuum) in the Buchner flask containing the stirrer bar. Degas (water vacuum) for at least 30 min while stirring.
2. Inflate the glove chamber bag with nitrogen and place a magnetic stirrer inside. This should be placed nearby the laminar flow hood. Maintain gentle N_2 flow to purge all oxygen.
3. Add D-glucose (0.675 g) and BSA (2.5 g) to a 50 ml propylene tube in the laminar flow hood.
4. When degassed, add 25 ml of the ultrapure RO water from **step 1** to the 50 ml tube to dissolve the D-glucose/BSA, close the lid and mix the solution by gently tapping and slowly inverting the tube up and down. Final concentration should be 150 mM D-glucose/100 mg/ml BSA (*see* **Note 9**).

5. Move the mixture promptly to the glove chamber bag and remove the lid. Leave it under nitrogen atmosphere to purge oxygen.
6. Prepare the incubation buffer by adding 1M Na2HPO4 (50 ml) and 2 M NaOH (26 ml) to the Buchner flask containing the stirrer bar in the laminar flow hood. pH should be 12.43-12.29.
7. Degas the incubation buffer for at least 30 mins while stirring.
8. Transfer the buffer to the glove chamber bag.
9. Start the 50 °C incubator with gentle nitrogen flow.
10. Filter the D-glucose/BSA solution and the buffer through a 0.22 mm filter into polypropylene tubes under nitrogen atmosphere.
11. Mix the buffer with the D-glucose/BSA solution (1:1) into a polypropylene tube. Final concentration 50 mg/ml BSA, 75 mM glucose and pH 11.
12. Keep the tube with the lid loose under gentle nitrogen flow in the glove chamber.
13. After 30 min put the lid on firmly.
14. Open the latex condom in the glove chamber, and place the tube inside. Seal the tube with the condom.
15. Place the tube in the pre-warmed 50 °C incubator. Incubate for 2 weeks under nitrogen atmosphere.
16. After 2 weeks the AGE-BSA solution should be brown. Remove to a laminar flow hood and cool to room temperature.
17. Decant and filter the solution through a 0.22 μm filter into polypropylene tubes, transfer solution to pre-wetted dialysis tubing and dialyze against 15,000 volumes of PBS at 4 °C.
18. After dialysis, filter the solution through 0.22 μm filter and aliquot desired volumes into polypropylene tubes (*see* **Note 10**). Determine the final protein concentration.
19. At this stage the AGE-BSA can be frozen at −70 °C (*see* **Note 11**).

3.2 FITC Labelling of AGE-BSA

1. FITC-AGE-BSA is prepared by incubating FITC and AGE-BSA at a 1:5 dye–protein ratio in 0.1 M $Na_2CO_3/NaHCO_3$ pH 9.5 buffer at room temperature overnight, light protected. Typically after **step 11** of Subheading 3.1 there is approximately 20 mg/mL AGE-BSA, so a 3 mL incubation mix would consist of FITC (10.5 mg) dissolved 1 M Na_2CO_3/ $NaHCO_3$ pH 9.5 buffer (0.3 mL) plus AGE-BSA (54 mg in 2.7 mL PBS).

2. After the overnight incubation, dialyze (10,000 MWCO) the reaction mix against 35,000 volumes of PBS at 4 °C (*see* **Note 12**) to remove unbound FITC.
3. Carefully decant and measure the volume of the dialyzed FITC-AGE-BSA. With this data calculate the protein concentration based on what was added in **step 1** (*see* **Note 13**). Alternatively a new protein assay can be performed.
4. Sterile filter the dialyzed FITC-AGE-BSA using a 0.22 μm filter into sterile polypropylene tubes and store at <–70 °C until use.

3.3 Characterisation of AGE-BSA and FITC-AGE-BSA

3.3.1 UV-VIS absorption Spectroscopy

1. UV-VIS absorption spectra of BSA, AGE-BSA and FITC-AGE-BSA is obtained by diluting 1 mg/ml stock solutions to appropriate concentrations (0.1–0.5 mg/ml) in PBS and measured on a Varian Cary 500 UV-VIS-NIR Spectrophotometer.
2. Prior to acquiring measurements, obtain a baseline correction by running a blank sample consisting of PBS.
3. Transfer 2 ml of BSA, AGE-BSA and FITC-AGE-BSA solution into 10 mm Cuvette cells.
4. Acquire measurements between 250–800 nm.
5. Absorption peaks at ~280 nm are observed for BSA (strong), AGE-BSA (medium) and FITC-AGE-BSA (low) due to absorption by aromatic amino acids. Absorption peaks at ~320 nm are observed for AGE-BSA (medium) and FITC-AGE-BSA (low) and are absent in BSA. An absorption peak at ~490 nm is only evident for FITC-AGE-BSA (strong). Representative absorption spectra are depicted in Figure 2a.

3.3.2 Fluorescence Emission Spectroscopy

1. Fluorescence emission spectra of BSA, AGE-BSA and FITC-AGE-BSA are obtained by adding a few drops of the solutions used in section 3.3.1 to a 10 mm cuvette containing 2

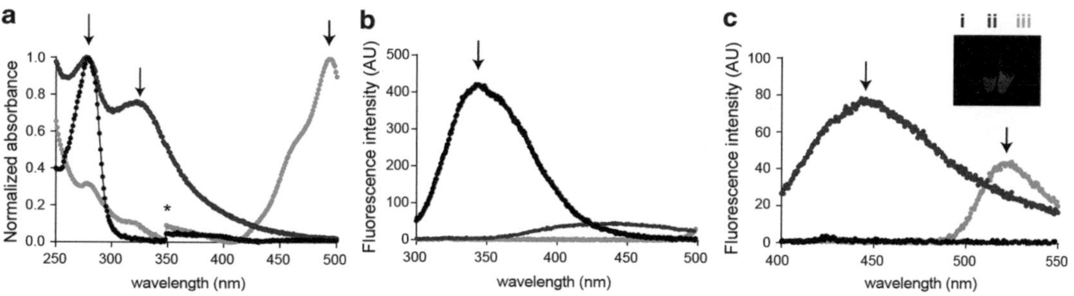

Fig. 2 (**a**) UV/VIS absorption spectroscopy of BSA (*black*), AGE-BSA (*blue*) and FITC-AGE-BSA (*green*). *lamp source changeover at 350 nm. Fluorescence emission spectra of BSA, AGE-BSA and FITC-AGE-BSA with an excitation wavelength of (**b**) 280 nm and (**c**) 370 nm. Inset: Image of 1 mg/ml solution of i. BSA, ii. AGE-BSA and iii. FITC-AGE-BSA under UV excitation (355 nm). Arrows indicate characteristic absorption/emission peaks as described in section 3.3.1 and 3.3.2

ml of PBS measured on a Perkin-Elmer LS50 Luminescence Spectrophotometer.

2. Acquire measurements between 250 and 500 nm with an excitation wavelength of 280 nm. Only BSA emits at 340 nm, which is indicative of fluorescence emission by aromatic amino acid residues. A representative emission spectrum is depicted in Figure 2b.
3. Acquire measurements between 400–600 nm with an excitation wavelength of 370 nm. BSA does not emit between 400–600 nm. A characteristic, broad emission peak at ~440 nm is observed for AGE-BSA, which is absent in both BSA and FITC-AGE-BSA. A peak at ~520 nm observed for FITC-AGE-BSA is attributed to the presence of FITC. A representative emission spectrum is depicted in Figure 2c.

3.4 FITC-AGE-BSA Intravenous Injection

1. Place mice under a heating lamp to dilate the tail vein.
2. Fill a 1 mL syringe with well-mixed FITC-AGE-BSA solution.
3. Weigh each mouse, wipe tail with 70 % ethanol, and place in the immobilization apparatus.
4. Inject FITC-AGE-BSA i.v. at 3.75 mg/kg by giving 200 μL, per 20 g of mouse into the lateral tail vein.
5. Allow the FITC-AGE-BSA solution to circulate for 60 min (*see* **Note 14**).

3.5 Sampling Endosteal Bone Marrow

1. Euthanize mice by cervical dislocation and dissect iliac crests, femurs, and tibiae.
2. Remove the epiphyseal and methaphyseal region of femurs and tibiae using the scalpel blade. Store these bone fragments in 10 mL PBS-2 %Se in a 50 mL Falcon tube (*see* **Note 15**).
3. Flush central marrow of the bones harvested (**step 1**) using the 23-gauge needle for the iliac crests and the 21-gauge needle for the femurs and tibiae, then store the flushed bones in same tube as above (**step 2**) (*see* **Note 16**).
4. Decant flushed bones and bone fragments into a sterile mortar.
5. Grind bones with the pestle until the marrow cavity is open to expose it to enzymatic digestion. Be careful not to pulverize the bones.
6. Thoroughly mix cell and crushed bone suspension by pipetting the supernatant up and down, then remove the cell supernatant and filter through a 40 μm nylon cell strainer into a 50 mL conical tube.
7. Rinse with PBS-2 %Se and if necessary crush any remaining intact bones.

8. Collect and filter the supernatant as indicated in **step 6** and fill tube to 50 mL with PBS-2 %Se and set aside on ice until **step 13**.
9. Transfer the crushed bone fragments into a new 50 mL conical tube containing the collagenase I/dispase II enzymatic suspension (1 mL enzyme per bone fragments of one mouse) and shake at 37 °C in an orbital shaker, 750 rpm for 5 min.
10. Add 20 mL PBS to the digested bone fragments and shake vigorously for 20 s.
11. Filter the cell suspension through a 40 μm nylon cell strainer into another 50 mL tube.
12. Repeat **steps 10** and **11** and filter cells into the same 50 mL conical tube. Top up the tube to 50 mL with PBS-2 %Se.
13. Centrifuge the cell suspension tubes (from **steps 8** to **12**) at 400 × *g* for 5 min at 4 °C.
14. Discard supernatant, resuspend and pool cell pellets from both tubes in 10 mL PBS-2 %Se. Perform a cell count and store cells on ice for BMSEC pre-enrichment by immunomagnetic separation.

3.6 BMSEC Pre-enrichment by Immunomagnetic Cell Separation

3.6.1 Immuno-labelling Cells with a Cocktail of Lineage Antibodies

1. Centrifuge cells at 400 × *g* for 5 min at 4 °C.
2. Stain cells at 1×10^8 cells/mL in the cocktail of lineage markers on ice for 20 min.
3. Wash cells by filling the 50 mL tube with PBS-2 %Se, remove an aliquot to perform a cell count, then centrifuging at 400 × *g* for 5 min at 4 °C to remove unbound antibodies.

3.6.2 Immunomagnetic Cell Separation

1. Resuspend cells at 1×10^8 cells/mL in PBS-2 mM EDTA-0.1 % BSA and transfer into 5 mL sterile polypropylene tube. Set aside on ice until **step 8**.
2. Resuspend Dynabeads.
3. Calculate the volume of Dynabeads needed based on the cell number. The optimal Dynabeads–cell ratio used in this protocol has been established as 0.5:1 repeated in two steps.
4. Dispense the volume of Dynabeads suspension required for both steps into a 5 mL polypropylene tube.
5. Remove azide in the Dynabeads by adding 1 mL PBS-2 mM EDTA-0.1 % BSA and mixing well. Place the tube in the magnet for 1 min, remove and discard supernatant.
6. Repeat **step 5**.
7. Resuspend aliquot of the washed Dynabeads in 500 μL of PBS-2 mM EDTA-0.1 % BSA.
8. Add the 250 μL of Dynabeads to the cells and mix well.

9. Incubate for 5 min at 4 °C with gentle tilting and rotation.
10. Place the mixture on the magnet for 2 min.
11. Without removing the tube from the magnet, transfer the supernatant containing the unbound cells to the tube containing the remaining aliquot of Dynabeads.
12. In order to collect any residual unbound cells, rinse the bead-bound cells with 1 mL PBS-2 mM EDTA-0.1 % BSA and place in magnet for 1 min.
13. Transfer the supernatant to the tube from **step 11**.
14. Incubate cells–bead mix for 10 min at 4 °C with gentle tilting and rotation.
15. Repeat **step 10**.
16. Transfer the supernatant to a new 14 mL polypropylene tube.
17. Make up the volume of the negative cell suspension (unbound cells) to 10 mL with PBS-2 %Se and perform a cell count.

3.7 BMSEC Fluorescence Activated Cell Separation

3.7.1 BMSEC Labelling

1. Pellet cells by centrifuging at 400 × *g* for 5 min at 4 °C.
2. Stain cells at 1×10^8 cells/mL in an optimally pre-titered antibody cocktail of rat anti-mouse CD31-APC and CD102-PE and incubate light protected on ice for 20 min.
3. Wash cells afterwards in 3 mL PBS-2 %Se by centrifuging at 400 × *g* for 5 min at 4 °C to remove unbound antibody. Discard the supernatant
4. Resuspend cells at 30–40 × 10^6 cells/mL in PBS-2 %Se and filter the cell suspension through a cell strainer into a new 5 mL polypropylene tube prior to fluorescence activated-cell sorting (*see* **Note 17**). Place on ice until sorted.

3.7.2 BMSEC Sorting

1. To set up the BMSEC sorting by flow cytometry, the following samples are required (*see* **Note 18**).
   (a) 0.5–1 × 10^6 unstained cells to set voltage for Forward Scatter, Side Scatter, FITC, APC, and PE.
   (b) Individual tubes containing 0.5–1 × 10^6 cells stained with FITC, APC, and PE for compensation controls (*see* **Note 19**).
2. Run the cell samples stained with the cocktail of endothelial cell antibodies and sequentially gate through FSC-H versus FSC-A, SSC-A versus FSC-A, SSC-A versus FITC. FITC$^+$ cells can be selected and sorted through APC and PE histograms (Fig. 1b–d).
3. Sort cells at predetermined optimal input speed and collect into 5 mL polypropylene tubes or 1.5 mL microtubes in culture medium or PBS-2 %Se depending on the functional assay requirement.

4 Notes

1. Dissolving Na2HPO4 at this concentration is difficult. We recommend dissolving it in warm water (50–60°), which will take approximately 5 min. Na2HPO4 will crystallize within hours.
2. An oversize stopper will cover the flask mouth and ease breaking the vacuum, with the added benefit of no contact with the inner surface of the flask.
3. The key to AGE production is cleanliness. Being sterile is preferable, but as this is technically very difficult and very little will grow in the incubation conditions of pH 11 and at 50 °C, clean non-sterile is acceptable.
4. The purpose of the laminar flow hood is to keep all reagents, glassware, and the incubation mix clean and dust-free.
5. An inflatable glove bag (I2R, Instruments for Research and Industry, PA, USA, model X-17-17) is used to work under nitrogen and purge remaining oxygen from the reaction mix.
6. Thermo Scientific's "Slide-A-Lyzer" which comes in various volume capacities can be used.
7. This osmolarity is appropriate for murine cells and results in better cell recovery.
8. CD31 (PECAM-1) and CD102 (ICAM-2) are well-known endothelial markers previously reported to be expressed in the BM sinusoidal lining [7, 15]. In addition, other antibodies of interest can be used. For all antibodies, the equivalent concentration of the appropriately conjugated isotype should also be used.
9. Dissolving the BSA may take a significant period of time.
10. This should be done aseptically into sterile tubes. Typically, aliquot 1 mL into 1.5 mL Eppendorf tubes.
11. Freezing at −20 °C will cause the AGE-BSA to aggregate.
12. We dialyze in Slide-A-Lyser cassettes, with three changes of PBS, at 1 h, then at 8 h, then overnight. Of note, a lot of color will come out in the PBS.
13. Do not expect to see any change in volume at this step, i.e., 3 mL is put into the Slide-A-Lyser and 3 mL collected.
14. This time frame is chosen to ensure a proper FITC-AGE-BSA uptake by BMSEC.
15. Volumes described are for marrow harvested from a single mouse.
16. We describe the isolation of endosteal BMSEC since they are located close to hemopoietic stem cells with higher hemopoi-

etic reconstitution ability [5]. In addition, BM can be separated as enhanced or as central fractions [15, 16]. For extended methods of cleaning and flushing bones refer to [15, 16].

17. 30–40 × 10^6 cells/mL is the concentration considered to obtain optimal yields when performing a two-step sorting strategy (enrichment, then purity) on a Cytopeia Influx cell sorter.
18. For convenience and saving enriched cell samples, use unfractionated BM single cells for compensation controls.
19. CD45-conjugated antibodies are used for compensation as CD45 is highly expressed on unfractionated BM cells.

Acknowledgements

P.M. is supported by a grant from the Tromso Research Foundation/Trond Mohn and from the Northern Norway Regional Health Authority Post-Doctoral Fellowship (HelseNord, SFP1000-11 ID5432). AO has a Northern Norway Regional Health Authority Post-Doctoral Fellowship (HelseNord, SFP1000-11 ID5432). S.N. has an ARC Future Fellowship and a Science and Industry Endowment Fund (SIEF) project grant. The authors would like to thank Brenda Williams and Shen Heazlewood for critically reviewing the chapter.

References

1. Schofield R (1978) The relationship between the spleen colony-forming cell and the haemopoietic stem cell. Blood Cells 4:7–25
2. Gong JK (1978) Endosteal marrow: a rich source of hematopoietic stem cells. Science 199:1443–1445
3. Nilsson SK, Johnston HM, Coverdale JA (2001) Spatial localization of transplanted hemopoietic stem cells: inferences for the localization of stem cell niches. Blood 97: 2293–2299
4. Kiel MJ, Yilmaz OH, Iwashita T et al (2005) SLAM family receptors distinguish hematopoietic stem and progenitor cells and reveal endothelial niches for stem cells. Cell 121: 1109–1121
5. Grassinger J, Haylock DN, Williams B et al (2010) Phenotypically identical hemopoietic stem cells isolated from different regions of bone marrow have different biologic potential. Blood 116:3185–3196
6. Haylock DN, Williams B, Johnston HM et al (2007) Hemopoietic stem cells with higher hemopoietic potential reside at the bone marrow endosteum. Stem Cells 25:1062–1069
7. Ellis SL, Grassinger J, Jones A et al (2011) The relationship between bone, hemopoietic stem cells, and vasculature. Blood 118: 1516–1524
8. Butler JM, Nolan DJ, Vertes EL et al (2010) Endothelial cells are essential for the self-renewal and repopulation of Notch-dependent hematopoietic stem cells. Cell Stem Cell 6: 251–264
9. Ding L, Saunders TL, Enikolopov G et al (2012) Endothelial and perivascular cells maintain haematopoietic stem cells. Nature 481:457–462
10. Hooper AT, Butler JM, Nolan DJ et al (2009) Engraftment and reconstitution of hematopoiesis is dependent on VEGFR2-mediated regeneration of sinusoidal endothelial cells. Cell Stem Cell 4:263–274
11. Qian H, Johansson S, McCourt P et al (2009) Stabilins are expressed in bone marrow sinusoidal endothelial cells and mediate scavenging and cell adhesive functions. Biochem Biophys Res Commun 383:336–339
12. Hansen B, Longati P, Elvevoid K et al (2005) Stabilin-1 and stabilin-2 are both directed into

the early endocytic pathway in hepatic sinusoidal endothelium via interactions with clathrin/AP-2, independent of ligand binding. Exp Cell Res 303:160–173

13. Ito Y, Sorensen KK, Bethea NW et al (2007) Age-related changes in the hepatic microcirculation in mice. Exp Gerontol 42:789–797
14. Sorensen KK, McCourt P, Berg T et al (2012) The scavenger endothelial cell—a new player in homeostasis and immunity. Am J Physiol Regul Integr Comp Physiol 303: R1217–R1230
15. Grassinger J, Nilsson SK (2011) Methods to analyze the homing efficiency and spatial distribution of hematopoietic stem and progenitor cells and their relationship to the bone marrow endosteum and vascular endothelium. Methods Mol Biol 750:197–214
16. Williams B, Nilsson S (2009) Investigating the interactions between haemopoietic stem cells and their niche: methods for the analysis of stem cell homing and distribution within the marrow following transplantation. Methods Mol Biol 482:93–107

Chapter 7

Analysis of Circadian Rhythms in Embryonic Stem Cells

Jiffin K. Paulose, Edmund B. Rucker III, and Vincent M. Cassone

Abstract

Recent attention on the early development of circadian rhythms has yielded several avenues of potential study regarding molecular and physiological rhythms in embryonic stem cells (ESCs) and their derivatives. While general guidelines of experimental design are—as always—applicable, there are certain idiosyncrasies with respect to experiments involving circadian rhythms that will be addressed. ESCs provide a number of challenges to the circadian biologist: growth rates are normally much higher than in established cell culture systems, the cells' innate drive towards differentiation and the lack of known synchronizing input pathways are a few examples. Some of these challenges can be addressed post hoc, such as normalization to total RNA or protein for transcript abundance studies. Most others, as outlined here, require special handling of the samples before and during experimentation in order to preserve any potential circadian oscillation that is present. Failure to do so may result in a disruption of endogenous oscillation(s) or, potentially worse, generation of an artificial oscillation that has no biological basis. This chapter begins with cultured ESCs, derived from primary blastocysts or in the form of cell lines, and outlines two methods of measuring circadian rhythms: the 2DG method of measuring glucose uptake (Sokoloff et al. J Neurochem 28: 897–916, 1977) and real-time measurement of molecular rhythms using transgenic bioluminescence (Yoo et al. Proc Natl Acad Sci U S A 101:5339–5346, 2004).

Key words Embryonic stem cells, Circadian, Clock genes, ESCs, 2-Deoxyglucose, Real-time bioluminescence

1 Introduction

1.1 Circadian Rhythms

Nearly every organism on the planet has evolved an endogenous mechanism to anticipate the myriad environmental changes that coincide with earth's rotation about its own axis, termed *circadian rhythms* [1]. These rhythms are expressed at every biological level: from gene expression to overt behavior [2]. Regardless of their modality, all circadian rhythms share three defining characteristics: (1) periodicities of approximately 24 h under constant conditions; termed *free-running* [3], (2) the ability to synchronize to an external stimulus ("zeitgeber" from the German "time-giver") [4], and (3) temperature compensation; that is to say an insensitivity of the endogenous period to changes in temperature within physiological

Ivan N. Rich (ed.), *Stem Cell Protocols*, Methods in Molecular Biology, vol. 1235,
DOI 10.1007/978-1-4939-1785-3_7, © Springer Science+Business Media New York 2015

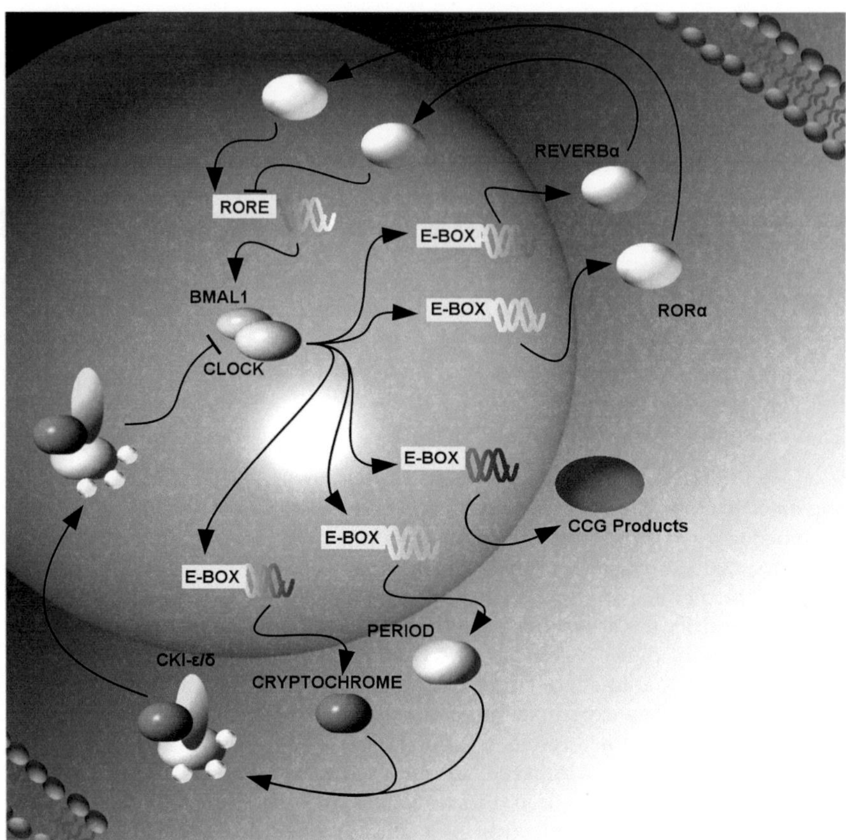

Fig. 1 Model of the molecular clock

range of the organism [5]. These characteristics are expressed within the laboratory and must therefore be considered when performing experiments dealing with circadian rhythms.

Similarly, all circadian mechanisms are regulated in a hierarchical fashion governed by, in the most general sense, a three component system. First, an input pathway provides a means for the cell/tissue/organism to receive temporal information. The classical example is photoreceptors in the retinae. That environmental signal is then relayed to the second component, the oscillator itself. In terms of mammalian cell culture systems, this could be the molecular clockworks (Fig. 1). Here, time of day information is translated to the expression and activity of the various components that act in a transcription-translation feedback loop. The "positive limb" of effector transcription factors, BMAL1 and CLOCK, activate transcription of E-box-containing genes, including those genes that encode the "negative limb" repressors PERIOD (PER1, PER2, and PER3) and CRYPTOCHROME (CRY1 and CRY2). *Bmal1* transcription is regulated at Retinoic acid-related Orphan Receptor Elements (ROREs) on its promoter by the orphan nuclear receptor proteins RORα and REVERBα which activate and repress *bmal1* expression, respectively [6]. Finally, the oscillator relays its own state to downstream outputs, including clock-controlled genes (CCGs).

Two of the three fundamental properties of circadian rhythms, dealing with entrainment and temperature compensation, require special consideration in that ignorance of either may lead to inaccurate data. While light is considered the dominant zeitgeber for circadian rhythms, it is not the only stimulus capable of synchronizing. Indeed, when considered in the context of mammalian stem cell culture, light becomes irrelevant due to the lack of photoreceptive machinery in ESCs. With respect to cell culture, we and others have found that passaging of cells [7] or even medium exchange [8] can synchronize rhythms in cell culture. Similar to nutrient exchange, temperature fluctuations are also capable of synchronizing rhythms in cultured cells [9], so care should be taken that cultures are maintained at 37 °C and time spent outside the incubator—presumably at lower temperatures—is minimal. Also, it is vital to consider the time of day when media changes are performed in order to avoid (or permit) imposing a zeitgeber to one's preparation.

1.2 2-Deoxyglucose Method

Observations of rhythmic glucose utilization using the 2DG method predate the formal recognition of the suprachiasmatic nucleus as the master pacemaker in mammals [10]. The designers of this method, Sokoloff and coworkers, had provided several lines of evidence that radioactively labeled 2-deoxyglucose could enter tissue using the same transport mechanisms as endogenous glucose and be used as a direct measure of metabolic activity in cortical tissues before formalizing the method in 1977 [11]. Since then, we and others have used 2DG utilization as a marker of circadian clock output in vivo and in cell culture [7, 12, 13]. While the theory is comprehensively explained in [11], it can be briefly stated that the glucose analog 2-deoxyglucose, after conversion by hexokinase (in competition with glucose) to 2-deoxyglucose-6-phosphate, is incapable of further enzymatic processing in the glycolysis pathway. The empirical value of the compound, therefore, is that it is biochemically indistinguishable from glucose until *after* the first step of glycolysis. Therefore, any uptake of the compound by the cell can be considered analogous to that of glucose itself. When investigating the utilization of glucose in cell culture, the concentration of 2DG used should be sufficiently low; otherwise the compound may interfere with glycolysis. Also, inoculation of the cells should be done via complete medium exchange, to ensure that the availability of 2DG is homogeneous across the culture. As described below, cell lysis must be preceded by multiple saline washes to ensure that the measured radioactivity is from the intracellular space and not from the medium. Finally, care should be taken when converting scintillation counts to concentrations of 2DG, as efficiency constants may vary from one machine to another.

1.3 Real-Time Bioluminescence

The use of luciferase-mediated bioluminescence in circadian rhythms research predates the discovery of molecular rhythms by some degree, as it was initially discovered as an output of the *Gonyaulax* clock [14]. Since then, advances in transgenic technology have allowed for the use of luciferase as a real-time reporter of clock gene expression (using luciferase transgene driven by *mPer1* promoter sequence [15]) and protein abundance (mPER2::LUCIFERASE protein fusions [16]). In the few studies investigating circadian rhythms in stem cells, stable transfection of reporter constructs into cell lines [17], adenovirus-mediated infection of cell lines [18], and primary stem cell cultures derived from transgenic mice [7] have all been successfully used.

Regardless of the transfection method, the biochemistry remains the same. When and as available in the cell, LUCIFERASE will catalyze the conversion of luciferin (provided in the culture media) to oxyluciferin in an ATP- and O_2-dependent, two-step process. In the laboratory, the light produced by the reaction is quite dim, requiring the use of photomultiplier-based optics to amplify the signal. Naturally this type of setup requires as little noise (i.e., light) as possible, so care should be taken that the culture apparatus and surrounding environmental chamber are shielded from light as much as possible.

Bioluminescence reporters provide several advantages to their fluorescent counterparts. One major advantage is that bioluminescence measurements require no exogenous light, whereas fluorescence reporters must have some wavelength of light to excite the fluorophore. Again, this can be seen as a non-issue when dealing with embryonic stem cells; however, there is no evidence that photobleaching over extended periods of time will not have deleterious effects on the cells.

Bioluminescence reporters also allow for automated observation with higher temporal resolution. Current software-driven interfaces, such as the Lumicycle (Actimetrics, IL), can measure multiple samples simultaneously as often as once per minute. However, this technique is not without its challenges. Cultures are kept at 37 °C with as little humidity as possible due to the sensitive electronics involved. Culture plates are sealed with glass coverslips to reduce evaporation of the culture medium, which also must be buffered using HEPES in order to minimize accumulation of toxic CO_2 in the cultures. It has been our experience that different equipment and laboratory spaces have different environmental characteristics. While the provided concentrations of buffering agents is widely used, it may be necessary to experiment with a range of buffer concentrations to find one that is suitable.

With respect to embryonic stem cells, the concentration of cells during plating should also be considered with care. Cellular senescence is a concern with low density cultures, and spontaneous differentiation can occur when cultures become too dense [19].

2 Materials

1. Embryonic Stem cell DMEM (ESDMEM): High Glucose, without Phenol Red or Sodium Bicarbonate, but supplemented with:Sodium Bicarbonate (2.2 g/L), MEM Nonessential Amino Acids (0.1 mM), l-glutamine (2 mM), 2-Mercaptoethanol (0.1 mM), penicillin (50 U/mL)/streptomycin (50 μg/mL) antibiotic (1 mM), and leukemia inhibitory factor (LIF) (500–1,000 U/mL).
2. 6-well cell culture plates: gelatin-coated.
3. Phosphate-Buffered Saline (PBS): Sterile.
4. TRIzol.
5. ^{14}C-labeled 2-deoxyglucose.
6. Scintillation vials.
7. Scintillation fluid.
8. Scintillation counter.
9. Recording Medium: ESDMEM modified as follows:Sodium Bicarbonate (0.35 g/L), HEPES (10 mM), d-Luciferin (0.1 mM).
10. 35 mm cell culture dishes: gelatin-coated.
11. 40 mm round glass coverslips.
12. Vacuum grease: sterilized by autoclave.
13. Photomultiplier-based bioluminescence recorder (e.g., Lumicycle, Actimetrics, IL).

3 Methods

3.1 2DG Uptake Protocol

1. ESCs should be at passaged three or more times under feeder-free conditions and seeded at an approximate density of 1×10^5 cells/plate into a sufficient number of 6-well plates to conduct the time series experiment (*see* **Notes 1–3).**
2. Synchronize ESCs, either via established chemical methods or by centrifugation into the final passage before the first time point. One hour prior to each time point, incubate the cells with 2DG by exchanging the entire volume of media for medium supplemented with 2DG.

3. At each time point, wash the cells twice with sterile PBS to remove any traces of 2DG.
4. Add 1 mL of cold TRIzol to the cells and incubate for 2 min to ensure complete cell lysis.
5. Scrape the bottom of each well and collect the lysate (*see* **Note 4**).
6. To measure 2DG uptake, 0.1 mL of the cell lysate will be used. The remainder can be used for downstream applications (e.g., qPCR).
7. Calculate the molar quantity of 2DG from the blank-subtracted scintillation counts using the specific activity (usually reported in μCi/mmol) of the radioisotope and normalize against total RNA, total protein, or cell number as suitable.

3.2 Real-Time Bioluminescence Recording

1. Passage cells into gelatin-coated 35 mm culture dishes and incubate for 24 h for cells to attach.
2. Synchronize cells by established chemical methods, or use time of centrifugation/passage as time of synchronization.
3. Exchange media for recording media and seal the coverslip to the top of the dish with vacuum grease. (*see* **Note 5**).
4. Place dishes in recording apparatus and begin recording.

4 Notes

1. It is vital that ESC cultures do not include any fibroblasts that may be present from the initial derivation of the cell line, as fibroblasts will also take up 2DG in a way that is indistinguishable from the ESCs.
2. When feeding cells prior to a time-series experiment, daily media changes should not be performed at the same time of day. A 24 h feeding cycle may impose an artificial rhythm onto the cells that will mask any endogenous rhythm that is present.
3. Some prior experimentation may be necessary to determine the concentration of cells to be plated during the final passage before beginning a time-series experiment. A high concentration of cells may lead to overgrowth and induce differentiation before the time-series is concluded. At the same time, a low concentration of cells may lead to senescence. We have been successful with a cell concentration of 1×10^5 cells per well in a 6-well plate.
4. Cell scrapers can be purchased for this step, but are not necessary. Tilting the plate forward after incubating in TRIzol will help to collect all of the cell lysate.

5. A 3 cc syringe can be filled with sterile vacuum grease to lay down a bead of grease along the lip of the culture plate. The coverslip can then be gently pressed down to create the air-tight seal. Coverslips should be handled by the edge to avoid getting residue on the top surface as this may occlude the light path between the cells and the photodetector.

References

1. Pittendrigh CS (1993) Temporal organization: reflections of a Darwinian clock-watcher. Annu Rev Physiol 55:16–54
2. Bell-Pedersen D, Cassone VM, Earnest DJ et al (2005) Circadian rhythms from multiple oscillators: lessons from diverse organisms. Nat Rev Genet 6:544–556
3. Pittendrigh CS (1960) Circadian rhythms and the circadian organization of living systems. Cold Spring Harb Symp Quant Biol 25: 159–184
4. Pittendrigh CS (1981) Circadian systems: entrainment. In: Aschoff J (ed) Biol. Rhythm. Springer, Boston, pp 95–124
5. Pittendrigh CS (1954) On temperature independence in the clock system controlling emergence time in Drosophila. Proc Natl Acad Sci U S A 40:1018–1029
6. Buhr ED, Takahashi JS (2013) Molecular components of the Mammalian circadian clock. Handb Exp Pharmacol (217):3–27. doi:10.1007/978-3-642-25950-0_1
7. Paulose JK, Rucker EB, Cassone VM (2012) Toward the beginning of time: circadian rhythms in metabolism precede rhythms in clock gene expression in mouse embryonic stem cells. PLoS One 7:e49555. doi:10.1371/journal.pone.0049555
8. Guido ME, Garbarino-Pico E, Contin MA et al (2004) Circadian regulation of phospholipid biosynthesis in chick retinal ganglion cells. ARVO Meet Abstr 45:3666
9. Sládek M, Sumová A (2013) Entrainment of spontaneously hypertensive rat fibroblasts by temperature cycles. PLoS One 8:e77010. doi:10.1371/journal.pone.0077010
10. Schwartz WJ, Gainer H (1977) Suprachiasmatic nucleus: use of 14C-labeled deoxyglucose uptake as a functional marker. Science 197: 1089–1091
11. Sokoloff L, Reivich M, Kennedy C et al (1977) The [14C]deoxyglucose method for the measurement of local cerebral glucose utilization: theory, procedure, and normal values in the conscious and anesthetized albino rat. J Neurochem 28:897–916
12. Earnest DJ, Liang FQ, Ratcliff M et al (1999) Immortal time: circadian clock properties of rat suprachiasmatic cell lines. Science 283: 693–695
13. Paulose JK, Peters JL, Karaganis SP et al (2009) Pineal melatonin acts as a circadian zeitgeber and growth factor in chick astrocytes. J Pineal Res 46:286–294
14. Hastings JW, Sweeney BM (1958) A persistent diurnal rhythm of luminescence in Gonyaulax polyedra. Biol Bull 115:440–458
15. Yamazaki S, Numano R, Abe M et al (2000) Resetting central and peripheral circadian oscillators in transgenic rats. Science 288: 682–685
16. Yoo S-HH, Yamazaki S, Lowrey PL et al (2004) PERIOD2::LUCIFERASE real-time reporting of circadian dynamics reveals persistent circadian oscillations in mouse peripheral tissues. Proc Natl Acad Sci U S A 101: 5339–5346
17. Yagita K, Horie K, Koinuma S et al (2010) Development of the circadian oscillator during differentiation of mouse embryonic stem cells in vitro. Proc Natl Acad Sci U S A 107: 3846–3851
18. Kowalska E, Moriggi E, Bauer C et al (2010) The circadian clock starts ticking at a developmentally early stage. J Biol Rhythms 25: 442–449

Chapter 8

Measuring Stem Cell Circadian Rhythm

William Hrushesky and Ivan N. Rich

Abstract

Circadian rhythms are biological rhythms that occur within a 24-h time cycle. Sleep is a prime example of a circadian rhythm and with it melatonin production. Stem cell systems also demonstrate circadian rhythms. This is particularly the case for the proliferating cells within the system. In fact, all proliferating cell populations exhibit their own circadian rhythm, which has important implications for disease and the treatment of disease. Stem cell chronobiology is particularly important because the treatment of cancer can be significantly affected by the time of day a drug is administered. This protocol provides a basis for measuring hematopoietic stem cell circadian rhythm for future stem cell chronotherapeutic applications.

Key words Stem cell proliferation, Circadian rhythm, Chronobiology, Chronotherapy, Hematopoiesis

1 Introduction

Biological rhythms are a fundamental and integral part of all life forms. Biological rhythms can be annual, monthly, and daily. They can be found in virtually all systems from organismic to the molecular level [1, 2]. Circadian rhythms are daily rhythms that take the form of a sine wave with a "high" or "active" and "low" or "quiet" periods over the 24-h clock. The sleep and wake cycle is the best example. To ensure biological economy, many molecular, cellular, and organismic processes are temporally coordinated so that groups of genes and their products are expressed at different critical times of the day, and various cellular and organ functions are likewise coordinated in circadian time. In some diseases, there is a disturbance of the circadian clock. Internal des-synchronization is a hallmark of profound circadian disruption, but whether an abnormality in molecular and cellular circadian time keeping could contribute to or even cause certain diseases is still unknown.

Definitive stem cell systems can be divided into continuously proliferating or partially proliferating systems. All cells that are capable of proliferating do so in a circadian manner; that is, there

Ivan N. Rich (ed.), *Stem Cell Protocols*, Methods in Molecular Biology, vol. 1235,
DOI 10.1007/978-1-4939-1785-3_8, © Springer Science+Business Media New York 2015

are times of the day when proliferation is greatest and other times where it is least. Virtually all the hematopoietic stem and progenitor cells exhibit a predictable circadian variation in rodents [3–7], dogs [8, 9], and humans [10–12]. Cells of the gut mucosa exhibit circadian organization from the tongue to the rectum [13] as do cells of the skin [14, 15] and the corneal epithelium of the eye [13]. For human bone marrow and gastrointestinal tissues, the proportion of cells in S-phase of the cell cycle is synchronized within the day with substantially more DNA synthesis occurring in the morning hours than in the evening or nighttime hours. This overlap of cytokinetic phasing between the two most important chemotherapy toxicity target tissues implies that S-phase specific cytotoxic agents that damage the gut and/or bone marrow might be expected to be less toxic if given during the nighttime hours. Indeed, it has been demonstrated that in vitro assays can be used to predict the best time of day to administer anti-neoplastic drugs and this prediction correlates with the chronotherapeutic administration in the patient [16].

The protocol below provides the basis for measuring stem cell circadian rhythms that might be useful in preclinical and clinical applications. Stem cell proliferation ability provides the basis for determining circadian rhythm. This chapter focuses on hematopoietic stem cell circadian rhythm using absorbance, fluorescence, or bioluminescence instrument-based assays.

2 Materials

2.1 Human Donors

To measure hematopoietic stem cell circadian rhythm, peripheral blood is used. This is taken from the arm vein every 4 h over a 24-h period. An Internal Review Board (IRB) must approve all studies involving human donors, and all donors and/or patients must sign a consent form approved by this IRB. The donor or patient should wear a wrist accelerometer, known as a "actigraph" to monitor sleep and activity cycles over a 7-day period prior to the study. Information from the actigraph can be uploaded to a computer after the study. In addition, cortisol and melatonin are circadian rhythm markers that can be measured in the serum using an enzyme immunoassay. For all peripheral blood samples, a Vacutainer containing heparin should be used and a mononuclear cell (MNC) suspension should to be prepared by density gradient centrifugation (DGC).

2.2 Animals

Animals are usually acclimatized to a light–dark cycle prior to starting any circadian rhythm studies. Bone marrow is used to determine hematopoietic stem cell circadian rhythm. It is best to perform these studies on individual animals, rather than pooling cells from several animals at each time point. For mice, fractionation of bone marrow by DGC is not usually necessary. However, for rats, dogs, and marrow from other species, a DGC should be performed to remove red blood cells (RBC) and other mature functional cells from the sample.

2.3 Preparation of Mononuclear Cell Suspension (See Note 1)

1. Biohazard hood for performing sterile culture.
2. PipetteAid, motorized pipette dispenser.
3. NycoPrep 1.077 for human cells or 1.077A (Axis-Shield) for animal cells. Alternatively Ficoll (GE Healthcare) can be used as a DGC medium.
4. Sterile conical tubes with screw caps, 50 and 15 mL (Becton Dickenson).
5. Dulbecco's Phosphate Buffered Saline (dPBS, Gibco).
6. Iscove's Modified Dulbecco's Medium (IMDM, Gibco).
7. Sterile 5, 10, 25, and/or 50 mL plastic pipettes.

2.4 Cryopreservation of Sample Aliquots

1. Sterile cryopreservation vials 1.5–2 mL.
2. Cryopreservation reagent: IMDM + 10 % FBS + 10 % DMSO.
3. Single channel manual or electronic pipettes (0.1–1 mL) and sterile pipette tips.
4. CoolCells (Biocision) in which cells can be frozen.
5. Freezer at −80 °C.
6. As an alternative to [4] an automated cryopreservation rate-freezer can be used.
7. Liquid nitrogen (LN2).

2.5 Cell Thawing

1. Cell thawing medium: IMDM + 10 % FBS.
2. Water bath at 37 °C.
3. Sterile 10 mL plastic tubes with either a screw cap or snap-on cap.
4. DNase (Sigma-Aldrich).
5. Laboratory centrifuge.

2.6 Preparation of Sample for Testing

1. Hemocytometer or automated cell counting instrument for counting MNCs.
2. Hemocytometer, automated cell counter or flow cytometer for measuring dye exclusion viability, e.g., by trypan blue, 7-aminoactinomycin D (7-AAD).
3. Sterile 5 and/or 10 mL plastic tubes for adjusting and preparing the cell suspensions.

2.7 Detecting and Measuring Hematopoietic Stem Cell Population Proliferation in Culture (See Note 2)

1. Stem cell proliferation assay (HemoLIGHT-96 Research, HemoFLUOR-96 Research or HALO-96 Research, HemoGenix).
2. Calibrated, manual or electronic 8-channel or 12-channel pipette to dispense 0.1 mL with sterile tips.
3. Reagent reservoir for an 8- or 12-channel pipette.
4. Fully humidified incubator at 37 °C gassed with 5 % CO_2 and, if possible, 5 % O_2.

2.8 Analysis of Results

Software capable of cosinor time-series analysis is required. Specialized software for circadian rhythm studies can be obtained from Expert Soft Technologie, France. Alternatively, software such as OriginLab's Origin software or SPSS TableCurve 2D can also be used.

3 Methods

Human peripheral blood (10–20 mL is usually sufficient) is collected in Vacutainers or similar closed blood collection system containing an anticoagulant. Bone marrow is collected from mice and rats by flushing out the marrow with IMDM. With the exception of mice, bone marrow cells from animals usually have to be fractionated using a DGC procedure to remove red blood cells, granulocytes, and platelets. Cells and/or tissues are collected every 4 h over a 24-h period. After each collection Subheadings 3.1 and 3.2 will be performed. Once the cells from all time points have been collected and cryopreserved, samples from all time points are tested and analyzed simultaneously. A diagrammatic representation of the procedure is shown in Fig. 1.

Unless specifically stated, all of the following procedures must be performed under sterile conditions.

3.1 Preparation of the Mononuclear Cell Fraction from Human Peripheral Blood or Animal Bone Marrow Using Density Gradient Centrifugation (See Note 3)

1. Dilute the sample to be fractionated with an equal volume of dPBS and mix gently by inversion. Canine bone marrow should be diluted 1:2 with dPBS.
2. For samples 3 mL or less, use a 15 mL conical plastic tube for separation. For samples greater than 3 mL, use a 50 mL conical, plastic tube for separation.
3. For samples of 3 mL diluted to 6 mL with dPBS, dispense 5 mL of the density gradient reagent into the tube. For samples greater than 3 mL, dispense 15 mL of the density gradient reagent into a 50 mL tube.
4. Using a sterile, serological pipette, dispense the diluted sample gently on top of the density gradient reagent by holding the tube at approx. 45° and using a PipetteAid on slow delivery.

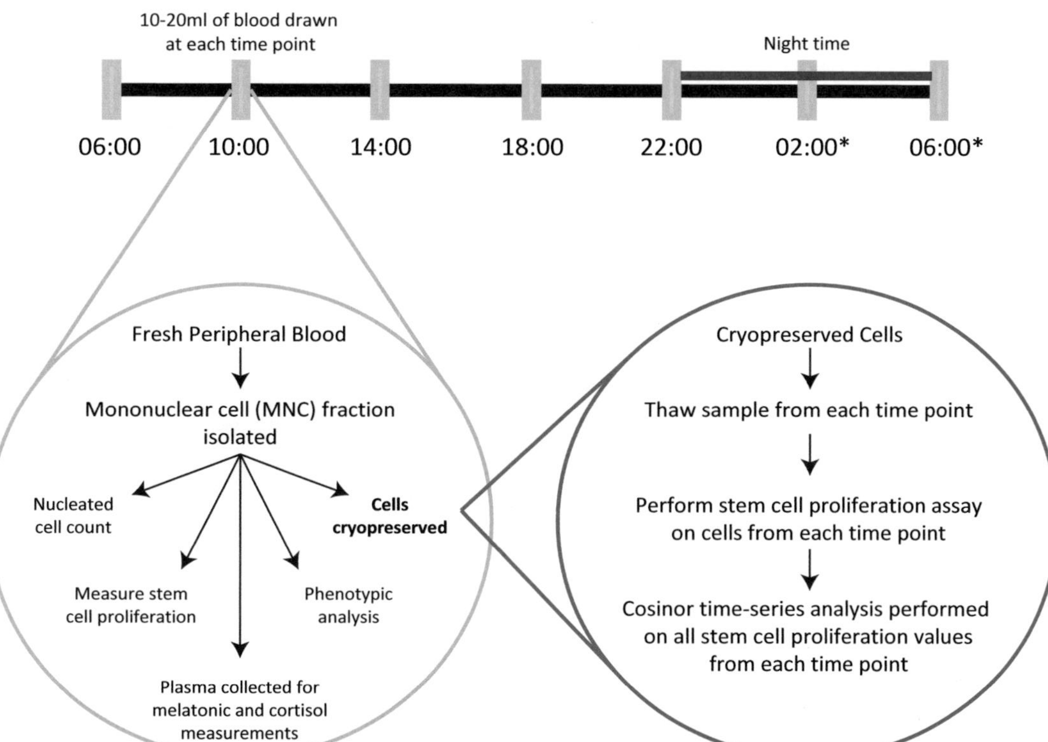

Fig. 1 Diagrammatic summary of the protocol to determine human hematopoietic stem cell circadian rhythm

5. Centrifuge for 10 min at 1,000 × *g* or 20 min at 600 × *g* at room temperature with NO brake.
6. After centrifugation, remove the tube(s) gently and carefully aspirate the top layer above the MNC interface leaving approximately 5 mm above the interface. Discard the supernatant.
7. Harvest the MNCs from the interface and transfer the cells to another sterile tube. It is best to harvest the cells using a manual 1 mL, single channel pipette. Do not remove cells below the interface.
8. Add approx. 10–20 mL dPBS, mix gently, and centrifuge the cells for 10 min at 300 × *g* at room temperature.
9. Aspirate the supernatant after centrifugation taking care not to aspirate the cell pellet.
10. Add 1–2 mL of IMDM and resuspend the cells, breaking up any clumps using a 1 mL manual pipette.
11. Perform a nucleated cell count and viability. The cell viability must be greater than 85 %. Using cells with a viability lower than 85 % will produce results with low proliferation ability.

3.2 Cryopreservation of Samples After Collection (See Note 4)

1. Prepare a cryopreservation reagent fresh and keep the container on ice or at 4 °C.
2. After determining the nucleated cell count, calculate the volume of the cell concentration that will produce 5×10^6 MNC/mL. Keep the cells on ice or at 4 °C.
3. Determine the number of cryopreservation vials required to freeze a certain number of cell suspension aliquots and label each vial.
4. Calculate the amount of cryopreservation reagent needed to make up 1 mL in each vial and dispense this amount of reagent into the vials.
5. Dispense the required volume of cell suspension into each vial.
6. Cap each vial securely and mix by inversion.
7. Place the vials in the CoolCell and transfer the CoolCell to a −80 °C freezer and leave overnight.
8. Alternatively, cryopreserve the cells using an automated rate freezer.
9. After freezing, transfer the vials to a LN2 container for storage.

3.3 Cell Thawing (See Note 5)

When the cells are thawed, granulocytes and other cell components will rupture and release DNA. Large amounts of released DNA will clump together encasing valuable stem cells. If the cell preparation originally cryopreserved was a MNC or similar fraction, the chances of clumping will be low. However to reduce or alleviate the possibility of clumping during cell thawing, it is recommended that DNase be added to the cell suspension.

1. Thaw the contents of the vials in a 37 °C water bath, by swirling the vial for approx. 1 min.
2. When a small ball of ice still remains in the vial (1–2 min), remove the vial from the water bath, sterilize the outside of the vial by spraying with 70 % ethanol, and carefully unscrew the vial lid.
3. It is possible that clumping can occur at this stage, in which case, add DNase to the total volume in the vial to achieve a concentration of 6 μg/mL before proceeding to the next step.
4. Using a 1 mL pipette, gently mix the contents of the vial and transfer to a 50 mL tube containing 20 mL of thaw medium. Up to three vials of the same cells can be added to this 20 mL of thaw medium. However, clumping can also occur at this stage. In this case, DNase at a final concentration of 6 μg/ mL should be added before proceeding to the next step.
5. Gently mix the cells by swirling the contents of the tube. Do not use repeat pipetting to mix the cells. This could cause

further rupture of cells and the release of DNA resulting in increased clumping.

6. Centrifuge the cells at 300 ×*g* for 10 min at room temperature and discard the supernatant after centrifugation.
7. Resuspend the cells in 1 mL of culture medium. If necessary, add 6 μg/mL DNase.
8. Perform a nucleated cell count and viability assessment.
9. Repeat this cell thawing process for one or more vials from all 7 time points (0, 4, 8, 12, 16, 20, and 24 h).

3.4 Assay Setup to Determine Stem Cell Circadian Rhythm (See Note 6)

The type of stem cell proliferation assay used will depend on the 96-well plate reader available. Plate readers are available as multimode readers or single readers. Multimode plate readers will have the capability of measuring absorbance, fluorescence, and even luminescence. Single mode plate readers detect only one type of signal (absorbance, fluorescence, or luminescence), but are generally more sensitive and more costly than multimode readers. Therefore, the type of signal detection system used will depend on the instrument available.

For absorbance, fluorescence, or bioluminescence readouts, HemoLIGHT-96 Research, HemoFLUOR-96 Research, or HALO-96 Research are available, respectively. These are cell proliferation assays that can be used to detect up to eight different hematopoietic stem cell populations from human, nonhuman primate, horse, pig, sheep, dog, rat, and mouse. The different stem cell populations detected are dependent upon the growth factor/cytokine cocktail used.

Although the circadian rhythms of multiple stem cell populations can be detected and measured, two are of importance. The first is designated CFC-GEMM for Colony-Forming Cell—Granulocyte, Erythroid, Macrophage, Megakaryocyte. As its name implies, it is a primitive hematopoietic stem cell population that can produce cells from all three hematopoietic lineages. The second is designated HPP-SP or High Proliferative Potential—Stem and Progenitor cell. This stem cell population is more primitive than the CFC-GEMM and can produce cells of both the lymphopoietic and hematopoietic lineages.

The assay kits to determine these stem cell populations contain Master Mixes consisting of all the reagents necessary to initiate and maintain stem cell proliferation and growth. To set up the assay for one or more stem cell populations, the following procedure is used:

1. Remove the Master Mix from the kit stored at −20 °C and thaw the contents at room temperature (RT) or in a 37 °C water bath.
2. During this time, establish the total number of samples to be tested. If only a single stem cell population is examined, then

the total number of samples, and therefore sterile 5 mL tubes required, will be 7. Label each of the tubes.

3. To obtain a baseline (background control), each sample can also be cultured using a Master Mix that does not contain any growth factor/cytokine cocktail. The same number of sterile 5 mL tubes will be required and labeled.
4. Using calibrated pipettes or better still, electronic pipettes, accurately dispense 0.9 mL of the Master Mix into each sterile tube (*see* **Note 7**).
5. For each of the time point samples to be tested, adjust the cell concentration using IMDM so that the working concentration of each sample is 500,000 cells/mL. Mix the cell suspension.
6. Accurately dispense 0.1 mL of each cell suspension into each tube containing the stem cell Master Mix and NGF Master Mix. This is now the Culture Master Mix. The total volume of each tube will be 1 mL. This is sufficient to produce up to 8 replicate wells containing 0.1 mL of Culture Master Mix.
7. Place the lid on each tube and mix gently by vortexing.
8. Remove a sterile 96-well plate from kit.
9. The following dispensing should be performed using the repeat dispensing function available with an electronic pipette. If this is not available, accurate dispensing using a single channel, manual pipette is required.
10. Dispense 0.1 mL into the middle of each of 8 replicate wells in each column of the plate. For example, column 1 (A–H) will contain the 0 hr sample. Column 2 (A–H) will contain the 4 h sample and so on.
11. After all the samples have been dispense into the wells of the 96-well plate, replace the lid and transfer the plate(s) to the incubator.
12. With the exception of nonhuman primate cells, incubate all animal cell species for 4 days. This can be extended to 5 days for greater sensitivity.
13. For nonhuman primate cells and human cells, incubate for 5 days, although this can be extended to 7 days.

3.5 Sample Processing and Assay Readout (See Note 8)

With the exception of Subheading 3.5.1, **step 3**, all further steps to determine hematopoietic stem cell proliferation for circadian rhythm can be performed under non-sterile conditions.

3.5.1 Absorbance or Fluorescence Readout

1. Remove the absorbance or fluorescence enumeration reagent from the freezer and thaw either in a beaker of water or at room temperature. Mix the reagent by gentle inversion. Do not shake the bottle.

2. After the culture incubation period has elapsed, remove the 96-well plate(s) from the incubator and allow the plate(s) to attain room temperature.
3. If less than a whole plate has been used, it is possible to maintain the unused wells sterile for later use. To do this, transfer the plate to a hood, remove the lid and place the sterile adhesive foil that is included with the assay kit, over the whole of the plate. With a sharp knife, cut away the foil of the wells that are to be process and peel the foil away so that only the unused wells remain with the foil covering them.
4. When the enumeration reagent has thawed completely and attained room temperature, mix gently by inversion and pour the reagent into a reservoir.
5. Using an 8-channel pipette, attach the tips to the pipette and withdraw 0.1 mL of the enumeration reagent into each tip.
6. Dispense the reagent into the wells of the first column of the plate and mix the contents without causing bubbles in the well.
7. If a complete row of 12-wells is to be processed, use the 12-channel pipette and follow the **steps 5** and **6**.
8. Once the reagent has been dispensed into all wells start a timer, cover the plate with the lid and transfer the plate back to the incubator.
9. For an absorbance reading, the soluble yellow formazan product will develop between 1 and 4 h. Usually 3 h is optimal. Read the absorbance in a plate reader with a 490 nm filter.
10. For a fluorescence reading, the reaction will develop between 30 min and 3 h, with an optimum at 2 h. Read the fluorescence using a 380–400 nm excitation filter and a 505 nm emission filter.
11. Although the mean, standard deviation and percent coefficient of variation (%CV) can be calculated, the values for individual wells will be used to determine the circadian rhythm parameters by cosinor analysis.

3.5.2 Bioluminescence Readout [17] (See ***Notes 9*** *and* ***10****)*

1. Remove the ATP standard, controls, and ATP Enumeration Reagent (ATP-ER) that are included with the assay kit, from the freezer and thaw at room temperature.
2. Remove a non-sterile 96-well plate and allow it to attain room temperature. The non-sterile 96-well plate will be used to calibrate and standardize the assay, prior to measuring the samples.
3. Once the reagents have thawed, setup the ATP standard serial dilution curve as described in the assay manual.
4. The ATP standard curve allows the output of the luminescence plate reader in Relative Luminescence Units (RLU) to

be converted into standardized ATP concentrations. The ATP controls must lie on the ATP standard curve and are used to calibrate the assay. In this way, the ATP bioluminescence assay is fully calibrated and standardized. Performing the ATP standard curve and controls also provides information to demonstrate that the assay is working correctly prior to measuring the samples.

5. After ensuring that the ATP standard curve and control parameters agree with those in the assay manual, the samples can be processed and measured.
6. Using an 8- or 12-channel pipette, affix the pipette tips and transfer 0.1 mL of the ATP-ER from the reservoir into each replicate well.
7. Mix the contents of the wells with the same pipette tips making sure that no bubbles are formed. It is important that the mixing is thorough since the reagent contains a lysis buffer that is used to release the intracellular ATP so that it can react with the luciferin-luciferase reagent.
8. After the ATP-ER has been added to all wells, transfer the plate to the luminescence plate reader and incubate the plate in the dark for 10 min before reading the luminescence.
9. If available, use the instrument software to automatically convert the output of the luminometer in Relative Luminescence Units or RLU values to ATP concentrations (μM) using the ATP standard curve and calculate the mean, standard deviation and %CV of the replicate wells.
10. If the instrument software cannot perform the interpolation function from RLU to ATP concentrations automatically, the results will have to be exported to an Excel file. Microsoft Excel does have the mathematical functions to perform the interpolation from RLU to ATP concentration. However, third-part software programs are also available that can perform this task, e.g., SPSS TableCurve 2D and GraphPad Prism.
11. By standardizing the assay using the ATP standard curve and controls, it is possible to compare results between different human and animal donors over time.

3.5.3 Evaluating the Hematopoietic Stem Cell Circadian Rhythm

1. It is possible to plot the absorbance, fluorescence, or ATP concentrations after stem cell culture for each time point over the 24-h period. An indication of the oscillation during the 24-h period will be observed.
2. However, to obtain the parameters of the circadian rhythm and compare these parameters to other donors, it is necessary

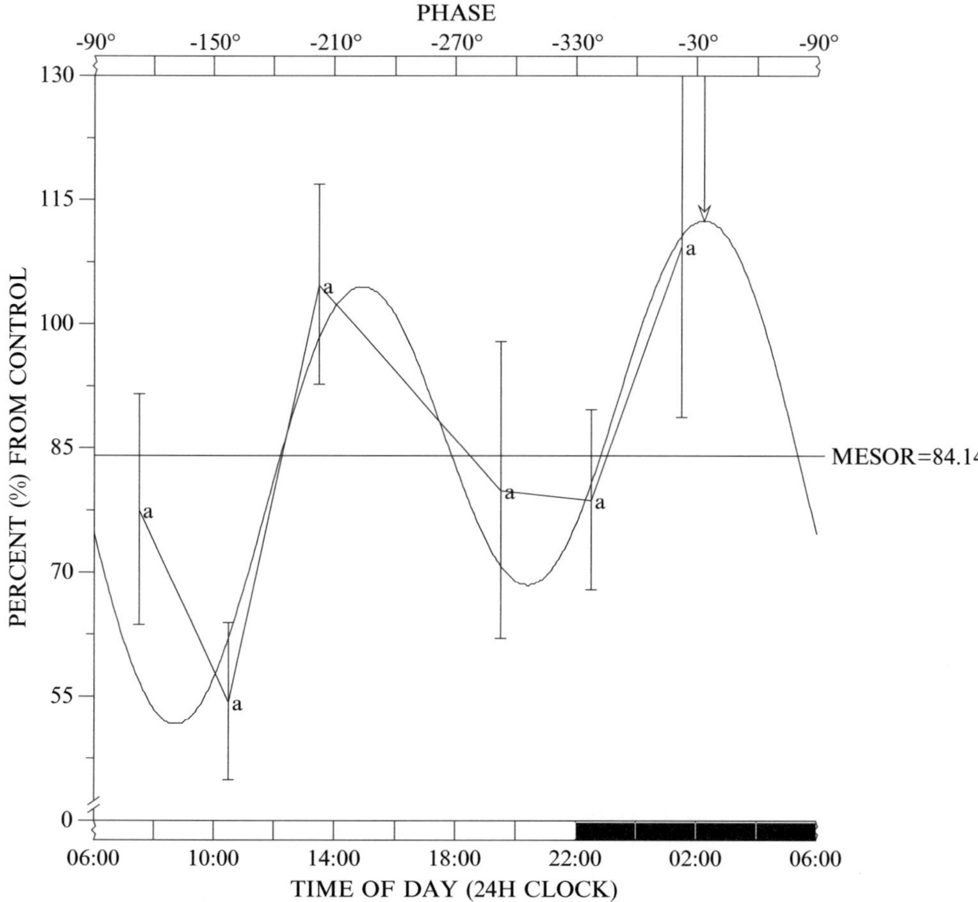

Fig. 2 The Circadian Rhythm for Human Hematopoietic Stem Cell (CFC-GEMM) proliferation ability. These results were obtained using HALO-96 Research. The ATP concentrations from individual wells were converted to percent from control (0 h time point). Using the cosinor time-series software from Expert Soft Technologie, the mean and standard deviations are plotted (slightly offset) and the circadian rhythm fitted by the least-squares method to the data. The *bottom X*-axis shows the time of day as a 24-h clock and the dark or nighttime period. The "phase" on the *upper X*-axis represents the distance between oscillations. The term "MESOR" represents the Midline Estimating Statistic of Rhythm and is the value midway between the highest and lowest values of the cosine function. The results show two peaks of stem cell proliferation, one in the afternoon and the other early in the morning. If, instead of single proliferation values detected at a specific cell dose (in this case 5,000 cells/well), the stem cell proliferation potential, that detects stem cell primitiveness, was plotted, a different circadian rhythm would be obtained. The proliferation potential is determined from the slope of a cell dose response performed for the stem cells at each time point. For more information visit the American Association of Chronobiology and Chronotherapeutics website (www.aamcc.net)

to subject the individual data points to time series or consinor analysis using specialized software (Subheading 2.8).

3. Figure 2 shows an example of the circadian rhythm of human, normal peripheral blood, CFC-GEMM.

4 Notes

1. Although both NycoPrep and Ficoll can be used, the latter is toxic to cells and should only be in contact with cells for a short period.
2. Culturing cells under low oxygen tension is advantageous because it reduces the production of free oxygen radicals that cause oxygen toxicity to the cells. Most, but not all cells, live under much lower oxygen tensions than present in the atmosphere (21 %). By culturing cells in an atmosphere containing oxygen tensions that approximate the tissue from which they originate, more superior in vitro plating efficiencies can be obtained. In the bone marrow, the partial oxygen tension is between 40 and 45 mmHg, which is approximately equivalent to 5 % oxygen [18, 19].
3. It is important that MNCs are used for the studies since this fraction contains hematopoietic stem cells as well as hematopoietic progenitor cells. Thus, if required, the circadian rhythm of the progenitor cells can also be determined. Using the MNC fraction means that few, if any, red blood cells, granulocytes and platelets, in other words, mature blood cells, are present in the culture. Red blood cell content greater than 10 % can lead to false positive results and therefore should be avoided. It is not advised to use cell lysis, since this does not produce a "clean" MNC fraction.
4. There is no standardized method of cryopreserving stem cells. In the cellular therapy field, automatic rate freezers are used to cryopreserve mobilized peripheral blood and umbilical cord blood as well as mesenchymal stem/stromal cells from different sources. This type of freezing procedure probably produces more consistent results, but it does not mean that it is the best to produce the highest yield of stem cells with the highest "quality". An automatic rate freezer is expensive. Using a CoolCell or similar container to freeze the cells could produce similar results at a fraction of the cost.
5. Like cryopreservation, cell thawing is also not a standardized procedure, but the method described works well. The addition of DNase to the thawing solution is an important aspect of the method. When nucleated cells, (e.g., granulocytes) die when the cells are thawed, they release DNA. If large numbers of nucleated cells are damaged during cryopreservation and subsequent thawing, the DNA released clumps together and encompasses viable cells in the suspension. This process can lead to significantly lower cell counts and a drastic reduction in stem cells. In some cases, even the addition of DNase will not prevent the formation of clumping. Therefore, the better the

of erythropoietic cells in vitro. Br J Haematol 52:579–588
19. Rich IN (1986) A role for the macrophage in normal hemopoiesis II Effect of varying physiological oxygen tensions on the release of hemopoietic growth factors from bone-marrow-derived macrophages in vitro. Exp Hematol 14:746–751
20. Bradley TR, Metcalf D (1966) The growth of mouse bone marrow cells in vitro. Aust J Exp Biol Med Sci 44:287–295
21. Pluznik DH, Sachs L (1966) The induction of colonies of normal "mast" cells by a substance in conditioned medium. Exp Cell Res 43(3): 553–559
22. Iscove NN, Sieber F, Winterhalter C (1974) Erythroid colony formation in cultures of mouse and human bone marrow: analysis of the requirement for erythropoietin by gel filtration and affinity chromatography on agarose-concanavalin A. J Cell Physiol 83: 309–320

Chapter 9

Cryopreservation of Human Pluripotent Stem Cells: A General Protocol

Takamichi Miyazaki and Hirofumi Suemori

Abstract

Cryopreservation is an essential technique to preserve stem cells, semipermanently sustaining their potentials. There are two main approaches of cryopreservation for human pluripotent stem cells (hPSCs). The first is the vitrification, which involves instantaneous freeze and thaw of hPSCs. The second is the conventional slow-cooling method and a rapid thaw. Both cryopreservation protocols have been standardized and optimized to yield high survivability of hPSCs.

Key words Cryopreservation, Vitrification, Slow-cooling, Single cell dissociation, Human pluripotent stem cells

1 Introduction

Human pluripotent stem cells (hPSCs), including human embryonic stem (ES) cells and human induced pluripotent stem (iPS) cells, have an infinite proliferative potential and capacity for differentiation into all cells of the three germ layers. Cryopreservation of hPSCs is a key procedure because it enables semipermanent preservation and easy transportation. Currently, hPSCs are cryopreserved using two different methods, vitrification and conventional slow-cooling. Vitrification is a method used for freezing unfertilized eggs [1] and modified to deal with a larger volume of primate ES cells and hPSCs [2]. The process of vitrification, involves suspending hPSC, grown as colonies, in a special cryoprotectant and instantaneously freezing them with liquid nitrogen. The thawing process after vitrification requires that the hPSCs are instantaneously thawed with warmed culture medium. Vitrification produces 20–90 % recovery of hPSC colonies, post-thawing [2, 3], but it requires an efficient and rapid operation to yield high survivability. For slow-cooling method, the cells are cryopreserved at rate of about 1 °C/min, using a cryoprotectant that generally contains dimethyl sulfoxide (DMSO). The manipulation is uncomplicated

Ivan N. Rich (ed.), *Stem Cell Protocols*, Methods in Molecular Biology, vol. 1235,
DOI 10.1007/978-1-4939-1785-3_9, © Springer Science+Business Media New York 2015

and appropriate for mass freezing, but slow-cooling has poor survivability (less than 10 %) for hPSC cryopreservation [4]. An optimized procedure for hPSCs has been reported to produce a high recovery [5]. The key to successful hPSCs cryopreservation is to use a single cell suspension, since freezing hPSCs as colonies damages the cells., Single hPSCs are resistant to the freeze–thaw process despite the same cryopreservation conditions. After thawing, recovery of single hPSCs in culture can be rapid, even without ROCK inhibitor treatment, by adjusting the cells density. The cells will return to the previous subculture state in a short period.

In this chapter, we describe both protocols of vitrification and an improved procedure for slow-cooling. The protocols can be applied to both feeder and non-feeder culture conditions.

2 Materials

2.1 Human Pluripotential Stem Cell (hPSC) Lines

The vitrification and slow-cooling methods can be applied to both feeder-dependent and feeder-free culture conditions. For seeding cells after thawing, MEF feeder layers or substrate-coated culture vessels can be prepared for routine maintenance of the cell lines. *See* **Note 1** for examples for the preparation of matrix coated plates. Although specific pretreatment of cells is not required, cells should be frozen at their sub-confluence.

2.2 Supplies

1. Dimethylsulfoxide (DMSO).
2. Acetamide.
3. Propylene glycol.
4. Dulbecco's Phosphate Buffered Saline without magnesium and calcium (D-PBS(−)).
5. Dulbecco's Modified Eagle's Medium (DMEM).
6. TrypLE Select dissociation medium (Life Technologies).
7. Ethylenediaminetetraacetic acid (EDTA·4Na, sodium salt).
8. Matrigel (BD BioSciences).
9. Laminin-521 (BioLamina).
10. iMatrix-511 (Iwai or Takara Bio).
11. mTeSR1 medium (Stem Cell Technologies).
12. TeSR2 medium (Stem Cell Technologies).

2.3 Reagents for Vitrification

1. DAP freezing medium (hPSC culture medium containing 2 M DMSO, 1 M Acetamide, and 3 M propylene glycol): Weigh 0.59 g of Acetamide and transfer to 15 mL conical tube containing 6.38 mL of ES/iPS cell culture medium. Add 1.42 mL of DMSO and 2.2 mL of Propylene glycol. Mix well and sterilize by filtering through 0.22 μm Millex-GV filter

(or equivalent). Aliquot and store at –80 °C. Avoid repeated freeze-thawing.

2. hPSC cell detachment solution, e.g., CTK solution (ReproCELL) (*see* **Note 2**).
3. Liquid nitrogen in appropriate reservoir (e.g., small styrofoam box) for flash freezing.

2.4 Reagents for Slow-Cooling

1. 2× concentrated freezing medium, e.g., hPSC culture medium containing 20 % DMSO (final concentration of DMSO is 10 %).
2. TrypLE select (Life technologies).
3. 0.2 % EDTA solution: Dissolve 0.2 g of EDTA·4Na into 100 mL of D-PBS(−). Sterilize by autoclave at 121 °C for 15 min (Equivalent is available).
4. Mr. Frosty™ Freezing Container (Thermo Scientific) (Equivalent is available).

2.5 Preparation of Culture Substrate-Coated Plates

1. Matrigel-Coated Plates: Prepare Matrigel in DMEM as described by the manufacturers and coat the culture vessels at 20 μg/cm^2 for 3 h at room temperature.
2. Laminin-521 Coated Plates: Prepare laminin-521 in D-PVS(−) and coat plates at 6 μg/cm^2 for 3 h at room temperature.
3. iMatrix-511 Coated Plates: iMatrix-511 is prepared in D-PBS(−) and plates are coated at 1.5 μg/cm^2 for 3 h at room temperature. Use the culture vessels removing supernatant.

3 Methods

3.1 Vitrification Method

The vitrification requires rapid freezing and thawing of hPSCs. It is therefore necessary to have all the necessary reagents and tools ready for the procedure. Cells in each dish should be frozen one by one to avoid long exposures of hPSCs to the cryoprotectant.

3.1.1 Freezing by Vitrification

The amount of each solution is adjusted to 60 mm culture dish.

1. Keep DAP freezing medium on ice and prepare liquid nitrogen in a proper container.
2. Remove culture supernatant from culture vessel.
3. Add 1 mL of detachment solution and incubate for 5 min at 37 °C until the edges of hPSC colonies detach (*see* **Note 3**).
4. Carefully rinse the culture surface once with 1 mL of hPSC culture medium to dilute and remove the detachment solution.
5. Add 5 mL of hPSC culture medium and detach hPSC colonies by gentle pipetting. Do not break up the colonies.

6. Transfer the hPSC suspension into a 15 mL centrifuge tube.
7. Centrifuge at 200 × *g* for 3 min.
8. Discard the supernatant after centrifugation.
9. Add 200 μL of DAP cryoprotectant, rapidly suspend cell pellet and transfer to a cryotube (*see* **Note 4**).
10. Immediately immerse the cryotube in liquid nitrogen (*see* **Note 5**).
11. Transfer the cryotube in liquid nitrogen tank or −150 °C freezer (*see* **Note 6**).

3.1.2 Thawing of Cells Frozen by Vitrification

1. Warm 3 mL of culture medium to 37 °C (*see* **Note 7**).
2. Keep the frozen cryotubes in liquid nitrogen until just prior to thawing (*see* **Note 8**).
3. Add the warmed culture medium to the cryotube and immediately thaw the frozen hPSC suspension by pipetting. Take care not to disaggregate the hPSC colonies.
4. Transfer the cell suspension into a 15 mL centrifuge tube containing 7 mL of cold hPSC culture medium (*see* **Note 9**).
5. Centrifuge at 200 × *g* for 3 min.
6. Discard the supernatant after centrifugation.
7. Suspend the cell pellet with an appropriate volume of hPSC culture medium.
8. Seed the cells into newly prepared culture vessels. Split ratio at seeding should be equal to that used prior to cryopreservation.
9. Incubate the culture vessels in CO_2 incubator.
10. Replace the hPSC culture medium every day until the hPSC grow to confluence.
11. Thereafter, subculture the hPSCs by routine passaging.

3.2 Slow-Cooling Method

To obtain high survival of hPSCs using the slow-cooling method, it is essential to cryopreserve hPSCs in a single cell suspension state. It is necessary to completely dissociate the hPSCs. Unlike vitrification, large numbers of sample vials can be processed (frozen or thawed) at the same time.

3.2.1 Cryopreservation by Slow-Cooling

The volume of each solution is adjusted for a 60 mm culture dish.

1. Cool the Mr. Frosty™ freezing container.
2. Cool the 2× concentrated freezing medium.
3. Remove culture supernatant from the cell cultures.
4. Gently wash the cells in the vessel with 5 mL of D-PBS(−).

5. Add 2 mL of EDTA solution into the culture vessel and incubate at room temperature until the hPSC colonies disaggregate (*see* **Note 10**).
6. Remove EDTA solution.
7. Add 1 mL of TrypLE select and remove immediately.
8. Incubate the culture vessel for just 1 min (*see* **Note 11**).
9. Add 1 mL of culture medium, resuspend the cells and transfer to a 15 mL centrifuge tube.
10. Add hPSC culture medium so that the total volume is 5 mL and centrifuge at 200 × *g* for 3 min.
11. Remove the supernatant.
12. Suspend the cell pellet with half of the volume of cold hPSC culture medium (e.g., 0.5 mL/tube).
13. Add 2× concentrated freezing medium dropwise, gently shaking the tube to gradually mix in the freezing medium.
14. Transfer the cell suspension to a cryotube and transfer this to the Mr. Frosty™ freezing container.
15. Transfer the freezing container to a –80 °C freezer and leave overnight.
16. Transfer the cryotubes to liquid nitrogen tank or –150 °C freezer.

3.2.2 Thawing of Cells Frozen by the Slow-Freezing Method

1. Warm the frozen cryotubes at 37 °C in water bath to thaw rapidly.
2. Transfer the cell suspension to a 15 mL centrifuge tube.
3. Add 9 mL of hPSC culture medium dropwise to the tube, gently shaking the tube to gradually mix the contents.
4. Centrifuge at 200 × *g* for 3 min.
5. Remove the supernatant.
6. Suspend cells with appropriate culture medium.
7. Perform a cell count.
8. Seed the cells with an appropriate volume of hPSC culture medium in newly prepared culture vessels at a density of 1×10^5 cells/cm^2 (*see* **Notes 12** and **13**).
9. Incubate culture vessels in CO_2 incubator.
10. Change the hPSC culture medium after 24 h. Then change the medium daily until hPSC grow to confluence (approximately 3 days post-thawing) (Fig. 1). Thereafter, subculture the hPSCs by routine passaging (*see* **Note 14**).

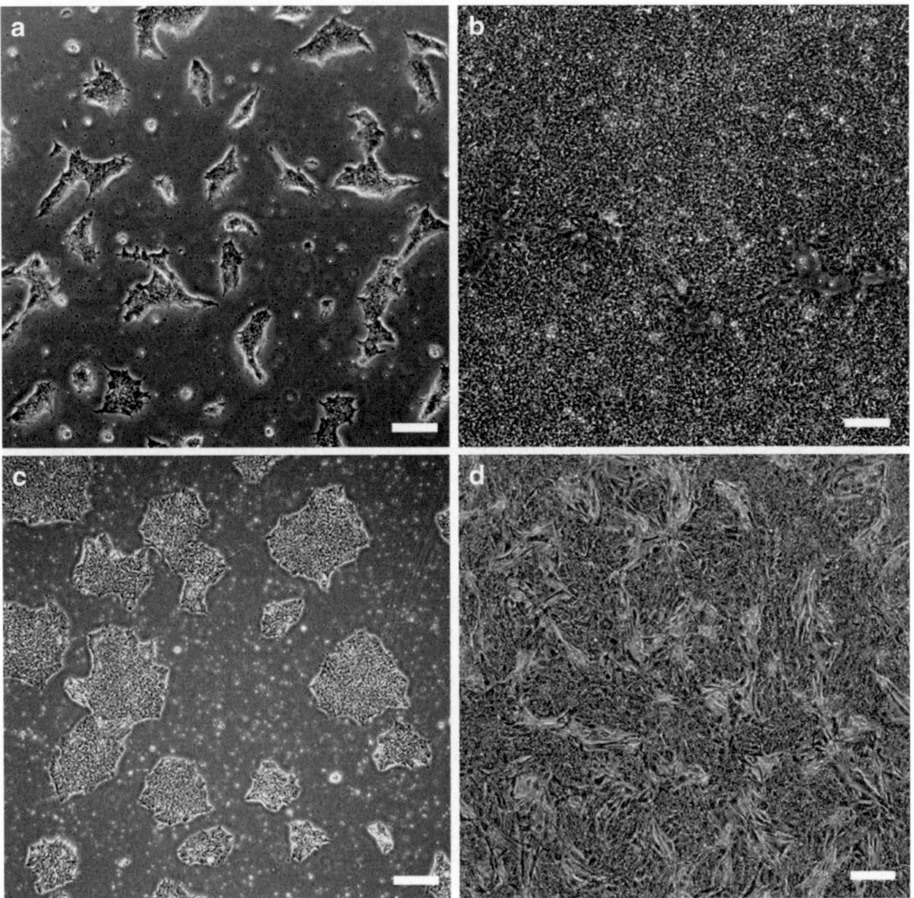

Fig. 1 Attachment and recovery of freeze-thawed, single hPSCs. (**a**, **b**) Phase contrast images of freeze-thawed, single hPSCs in mTeSR1 medium on Matrigel at 12 h (**a**) and 3 days (**b**) post-seeding at 3×10^5 cells/cm^2. (**c**) Phase contrast image of hPSC at subculture. (**d**) Phase contrast images of freeze-thawed single hPSCs on feeder-dish at 3 days post-seeding at 1×10^5 cells/cm^2. Scale bars: 200 μm. (**a–c**: reproduced from [5])

4 Notes

1. The culture substrates, laminin-521 and iMatrix-511 highly enhance the adhesion of single hPSCs. Single hPSCs can adhere to the Matrigel, but it is necessary to use a higher seeding density compared to the laminin or iMatrix-511 (*see* also **Note 12**). In addition to laminin-521, laminin-511 is also available.
2. Vitrification requires large colonies of hPSCs during the freeze–thaw process. A detachment solution should be used that allows removal of large colonies, but does not dissociate them into individual cells. Collagenase IV in DMEM also

works well, but dispase is not recommended even though it is used for subculture

3. In some instances, the hPSC colonies may not detach well from the surface. In this case, avoid excess treatment using the detachment solution. Instead, detach the colonies gently using a cell scraper.

4. and 5. The DAP cryoprotectant is harmful to hPSCs. **Steps 9–10** should be completed within 15 s.

6. When hPSCs are vitrified, they should be stored at temperatures below −150 °C. If stored at −80 °C, increased cellular damage can occur compared to the conventional slow-cooling method using 10 % DMSO.

7. It is essential that frozen hPSCs in DAP cryoprotectant be instantaneously thawed and diluted with hPSC culture medium. The hPSC culture medium must be warmed prior to use.

8. Exposure of frozen hPSCs, cryopreserved by vitrification, to room temperature causes damage to the hPSCs. Transfer the cryotubes in a liquid nitrogen container until just prior to use.

9. Once the hPSC suspension has been removed from liquid nitrogen, the cells should be thawed immediately. To avoid slow-thawing and multiple pipetting, quickly change the warmed culture medium to rapidly thaw the frozen hPSC suspension. Thawing and suspending in culture medium should be performed within 20 s.

10. and 11. Treatment with of TrypLE Select should be kept to a minimum. If it is difficult to detach hPSCs from culture surface, increase the incubation time with EDTA but do not change the treatment time with TrypLE Select.

12. The optimized seeding density, post-cryopreservation, varies depending on culture conditions as follows: 3×10^5 cells/cm^2 with mTeSR1 medium on Matrigel; 1×10^5 cells/cm^2 in TeSR2 medium on laminin-521 (or iMatrix-511); 1×10^5 cells/cm^2 on MEF feeder layer dishes.

13. If the survivability or viability of thawed, single hPSCs is low, increase the seeding density to match to the condition of individual hPSCs. Note that the survivability of thawed hPSCs when plated onto MEF feeder dishes is dependent upon the condition of MEF feeder cells. When the condition of the cells is low, add ROCK inhibitor Y-27632 at the time of seeding.

14. hPSCs cryopreserved using the slow-freezing method can rapidly attain confluency when cultured. Monitor the cells closely because confluency can also accelerate spontaneous differentiation.

Acknowledgements

This work was supported by New Energy and Industrial Technology Development Organization (NEDO), Japan; Ministry of Health, Labour and Welfare, Japan.

References

1. Fujioka T, Yasuchika K, Nakamura Y et al (2004) A simple and efficient cryopreservation method for primate embryonic stem cells. Int J Dev Biol 48:1149–1154
2. Suemori H, Yasuchika K, Hasegawa K et al (2006) Efficient establishment of human embryonic stem cell lines and long-term maintenance with stable karyotype by enzymatic bulk passage. Biochem Biophys Res Commun 345:926–932
3. Li T, Mai Q, Gao J et al (2010) Cryopreservation of human embryonic stem cells with a new bulk vitrification method. Biol Reprod 82: 848–853
4. Wagh V, Meganathan K, Jagtap S et al (2011) Effects of cryopreservation on the transcriptome of human embryonic stem cells after thawing and culturing. Stem Cell Rev 7: 506–517
5. Miyazaki T, Nakatsuji N, Suemori H (2013) Optimization of slow cooling cryopreservation for human pluripotent stem cells. Genesis 52(1):49–55

Chapter 10

Biological Differences Between Native and Cultured Mesenchymal Stem Cells: Implications for Therapies

Elena Jones and Richard Schäfer

Abstract

We describe the current knowledge of the surface marker phenotype of native bone marrow mesenchymal stem/stromal cells (MSCs) in humans and in mouse models, highlighting similarities in the MSC marker "signature" between the two species. The chapter proceeds to discuss the published literature pertaining to native MSC topography and their interactions with hematopoietic stem cells and their progeny, as well as with blood vessels and nerve endings. Additionally, the chapter describes phenotypic and functional "drifts" that occur in MSC preparations as they are taken out of their native bone marrow microenvironment and induced to proliferate in vitro (in the presence of animal or human serum). We propose that the understanding of the biology of MSCs in their native niches in the bone marrow could lead to future developments in the treatment of hematological diseases such as multiple myeloma. Additionally, this knowledge would assist in the development of more "natural" MSC culture conditions, best preserving MSC functionality including their homing potential in order to optimize MSC transplantation in the context of graft-versus-host and other diseases.

Key words Mesenchymal stem/stromal cells, MSCs, Bone marrow, Phenotype, Flow cytometry, Transcriptional profile

1 Introduction

Unlike hematopoietic stem cells (HSCs), a universally acceptable definition of mesenchymal stem/stromal cells (MSCs) has not been agreed upon [1, 2]. Certain essential criteria for MSCs were laid down several decades ago by Alexander Friedenstein and Maureen Owen; they postulated that these cells, at least in the bone marrow (BM), should be adherent, clonogenic, and multipotential towards a number of mesenchymal lineages including bone, cartilage, and the fibrous tissue [3–5]. Single MSC-derived clones from a human were capable of culture-expansion in fairly rudimentary conditions [3], but this was not the case for murine MSCs, which were easily overgrown by adherent macrophages and required more complex media to grow [6]. Later findings have documented significant chromosomal aberrations in murine

Ivan N. Rich (ed.), *Stem Cell Protocols*, Methods in Molecular Biology, vol. 1235,
DOI 10.1007/978-1-4939-1785-3_10, © Springer Science+Business Media New York 2015

long-term cultured MSCs [7, 8] suggesting that basic conditions commonly employed for human MSC culture were unsuitable for the maintenance of mouse MSCs.

Since MSCs are anchorage-dependent, adherent cells, their successful in vitro cultivation depends on at least two variables: (1) growth factors driving MSC proliferation and (2) attachment factors responsible from MSC adherence and spreading. The main growth factors required for human MSC proliferation are reasonably well understood; these include the members of the platelet-derived growth factor (PDGF) signalling pathway (primarily PDGF-BB), epidermal growth factor (EGF), transforming growth factor (TGF) beta, basic fibroblast growth factor (FGF), and insulin-like growth factor (IGF) [9–11]. Interestingly, the majority of these cytokines are released by platelets upon their activation [12]. Hence is not surprising that platelet lysates (PL) are becoming increasingly popular and begin to replace traditional fetal calf serum (FCS) for expanding human MSCs intended as therapy [13]. Fibronectin is the best studied attachment factor required for human MSC adherence to plastic [14, 15].

As mentioned above, culturing mouse MSCs is more difficult [16, 17]. Not only the best growth factor "cocktail" for mouse MSCs expansion remains unclear [18], the presence of other plastic-adherent cells such as macrophages appears inhibitory for MSC growth [6, 18–20]. Because of this, the study of native mouse MSCs have taken a somewhat different route compared to human MSCs. In many human studies as those described below, native MSCs are isolated using a candidate marker; subsequently, marker-positive cells are expanded and differentiated towards several mesenchymal lineages in vitro or in vivo. A limiting dilution assay is normally used to measure the frequency of "true" clonogenic and multipotential MSCs in the selected marker-positive fraction [21, 22]. Subsequently, the phenotype of freshly isolated cells is compared to that of their progeny at different stages of cultivation (i.e., during different passages, best measured as cumulative population doublings).

In a mouse system, due to culturing "challenges" outlined above, such methodology is limited [23]. The majority studies dedicated to finding the nature of murine MSCs have been performed using transgenic mice and gene tracing technologies [24–27]. Therefore, the data comparing the phenotype of native and cultured MSCs in the mouse system are comparatively limited.

2 Markers and Topographies of Native Human BM MSCs

2.1 Stro-1

The first marker of human MSCs proposed in 1995 was named Stro-1 [28] highlighting the "stromal," non-hematopoietic character of these cells (Table 1). The majority of BM Stro-1^{+} cells were

Table 1
Candidate surface markers of human BM MSCs

Marker	Extended phenotype	Changes in culture
Stro-1 [28]	PDGFR/CD140, EGFR/CD312, IGFR/CD220, NGFR/CD271 [11]	Decrease [28] Stable [62]
Low affinity NGFR/CD271 [30, 31, 81] (p75 in the mouse)[82]	CD90, CD13, CD10, CD105, Stro-1, low/negative for CD45 [31] CD146 [22, 34, 43–45] CD15, CD73, CD140b, CD144, CD200, MSCA-1/W8B2 [34, 43]	Decline [30, 31]
CD105 [47], CD73 [37, 46]	Negative for CD45 [37, 46]	Stable [31, 37, 43, 62]
CD106 [21, 83]	Stro-1 [21, 83]	Decline [44, 62] Increase [58] Expression depends on inflammatory environment [84]
CD146 [42, 85]	CD271 [22, 34, 43–45, 51]	Decline [44, 62] Increase [36] Expression depends on hypoxia levels [22]
SSEA-4 [50]	Negative for CD45 [50]	Increased [50]
GD2 [49]	CD105, CD73 [49]	Stable [49] Decline [86]
CD200 [51]	CD271 [43], Leptin Receptor/CD295, integrin alphaV/CD51 [51]	Increased by interferon-gamma [87]

shown to express a red cell lineage marker glycophorin A. However, all colony-forming cells were detected in the Stro-1$^+$glycophorin$^-$ fraction and an approximately 1 % of Stro-1$^+$glycophorin$^-$ cells formed colonies of fibroblasts in standard FCS-rich conditions [28]. The multipotentiality of Stro-1$^+$ cells was not reported in this initial study however their osteogenic potential was proven several years later [29]. Flow cytometry analysis revealed the specific presence of several growth factor receptors on the surface of Stro-1$^+$glycophorin$^-$ cells: PDGFR, IGFR, EGFR, and nerve growth factor (NGF) receptor [11]. The simultaneous addition of PDGF-BB and EGF resulted in the formation of the biggest colonies from Stro-1$^+$ cells under serum-deprived conditions [11]. In their initial study of Stro-1$^+$ cells, the same authors reported a gradual decline in Stro-1 expression following standard cultivation [28].

2.2 CD271

In 2002 two studies reported the suitability of using magnetic beads [30] or cell sorting [31] for the isolation of human BM MSCs based on low-affinity NGF receptor (CD271). The CD271-positive fraction was 45-fold enriched for colony-forming unit-fibroblast (CFU-F) [30] and was tri-potential following standard cultivation [31]; CD271 was rapidly lost following culture [30, 31]. CD271 binds NGF [11] and is involved in mediating IL-6 production by MSCs thus controlling MSC support of hematopoiesis [32].

In 2007 Buhring et al. used CD271-based gating strategy to discover several novel markers of native human BM MSCs (Table 1) [33]. The same strategy was later used by Battula et al. showing that a subset of CD271^{+}MSCA-1^{+} double-positive cells was most enriched for MSCs [34]. The most recent studies have employed CD271, either alone or in combination with these other markers, to select for native human BM MSCs. Churchman et al. have demonstrated a broad and highly specific mesenchymal lineage transcript expression in BM CD271-positive cells as well as a strong expression of CXCL12, the gene encoding stromal-derived factor 1 (SDF-1) involved in hematopoiesis [35]. Tormin et al. have eloquently shown that CD146 co-expression on CD271^{+} cells was associated with their topographic location: CD271^{+}CD146^{+} cells were positioned perivascularly, whereas CD271^{+}CD146^{-} cells were located in more hypoxic areas near the bone surface (the endosteal region) [22].

Another recently published study has documented the lack of CD44 on CD271^{+} native MSCs and an increase in its expression when MSCs began to grow [36]. The authors used a microarray-based approach to find more genes differentially expressed in native and cultured MSCs. The genes encoding surface molecules included: integrins A3, AE, B5, and A6, CD109, CD151, CD59, CD146, and CD248 (increased in culture) and VCAM1/CD106, CD271, CD36, and EPOR (decreased in culture) [36]. Importantly, these authors have a shown a dramatic downregulation of genes encoding cytokines and chemokines produced by MSCs, most notably CXCL12/SDF-1, consistent with a Churchman et al. study [35], a molecule to be further discussed in the next section. More recently, CD44-negativity of native BM MSCs was confirmed independently [37]; significantly, a different combination of markers was used to select for MSCs (CD45^{-}CD 90^{+}CD73^{+}CD105^{+}).

Besides multipotentiality, another important function of MSCs that increasingly attains significant therapeutic relevance is their considerable immunoregulatory capacity [38, 39]. Kuçi et al. reported in a recent in vitro study on remarkable differences of prostaglandin (PG) E2-mediated suppression of T cell proliferation amongst CD271^{+} BM-MSC clones identifying high PGE2-depending, low PGE2-depending, and non-PGE2-depending CD271^{+} MSC clones. Moreover, they observed, besides the typical spindle shaped cell morphology, CD271^{+} MSC clones with endothelial-like cell shapes [40].

2.3 Other Markers and Marker Combinations

In 2003 Gronthos et al. reported that cells positive for Stro-1 were able to generate clonal progeny with an in vivo bone forming capacity [41]. Furthermore, cells double-positive for Stro-1 with CD106 were more enriched for MSCs compared with Stro-1 alone [21]. The frequency of CFU-Fs in double-positive fractions, established by limiting dilution analysis, was as high as 30 % [21]. Furthermore, double-positive cells expressed high levels of transcripts associated with mesenchymal tripotentiality, such as osterix, osteocalcin, lipoprotein lipase and collagen X [21]. This study was amongst the first two reports by the same authors [42] to highlight the expression of integrin molecules on MSCs in vivo. As with many other molecules expressed on MSCs in vivo, the expression of both CD146 and CD106 were later shown to be modulated in cultured MSCs (Table 1). CD146 expression on native MSCs, in particular, was later corroborated in several independent studies [22, 34, 43–45].

In an alternative approach, two independent studies have used "classical" markers of cultured MSCs, i.e., CD73 and CD105, to enumerate and purify native MSCs in human bone marrow [46, 47]. Their attempts were successful; however, the use of single markers did not provide sufficient selectivity and the negative depletion of CD45 cells, in order to remove predominant hematopoietic lineage cells, was deemed beneficial [46] as these antigens are also expressed on non-MSC cell entities (CD73 on lymphocytes and neutrophils, CD105 on endothelial cells and macrophages) [48].

At approximately the same time, Martinez et al. proposed a new marker, a ganglioside GD2 to purify BM MSCs [49] whereas Gang et al. used another glycoprotein—SSEA-4—for successful MSC isolation from both human and mouse BM [50]. In both studies, a prior CD45 cell depletion was used to remove the bulk of hematopoietic lineage cells prior to native MSC selection. Another more recent study by Delorme et al. used CD200 to purify human BM MSCs [51].

The short format of this chapter does not allow a more comprehensive analysis of all studies aimed at isolating human MSCs; the list of potential candidate molecules remains large and growing [52]. Nevertheless, it is clear that the last two decades of intensive research in this field have identified a set of "overlapping" candidate markers; the most commonly used amongst which are CD271, Stro-1 and CD146 (Table 1). The expression of all three markers tends to change in standard FCS-containing culture (Table 1). Standard markers of cultured MSCs (CD105, CD73, and CD90) [53] tend to be stable; however, their strong cross-reaction with hematopoietic lineage cells does not make them particularly suitable, as single markers, for isolating rare BM MSCs admixed with much more numerous hematopoietic cells in the marrow [48].

3 Markers and Topographies of Native Murine BM MSCs

Initial studies have used depletion methodologies to isolate native BM MSCs from the mouse; cells belonging to different blood lineages (lymphoid, myeloid, etc.) were removed by "lineage depleting cocktails" [18]. In 2009, Morikawa et al. showed that mouse MSCs could be isolated based on the lineage negative, Sca-1$^+$PDGFRa$^+$ phenotype; ~4 % of sorted Sca-1$^+$PDGFRa$^+$ cells were colony-forming, ~67 % of which were tripotential [54]. These cells were "located in the arterial perivascular space near the inner surface of the cortical bone" [54]. As expected from their topography, Sca-1$^+$PDGFRa$^+$ cells produced abundant paracrine factors supporting hematopoiesis: CXCL12/SDF-1 and angiopoietin-1 (Ang1) [54]. Following extensive cultivation, their homing capacity to BM following systemic transplantation was lost [54]. In a separate study, the same group of authors have shown that topographically similar perivascular cells expressed Leptin Receptor/CD295 [55]; subsequently, the same authors have demonstrated that these perivascular cells, together with neighboring endothelial cells, formed a niche for hematopoietic stem cells (HSCs) [25].

At approximately the same time, Mendez-Ferrer et al. studied circadian rhythms of HSC release into the circulation [56]. They observed that the cyclical release of HSCs was controlled by Nestin$^+$CXCL12$^+$ stromal cells [56]. In the subsequent study, the same group have demonstrated the true MSC nature of these nestin-positive perivascular stromal cells, including the expression of ten MSC-related transcripts, in vivo [24]. Importantly, nestin-positive cells expressed a number of HSC maintenance genes (including CXCL12, Ang1, and interleukin-7 (IL7)), as well as CD106/VCAM, consistent with the Morikawa et al. study [54]. Most interestingly, nestin-positive cells were located in close proximity to nerve endings and expressed beta3-adrenergic receptors, which linked native MSC functionality with the sympathetic nervous system [24]. The most recent study from the Frenette's group has shown that nestin-positive cells featured a PDGFRa$^+$CD51$^+$ surface phenotype [27].

Knockdown mouse models provide a powerful tool to explore developmental origins of BM MSC. Using double-transgenic mouse systems, Komada et al. have recently shown that native murine BM MSCs formed a heterogeneous population of cells, some of which were neural-crest derived whereas the others were mesoderm-derived [26]. Chan et al. revealed further phenotypic heterogeneity of mouse BM MSCs in vivo, demonstrating the existence of the three subsets of stromal cells exhibiting differential expression of CD105, Thy1/CD90, and 6C3/ENPEP having differential capacities to support HSC maturation [57].

Table 2
Candidate surface markers of mouse BM MSCs

Marker	Extended phenotype	Changes in culture
Lineage-negative [18]	Sca-1, CD29, CD44, CD81, CD106 [18]	Stable [88]
Sca-1 [54]	PDGFRa, CD29, CD34, CD49e, CD105, CD133, CD140b, CD146 [16, 54]	Sca-1 and the extended phenotype is maintained [23] Increased in hypoxia [89]
PDGFRa (CD140a) [54]	Sca-1, CD29, CD49e, CD105, CD133, CD140b, CD146 [54]	PDGFa and the extended phenotype is maintained [23]
Leptin Receptor/CD295 [55]	ND	ND
Integrin alphaV/CD51 [57]	CD90/Thy1, CD105 (not completely overlapping) [57] PDGFRa, CD105,CD29, CD44, p75, Leptin R/CD295 [27]	ND
SSEA-4 [50]	CD45-negative [50]	Increased [50]

As seen from both Tables 1 and 2, there is an emerging commonality in the marker expression on the surface of native human and mouse BM MSCs. These overlapping markers include: CD105, PDGFR/CD140, CD106, CD51, Leptin receptor, and possibly CD271, CD146, and SSEA-4. Remarkably, both human $CD271^+$ cells [35, 36, 58] and mouse $nestin^+PDGFRa^+CD51^+$ cells express high levels of CXCL12/SDF-1 and other hematopoiesis-supporting molecules, lending further support to the idea that MSC activity and hematopoiesis-supporting activity are "co-segregated" in the BM [27]. Figure 1 illustrates the expression of Leptin receptor and SSEA-4 on the surface of human $CD271^+$ MSCs in vivo.

4 Differences in MSC Transcriptome and Receptome Between Native MSCs and MSCs Cultured in Standard Conditions

As outlined above, MSC "phenotypic drifts" are better studied in the human system, compared to the mouse system. Nevertheless, it is becoming clear that MSC hematopoiesis-supportive function is quite poorly maintained in standard MSC culture; this is evident by the dramatic loss CXCL12/SDF-1 and IL-7 expression in cultured MSCs compared to native MSCs, reported both in human [35–37] and mouse [27] studies. MSCs and their differentiated progeny form "niches" for HSC maintenance and maturation [24, 57] and the absence of hematopoietic lineage cells in vitro

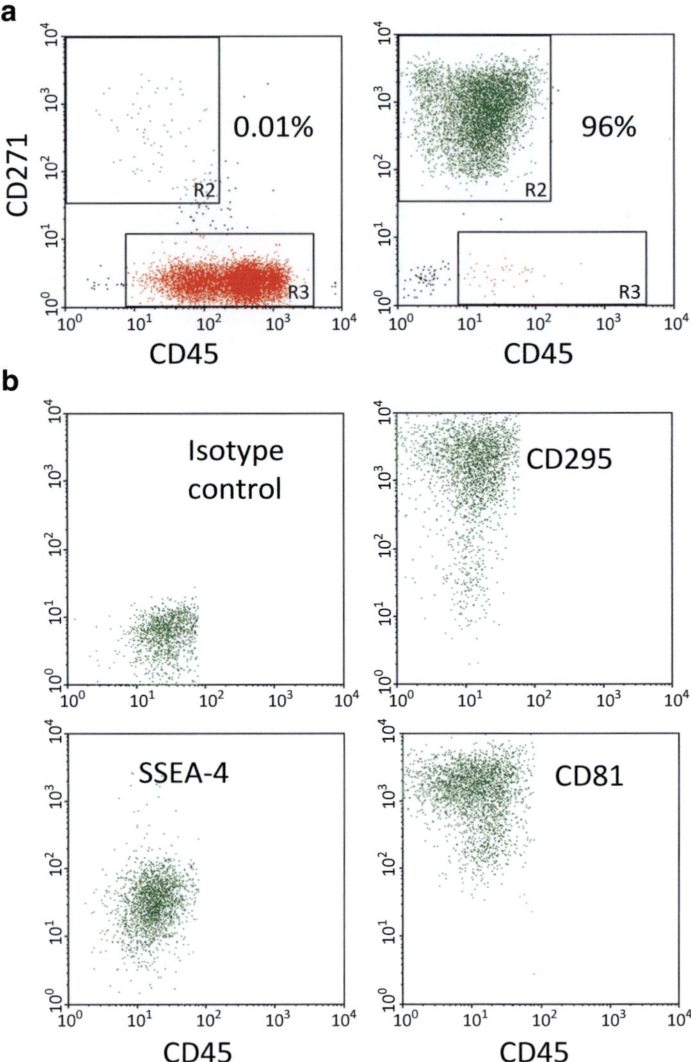

Fig. 1 The expression of Leptin R/CD295, SSEA-4, and the tetraspanin CD81 on human native BM MSCs. (**a**) Isolation of a pure population of MSCs by cell sorting for the CD45$^{-/low}$CD271^{+} phenotype (MSC purity is indicated in the *boxes*). *Left panel*—before sorting, *right panel*—after sorting. (**b**) Marker expression on pure MSCs: Isotype control, Leptin R/CD295, SSEA-4, and CD81

could be responsible for the notable loss of proliferative [58] and MSC hematopoiesis-supportive activity in standard culture conditions. Previous studies have reported the existence of MSC aggregates with megakaryocytes [59] and other maturing hematopoietic cells [60]. Megakaryocytes in particular, could control the "proliferation clock" in MSCs in vivo via generation of PDGFs thus affecting their proliferation [61] and osteogenesis [59].

Beyond the described changes in MSC hematopoiesis-supportive function, MSC surface phenotype tends to change following

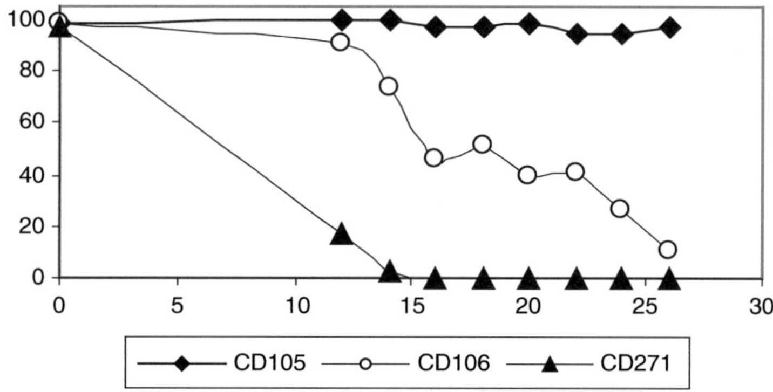

Fig. 2 The stable expression of CD105, a gradual decline in CD106, and a rapid loss of CD271 in long-term cultured MSCs compared to native MSCs. *X* axis—cumulative population doublings (CPDs), *Y* axis—% positive cells

standard cultivation. Both a decline and an increase in expression of some molecules have been documented (Table 1). Figure 2 illustrates the decline in CD271 and VCAM-1/CD106 (and the stable expression of CD105) in long-term cultured MSCs. The reasons for these dramatic phenotypic changes remain unclear but there are two mutually non-exclusive theories to explain that. According to one theory phenotypic "drifts" reflect major biochemical changes in MSCs as they adapt to in vitro culture conditions. For example, changes in CD146 expression could result from a highly oxygenated environment of standard cultivation [22] whereas changes in the MSC integrin profile could reflect adaptations to anchorage-dependent growth on plastic surfaces [62]. The quiescent status of native MSCs [21], in contrast to the dividing nature of cultured MSCs [58], could also bring major differences in their surface marker profile.

An alternative explanation is based on a documented evidence for morphological [63], phenotypic [34], and topographical [22] heterogeneity of MSCs in vivo. For example, immunohistochemical staining of human BM samples with markers that described MSC phenotypes in vitro revealed the presence of at least four non-hematopoietic stromal cell entities in the human BM in addition to perivascular cells [63]. These stromal cell entities were defined by morphology, marker expression and micro-anatomical localization (marrow, trabecular bone, and medullary cavity). According to this second theory, BM MSC isolation by plastic adherence and removal of non-adherent cells could lead on the one hand to a positive selection of MSC subpopulations that proliferate well at "standard" in vitro culture conditions at 21 % oxygen but on the other hand to a loss of rare subpopulations with possibly different phenotypes and functions (Fig. 3). Therefore, according to the second theory, a simple step of plastic adherence as well as current culture conditions could lead to the selection of some in vivo MSC subpopulations, but not the others.

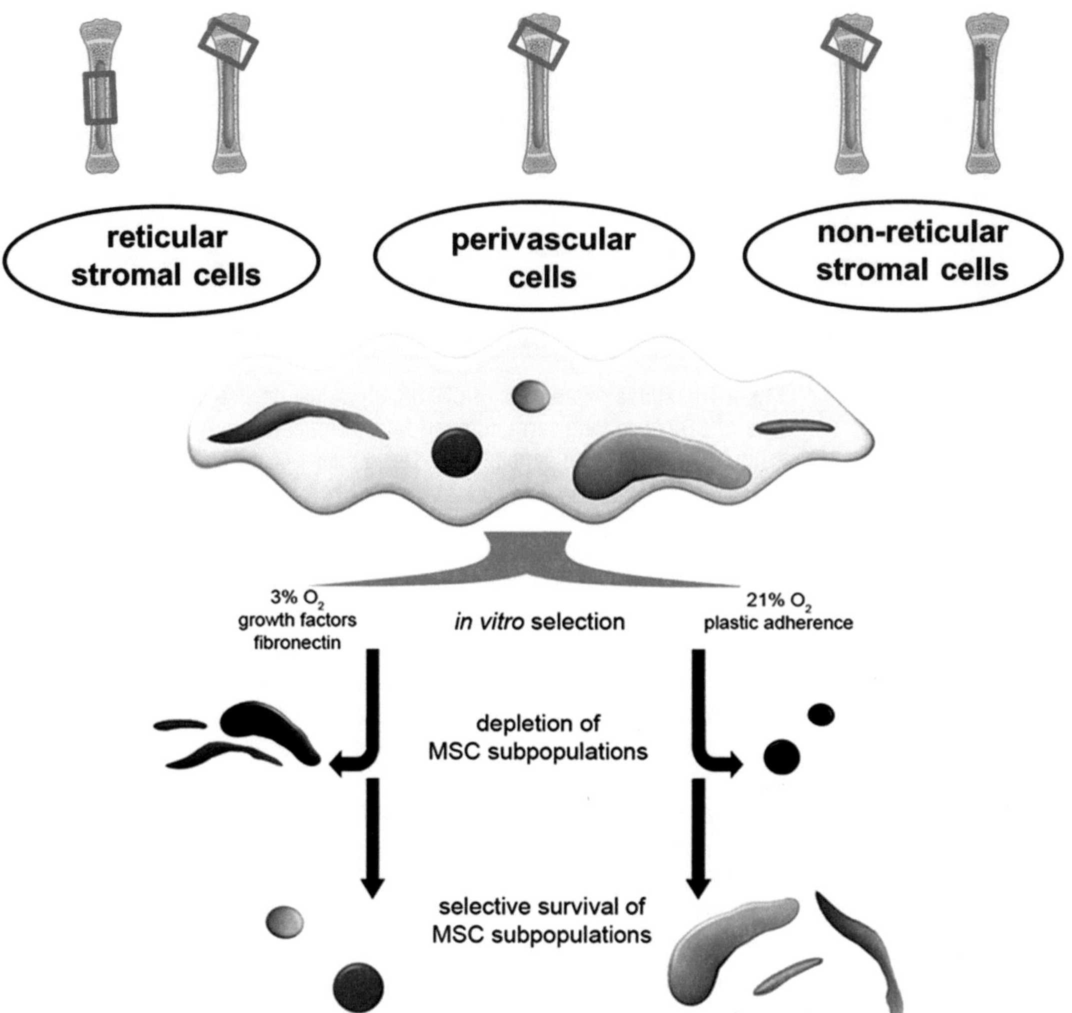

Fig. 3 Suggested concept of BM MSCs in vivo and in vitro. In the trabecular bone, reticular and non-reticular stromal cells as well as perivascular cells can be detected. Reticular stromal cells reside also in the medullary cavity and a subgroup of non-reticular stromal cells represents the bone lining cells [63]. BM MSC isolation by plastic adherence could lead on the one hand to a positive selection of MSC subpopulations that proliferate well at "standard" in vitro culture conditions at 21 % oxygen (*right panel*) but on the other hand to a loss of very rare subpopulations with possibly different phenotypes and functions that require more sophisticated isolation and culture conditions (*left panel*) [64, 67, 90, 91]

This view is supported by the observations that culturing BM plastic-adherent cells in low-oxygen conditions facilitated the growth of more potent MSC cultures [64]. Furthermore, the existence of non-plastic adherent MSCs have been proposed in two independent studies [65, 66]. A recent study based on in vitro analyses of multiple non-sorted human BM-MSC preparations addressed possible correlations of phenotypes to functional parameters [67]. Markers such as CD71, CD90, CD106, CD140b,

CD146, CD166, and CD274 were higher expressed in MSC preparations from younger donors. Moreover, CD10, CD29, CD44, CD71, CD106, CD119, CD146, CD166, CD271, and HLA class I correlated to BM-MSC clonogenic potential. CD119, CD146, and HLA class I correlated to the proliferation capacity of BM MSCs, whereas CD71, CD140b, and Galectin 1 correlated negatively to hepatocyte growth factor (HGF) secretion of BM MSCs [67]. It is reasonable to assume that the heterogeneity between and within BM MSC preparations observed in vitro reflects, at least in part, the in vivo situation.

5 Conclusions and Implications for Therapy

The therapeutic relevance of culture-expanded MSCs is clearly evident: MSCs have been and are actively used as cell therapy for graft-versus-host disease, Crohn disease, and other disease indications [2, 68, 69] as well as for engineering of tissues of musculoskeletal (bone, cartilage, tendon) and other systems [70, 71]. What would the knowledge of in vivo MSCs bring to modern Medicine?

One example would be creating a new knowledge on the interactions between MSCs and HSCs in the BM with a view of developing novel therapies for the treatment of hematological disease such as multiple myeloma (MM). As stated in previous sections, MSCs are intimately involved in controlling marrow hematopoiesis and in MM such a close cooperation between MSC and HSC is disturbed. As early as 1994, BM stromal cells from MM patients were noted to have an abnormal integrin profile [72]. A trend for increased CD271^{+} MSC numbers in MM compared to age-matched controls was recorded later [73] and confirmed more recently using the Stro-1 based methodology [74]. The authors have proposed that targeting MSCs and the modulation of the bone microenvironment in general could "alter the progression of the myeloma disease" [74]. On the other hand, functional deficits in MSCs have been documented in the myelodysplastic syndrome (MDS) disease spectrum: their premature senescence and reduced Ang1 production [75], an increased CXCL12 expression [76], and an aggravated genetic instability [77–79]. In order to study these abnormalities in more depth and to find new therapeutic solutions, an investigation of native MSCs is clearly warranted since the way MSCs from these patients are cultured "may significantly impact on results and is in fact likely to account for differences in conclusions drawn by different studies" [74]. The study of native MSCs in the disease of the musculoskeletal system, including osteoporosis and osteoarthritis, could also lead to the development of novel therapeutic agents targeting altered MSC differentiation pathways in bone leading to improved bone phenotypes [70].

The knowledge of the in vivo MSC surface marker "signature" could additionally help with the development of better expansion protocols "tailor-made" for specific disease indications or routes of administration. It has been proposed in several studies that the alteration in the integrin profile in cultured MSCs, compared to native MSCs, could be the reason responsible for the loss of their homing capacity following in vivo transplantation [25, 31, 51, 80]. As proposed in one study "loss of BM homing capacity of the cultured MSCs might be due to culture-induced changes in the expression of adhesion receptors including CD44 and CXCR4 on the cells, which might lead to unwanted entrapment in other organs" [36]. New media formulations could be therefore developed to "create" cultured MSCs with an adhesion molecule profile suitable for a particular mode of cell delivery.

In conclusion, the surface phenotype of native bone marrow MSCs is not identical to their culture-expanded progeny. Culture-induced changes on the MSC surface profile can be responsible for their reduced homing potential and hematopoiesis-supportive functions, due to the culture-induced loss of the appropriate surface receptors. Further study of native MSCs in their natural in vivo niches should facilitate the development of novel MSC expansion protocols. Furthermore, targeting native MSCs in situ could represent a novel means for the treatment of hematological diseases such as multiple myeloma.

References

1. Bianco P, Cao X, Frenette PS et al (2013) The meaning, the sense and the significance: translating the science of mesenchymal stem cells into medicine. Nat Med 19:35–42
2. Pittenger MF (2013) MSCs: science and trials. Nat Med 19:811–811
3. Friedens AJ, Deriglas UF, Kulagina NN et al (1974) Precursors for fibroblasts in different populations of hematopoietic cells as detected by in vitro colony assay method. Exp Hematol 2:83–92
4. Friedenstein AJ, Chailakhyan RK, Gerasimov UV (1987) Bone-marrow osteogenic stem-cells—in vitro cultivation and transplantation in diffusion-chambers. Cell Tissue Kinet 20: 263–272
5. Owen M, Friedenstein AJ (1988) Stromal stem-cells—marrow-derived osteogenic precursors. Ciba Found Symp 136:42–60
6. Friedenstein AJ (1976) Precursor cells of mechanocytes. Int Rev Cytol 47:327–359
7. Zhou YF, Bosch-Marce M, Okuyama H et al (2006) Spontaneous transformation of cultured mouse bone marrow-derived stromal cells. Cancer Res 66:10849–10854
8. Miura M, Miura Y, Padilla-Nash HM et al (2006) Accumulated chromosomal instability in murine bone marrow mesenchymal stem cells leads to malignant transformation. Stem Cells 24:1095–1103
9. Kuznetsov SA, Friedenstein AJ, Robey PG (1997) Factors required for bone marrow stromal fibroblast colony formation in vitro. Br J Haematol 97:561–570
10. Hirata J, Kaneko S, Nishimura J et al (1985) Effect of platelet-derived growth-factor and bone marrow-conditioned medium on the proliferation of human-bone marrow-derived fibroblastoid colony-forming cells. Acta Haematol 74:189–194
11. Gronthos S, Simmons PJ (1995) The growth-factor requirements of stro-1-positive human bone-marrow stromal precursors under serum-deprived conditions in-vitro. Blood 85: 929–940
12. Dhillon RS, Schwarz EM, Maloney MD (2012) Platelet-rich plasma therapy—future or trend? Arthritis Res Ther 14:219
13. Doucet C, Ernou I, Zhang YZ et al (2005) Platelet lysates promote mesenchymal stem cell

expansion: a safety substitute for animal serum in cell-based therapy applications. J Cell Physiol 205:228–236

14. Schwartz RE, Reyes M, Koodie L et al (2002) Multipotent adult progenitor cells from bone marrow differentiate into functional hepatocyte-like cells. J Clin Invest 109:1291–1302
15. Ogura N, Kawada M, Chang W-J et al (2004) Differentiation of the human mesenchymal stem cells derived from bone marrow and enhancement of cell attachment by fibronectin. J Oral Sci 46:207–213
16. Peister A, Mellad JA, Larson BL et al (2004) Adult stem cells from bone marrow (MSCs) isolated from different strains of inbred mice vary in surface epitopes, rates of proliferation, and differentiation potential. Blood 103:1662–1668
17. Krishinappa V, Boregowda SV, Phinney DG (2013) The peculiar biology of mouse mesenchymal stromal cells-oxygen is the key. Cytotherapy 15:536–541
18. Baddoo M, Hill K, Wilkinson R et al (2003) Characterization of mesenchymal stem cells isolated from murine bone marrow by negative selection. J Cell Biochem 89:1235–1249
19. Phinney DG, Kopen G, Isaacson RL et al (1999) Plastic adherent stromal cells from the bone marrow of commonly used strains of inbred mice: variations in yield, growth, and differentiation. J Cell Biochem 72:570–585
20. Xu S, De Becker A, Van Camp B et al (2010) An improved harvest and in vitro expansion protocol for murine bone marrow-derived mesenchymal stem cells. J Biomed Biotechnol. doi:10.1155/2010/105940
21. Gronthos S, Zannettino ACW, Hay SJ et al (2003) Molecular and cellular characterisation of highly purified stromal stem cells derived from human bone marrow. J Cell Sci 116: 1827–1835
22. Tormin A, Li O, Brune JC et al (2011) CD146 expression on primary non-hematopoietic bone marrow stem cells correlates to in situ localization. Blood 117:5067–5077
23. Houlihan DD, Mabuchi Y, Morikawa S et al (2012) Isolation of mouse mesenchymal stem cells on the basis of expression of Sca-1 and PDGFR-alpha. Nat Protoc 7:2103–2111
24. Mendez-Ferrer S, Michurina TV, Ferraro F et al (2010) Mesenchymal and haematopoietic stem cells form a unique bone marrow niche. Nature 466:829–834
25. Ding L, Morrison SJ (2013) Haematopoietic stem cells and early lymphoid progenitors occupy distinct bone marrow niches. Nature 495:231–235
26. Komada Y, Yamane T, Kadota D et al (2012) Origins and properties of dental, thymic, and bone marrow mesenchymal cells and their stem cells. PLoS One 7:e46436
27. Pinho S, Lacombe J, Hanoun M et al (2013) PDGFR alpha and CD51 mark human Nestin(+) sphere-forming mesenchymal stem cells capable of hematopoietic progenitor cell expansion. J Exp Med 210:1351–1367
28. Simmons PJ, Torok SB (1991) Identification of stromal cell precursors in human bone-marrow by a novel monoclonal-antibody, Stro-1. Blood 78:55–62
29. Gronthos S, Graves SE, Ohta S et al (1994) The Stro-1(+) fraction of adult human bone-marrow contains the osteogenic precursors. Blood 84:4164–4173
30. Quirici N, Soligo D, Bossolasco P et al (2002) Isolation of bone marrow mesenchymal stem cells by anti-nerve growth factor receptor antibodies. Exp Hematol 30:783–791
31. Jones EA, Kinsey SE, English A et al (2002) Isolation and characterization of bone marrow multipotential mesenchymal progenitor cells. Arthritis Rheum 46:3349–3360
32. Rezaee F, Rellick SL, Piedimonte G et al (2010) Neurotrophins regulate bone marrow stromal cell IL-6 expression through the MAPK pathway. PLoS One 15:e9690
33. Buhring H-J, Battula VL, Treml S et al (2007) Novel markers for the prospective isolation of human MSC. In: Kanz L, Weisel KC, Dick JE, Fibbe WE (Eds) Hematopoietic stem cells, vol 1106. Ann NY Acad Sci. pp 262–271
34. Battula VL, Treml S, Bareiss PM et al (2009) Isolation of functionally distinct mesenchymal stem cell subsets using antibodies against CD56, CD271, and mesenchymal stem cell antigen-1. Haematologica 94:173–184
35. Churchman SM, Ponchel F, Boxall SA et al (2012) Transcriptional profile of native CD271+ multipotential stromal cells: evidence for multiple fates, with prominent osteogenic and Wnt pathway signaling activity. Arthritis Rheum 64:2632–2643
36. Qian H, Le Blanc K, Sigvardsson M (2012) Primary mesenchymal stem and progenitor cells from bone marrow lack expression of CD44. J Biol Chem 287:25795–25807
37. Hall SRR, Jiang Y, Leary E et al (2013) Identification and isolation of small CD44-negative mesenchymal stem/progenitor cells from human bone marrow using elutriation and polychromatic flow cytometry. Stem Cells Transl Med 2:567–578
38. Barry F, Murphy M (2013) Mesenchymal stem cells in joint disease and repair. Nat Rev Rheumatol 9:584–594
39. Krampera M, Galipeau J, Shi Y et al (2013) Immunological characterization of multipotent

mesenchymal stromal cells—the International Society for Cellular Therapy (ISCT) working proposal. Cytotherapy 15:1054–1061

40. Kuçi Z, Seiberth J, Latifi-Pupovci H et al (2013) Clonal analysis of multipotent stromal cells derived from CD271+ bone marrow mononuclear cells: functional heterogeneity and different mechanisms of allosuppression. Haematologica 98:1609–1616
41. Gronthos S, Chen SQ, Wang CY et al (2003) Telomerase accelerates osteogenesis of bone marrow stromal stem cells by upregulation of CBFA1, osterix, and osteocalcin. J Bone Miner Res 18:716–722
42. Shi S, Gronthos S (2003) Perivascular niche of postnatal mesenchymal stem cells in human bone marrow and dental pulp. J Bone Miner Res 18:696–704
43. Buhring HJ, Treml S, Cerabona F et al (2009) Phenotypic characterization of distinct human bone marrow-derived MSC subsets. In: Kanz L, Weisel KC, Dick JE, Fibbe WE (Eds) Hematopoietic stem cells, vol 1176. Ann NY Acad Sci. pp 124–134
44. Jones E, English A, Churchman SM et al (2010) Large-scale extraction and characterization of CD271+ multipotential stromal cells from trabecular bone in health and osteoarthritis: implications for bone regeneration strategies based on uncultured or minimally cultured multipotential stromal cells. Arthritis Rheum 62:1944–1954
45. Maijenburg MW, Kleijer M, Vermeul K et al (2012) The composition of the mesenchymal stromal cell compartment in human bone marrow changes during development and aging. Haematologica 97:179–183
46. Veyrat-Masson R, Boiret-Dupre N, Rapatel C et al (2007) Mesenchymal content of fresh bone marrow: a proposed quality control method for cell therapy. Br J Haematol 139: 312–320
47. Aslan H, Zilberman Y, Kandel L et al (2006) Osteogenic differentiation of noncultured immunoisolated bone marrow-derived CD105+ cells. Stem Cells 24:1728–1737
48. Boxall SA, Jones E (2012) Markers for characterization of bone marrow multipotential stromal cells. Stem Cells Int 2012: 975871–975871
49. Martinez C, Hofmann TJ, Marino R et al (2007) Human bone marrow mesenchymal stromal cells express the neural ganglioside GD2: a novel surface marker for the identification of MSCs. Blood 109:4245–4248
50. Gang EJ, Bosnakovski D, Figueiredo CA et al (2007) SSEA-4 identifies mesenchymal stem cells from bone marrow. Blood 109:1743–1751
51. Delorme B, Ringe J, Gallay N et al (2008) Specific plasma membrane protein phenotype of culture-amplified and native human bone marrow mesenchymal stem cells. Blood 111: 2631–2635
52. Harichandan A, Buehring H-J (2011) Prospective isolation of human MSC. Best Pract Res Clin Haematol 24:25–36
53. Dominici M, Le Blanc K, Mueller I et al (2006) Minimal criteria for defining multipotent mesenchymal stromal cells. The International Society for Cellular Therapy position statement. Cytotherapy 8:315–317
54. Morikawa S, Mabuchi Y, Kubota Y et al (2009) Prospective identification, isolation, and systemic transplantation of multipotent mesenchymal stem cells in murine bone marrow. J Exp Med 206:2483–2496
55. Ding L, Saunders TL, Enikolopov G et al (2012) Endothelial and perivascular cells maintain haematopoietic stem cells. Nature 481:457–465
56. Mendez-Ferrer S, Lucas D, Battista M et al (2008) Haematopoietic stem cell release is regulated by circadian oscillations. Nature 452: 442–444
57. Chan CKF, Lindau P, Jiang W et al (2013) Clonal precursor of bone, cartilage, and hematopoietic niche stromal cells. Proc Natl Acad Sci U S A 110:12643–12648
58. Tormin A, Brune JC, Olsson E et al (2009) Characterization of bone marrow-derived mesenchymal stromal cells (MSC) based on gene expression profiling of functionally defined MSC subsets. Cytotherapy 11:114–128
59. Miao D, Murant S, Scutt N et al (2004) Megakaryocyte-bone marrow stromal cell aggregates demonstrate increased colony formation and alkaline phosphatase expression in vitro. Tissue Eng 10:807–817
60. Blazsek I, Chagraoui J, Peault B (2000) Ontogenic emergence of the hematon, a morphogenetic stromal unit that supports multipotential hematopoietic progenitors in mouse bone marrow. Blood 96:3763–3771
61. Castromalaspina H, Rabellino EM, Yen A et al (1981) Human megakaryocyte stimulation of proliferation of bone-marrow fibroblasts. Blood 57:781–787
62. Jung EM, Kwon O, Kwon K-S et al (2011) Evidences for correlation between the reduced VCAM-1 expression and hyaluronan synthesis during cellular senescence of human mesenchymal stem cells. Biochem Biophys Res Commun 404:463–469
63. Rasini V, Dominici M, Kluba T et al (2013) Mesenchymal stromal/stem cells markers in the human bone marrow. Cytotherapy 15: 292–306

64. D'Ippolito G, Diabira S, Howard GA et al (2004) Marrow-isolated adult multilineage inducible (MIAMI) cells, a unique population of postnatal young and old human cells with extensive expansion and differentiation potential. J Cell Sci 117:2971–2981
65. Baksh D, Zandstra PW, Davies JE (2007) A non-contact suspension culture approach to the culture of osteogenic cells derived from a CD49e(low) subpopulation of human bone marrow-derived cells. Biotechnol Bioeng 98:1195–1208
66. Otsuru S, Gordon PL, Shimono K et al (2012) Transplanted bone marrow mononuclear cells and MSCs impart clinical benefit to children with osteogenesis imperfecta through different mechanisms. Blood 120:1933–1941
67. Siegel G, Kluba T, Hermanutz-Klein U et al (2013) Phenotype, donor age and gender affect function of human bone marrow-derived mesenchymal stromal cells. BMC Med 11:146
68. Dryden GW (2009) Overview of stem cell therapy for Crohn's disease. Expert Opin Biol Ther 9:841–847
69. van Laar JM, Tyndall A (2006) Adult stem cells in the treatment of autoimmune diseases. Rheumatology 45:1187–1193
70. Jones E, Yang XB (2011) Mesenchymal stem cells and bone regeneration: current status. Injury 42:562–568
71. Smith JO, Aarvold A, Tayton ER et al (2011) Skeletal tissue regeneration: current approaches, challenges, and novel reconstructive strategies for an aging population. Tissue Eng B Rev 17:307–320
72. Gregoretti MG, Gottardi D, Ghia P (1994) Characterization of bone-marrow stromal cells from multiple-myeloma. Leuk Res 18: 675–682
73. Jones EA, English A, Kinsey SE et al (2006) Optimization of a flow cytometry-based protocol for detection and phenotypic characterization of multipotent mesenchymal stromal cells from human bone marrow. Cytometry B Clin Cytom 70:391–399
74. Noll JE, Williams SA, Tong CM et al (2013) Myeloma plasma cells alter the bone marrow microenvironment by stimulating the proliferation of mesenchymal stromal cells. Haematologica 99(1):163–171. doi:10.3324/haematol.2013.090977
75. Geyh S, Oz S, Cadeddu RP et al (2013) Insufficient stromal support in MDS results from molecular and functional deficits of mesenchymal stromal cells. Leukemia 27:1841–1851
76. Flores-Figueroa E, Varma S, Montgomery K et al (2012) Distinctive contact between CD34+ hematopoietic progenitors and CXCL12+ CD271+ mesenchymal stromal cells in benign and myelodysplastic bone marrow. Lab Invest 92:1330–1341
77. Flores-Figueroa E, Arana-Trejo RM, Gutierrez-Espindola G et al (2005) Mesenchymal stem cells in myelodysplastic syndromes: phenotypic and cytogenetic characterization. Leuk Res 29:215–224
78. Lopez-Villar O, Garcia JL, Sanchez-Guijo FM et al (2009) Both expanded and uncultured mesenchymal stem cells from MDS patients are genomically abnormal, showing a specific genetic profile for the 5q-syndrome. Leukemia 23:664–672
79. Song L-X, Guo J, He Q et al (2012) Bone marrow mesenchymal stem cells in myelodysplastic syndromes: cytogenetic characterization. Acta Haematol 128:170–177
80. Bensidhoum M, Chapel A, Francois S et al (2004) Homing of in vitro expanded Stro-1(−) or Stro-1(+) human mesenchymal stem cells into the NOD/SCID mouse and their role in supporting human CD34 cell engraftment. Blood 103:3313–3319
81. Kuci S, Kuci Z, Kreyenberg H et al (2010) CD271 antigen defines a subset of multipotent stromal cells with immunosuppressive and lymphohematopoietic engraftment-promoting properties. Haematologica 95:651–659
82. Kurth TB, Dell'Accio F, Crouch V et al (2011) Functional mesenchymal stem cell niches in adult mouse knee joint synovium in vivo. Arthritis Rheum 63:1289–1300
83. Gronthos S, Franklin DM, Leddy HA et al (2001) Surface protein characterization of human adipose tissue-derived stromal cells. J Cell Physiol 189:54–63
84. Laschober GT, Brunauer R, Jamnig A et al (2011) Age-specific changes of mesenchymal stem cells are paralleled by upregulation of CD106 expression as a response to an inflammatory environment. Rejuvenation Res 14: 119–131
85. Sacchetti B, Funari A, Michienzi S et al (2007) Self-renewing osteoprogenitors in bone marrow sinusoids can organize a hematopoietic microenvironment. Cell 131:324–336
86. Kozanoglu I, Boga C, Ozdogu H et al (2009) Human bone marrow mesenchymal cells express NG2: possible increase in discriminative ability of flow cytometry during mesenchymal stromal cell identification. Cytotherapy 11: 527–533
87. Najar M, Raicevic G, Jebbawi F et al (2012) Characterization and functionality of the CD200-CD200R system during mesenchymal

stromal cell interactions with T-lymphocytes. Immunol Lett 146:50–56

88. Sun S, Guo Z, Xiao X et al (2003) Isolation of mouse marrow mesenchymal progenitors by a novel and reliable method. Stem Cells 21: 527–535
89. Valorani MG, Germani A, Otto WR et al (2010) Hypoxia increases Sca-1/CD44 co-expression in murine mesenchymal stem cells and enhances their adipogenic differentiation potential. Cell Tissue Res 341:111–120
90. Roobrouck VD, Clavel C, Jacobs SA et al (2011) Differentiation potential of human postnatal mesenchymal stem cells, mesoangioblasts, and multipotent adult progenitor cells reflected in their transcriptome and partially influenced by the culture conditions. Stem Cells 29:871–882
91. Serafini M, Dylla SJ, Oki M et al (2007) Hematopoietic reconstitution by multipotent adult progenitor cells: precursors to long-term hematopoietic stem cells. J Exp Med 204: 129–139

Chapter 11

The Use of Multiparameter Flow Cytometry and Cell Sorting to Characterize Native Human Bone Marrow Mesenchymal Stem Cells (MSC)

Sally Boxall and Elena Jones

Abstract

This chapter describes a method for identification, phenotypic analysis, and cell sorting of rare mesenchymal stem cells (MSCs) from human bone marrow (BM) aspirates. The native BM MSC population is identified based on the CD45$^{-/low}$CD271^{+} phenotype. The method consists of three related procedures: Procedure 1 involves a microbead-based pre-enrichment step. Two other procedures describe direct flow cytometric analysis of MSCs following the isolation of the mononuclear cell (MNC) fraction (Procedure 2) or more rapidly, following a simple ammonium chloride-based red cell lysis (Procedure 3). Recently described multi-lineage transcript expression in the CD45$^{-/low}$CD271^{+} cells suggests that the native BM MSC fraction could be further subdivided into functionally distinct subpopulations. The present protocols are hoped to help MSC biologists to enter this exciting field of research and to take it forward towards a better understanding of MSC biology in vivo.

Key words Mesenchymal stem cells, MSCs, Flow cytometry, Cell sorting, Human bone marrow

1 Introduction

The presence of mesenchymal stem cells (MSCs) in the bone marrow (BM) was first demonstrated by expanding single adherent cells in vitro and testing their multipotentiality in vivo using a mouse diffusion chamber model [1, 2]. These and the following corroborative experiments have established that the frequency of MSCs in aspirated or flushed-out marrow was approximately 1 cell per 10,000 of mononuclear cells [3], which was ~100-fold lower than the frequency of their hematopoietic progenitor counterparts. Such a low percentage of native MSCs (in the region of 0.01–0.1 % of total cells) has posed a significant problem for flow cytometrists in the past, because early-generation FACS machines did not have the required data collection and processing power to enable rare-event analysis of millions of cells. Consequently, the original studies aimed to discover native MSC markers have utilized the so-called

Ivan N. Rich (ed.), *Stem Cell Protocols*, Methods in Molecular Biology, vol. 1235,
DOI 10.1007/978-1-4939-1785-3_11,

pre-enrichment methodologies whereby a candidate antibody, in conjunction with magnetic beads, was used to enrich the candidate cellular fraction [4–7] and to analyze whether it was indeed enriched for MSCs by performing functional colony assays. In the last 10 years, having assembled a panel of several candidate "positive" and "negative" markers [8] as well as having more powerful flow cytometry instrumentation, it has become possible to perform "direct" flow cytometry investigations to measure and sort native BM MSCs [9–11]. This chapter provides some procedures developed in our laboratory for the analysis and sorting of human BM MSCs either with or without the pre-enrichment step.

The identity of the "best" marker for MSC identification, in both human and mouse systems, remains disputed [8]. Whilst further work in this field could indeed identify more powerful and more selective markers, the purpose of this chapter is to illustrate a possibility of a direct study of native BM MSCs without resorting to culture expansion. In our laboratory we use the $CD45^{-/low}CD271^{+}$ phenotype to identify native human BM MSCs [7, 9, 12, 13]. Independent investigators have later shown that $CD45^{-/low}CD271^{+}$ MSCs could be analyzed without prior enrichment [14] and several independent groups have subsequently documented near-100 % purity of $CD45^{-/low}CD271^{+}$ MSCs to describe their transcriptional signature [13, 10, 11].

The need for a pre-enrichment step (Procedure 1) may now become obsolete. However, as shown below, it may be beneficial for allowing faster cell sorting, thus ensuring a better MSC viability post-sort. Direct analysis of MSCs can be performed either following the isolation of the mononuclear cell (MNC) fraction (Procedure 2) [15] or more rapidly, following a simple ammonium chloride-based lysis of red cells (Procedure 3) [16]. The latter also allows for direct volumetric enumeration of MSCs in a given volume of marrow sample [16]. The protocols and notes also highlight critical steps and potential sources of variability.

2 Materials

2.1 Native MSC Analysis Following Positive Selection Using MACs Columns

1. MACS buffer: PBS, 0.5 % BSA, 2 mM EDTA.
2. K_2 EDTA Vacuette (BD Biosciences).
3. Dulbecco's Phosphate Buffered Saline (PBS) (Invitrogen Life Technologies).
4. EDTA.
5. Lymphoprep™ (Axis-Shield).
6. Trypan Blue.
7. Acetic acid.
8. Anti-fibroblast Microbeads (Miltenyi Biotec).

9. CD271 Microbeads (Miltenyi Biotec).
10. MS columns (Miltenyi Biotec).
11. LS columns (Miltenyi Biotec).
12. Bovine Serum Albumin (BSA).
13. CD45 FITC (Clone T29/33) (Dako).
14. CD271 PE (Clone C40-1547) (BD Biosciences).
15. 7AAD (BD Biosciences).
16. IgG1 PE (BD Biosciences).
17. IgG1 FITC (AbD Serotec).
18. Dulbecco's Modified Eagle's Medium (DMEM) Low glucose (Invitrogen Life Technologies).
19. Fetal Bovine Serum (FBS).

2.2 Phenotypic Analysis of Native MSCs Population in the MNC Fraction Without Pre-enrichment

1. Acetic acid.
2. DNase1 (Sigma-Aldrich).
3. Bovine Serum Albumin (BSA).
4. CD45 PE-Cy7 (Clone HI30, BD Biosciences).
5. CD73 Pe-Cy5.5 (Clone AD2, BD Biosciences).
6. CD90 FITC (Clone F15-42-1, AbD Serotec).
7. CD271-APC (Clone ME20.4-1.H4, Miltenyi Biotec).
8. IgG1 PE (BD Biosciences).
9. IgG1 FITC (AbD Serotec).
10. IgG1 PE (AbD Serotec).
11. IgG1 PcP-Cy5.5 (BD Biosciences).
12. IgG1 PE-Cy7 (BD Biosciences).
13. IgG1 APC (Miltenyi Biotec).
14. FCR Block (Miltenyi Biotec).
15. DAPI (Sigma-Aldrich).

2.3 Direct Native MSC Analysis Following Ammonium Chloride Lysis

1. Ammonium Chloride Solution: 168 mM NH_2Cl, 10 mM $KHCO_3$, 1 mM EDTA, pH 8.0.

3 Methods

3.1 Native MSC Analysis Following Positive Selection Using MACs Columns

1. Collect bone marrow (BM) sample into a K_2 EDTA vacutainer and mix well. Dilute 1:1 with PBS, mixing gently (*see* **Note 1**).
2. Ensure that the Lymphoprep™ being used is at room temperature (approximately 20 °C) and add 20 mL to a 50 mL falcon

tube or 5 mL to a 15 mL falcon tube depending on the amount of sample being used. Layer the diluted sample on the top of the Lymphoprep. If using a 50 mL tube, layer up to 30 mL of diluted sample onto 20 mL of Lymphoprep. For more than 30 mL divide across multiple tubes. For less than 10 mL of diluted sample layer onto 5 mL of Lymphoprep in a 15 mL Falcon tube as this will give a more visible interface layer. Pipette slowly so as not to disturb the bottom Lymphoprep layer (*see* **Note 2**).

3. Centrifuge at 600 × *g* for 20 min (no brake).
4. After centrifuging, carefully remove the "cloudy" interface layer containing the mononuclear cells (MNCs) and transfer to a 50 mL tube (the sample can be pooled if multiple tubes were used).
5. Add at least an equal volume of PBS and mix gently to wash.
6. Centrifuge at 400 × *g* for 5 min (with brake on).
7. After centrifuging carefully discard the supernatant. Tap the pellet to resuspend (if the pellet is large this may take several minutes to break up all the clumps) and make up to 1 mL with MACS buffer. Depending on the cell concentration you may need to increase MACS buffer volume up to 2–5 mL.
8. Count cells on a hemocytometer using a standard trypan blue exclusion assay. Bone marrow samples usually require a dilution of around 1:20 to count, but if the cell count is too high either resuspend the sample in a larger volume and recount or make a larger dilution. If there are still large amounts of red cells present, use 4 % acetic acid to dilute and count the cells instead of trypan blue as this will lyse the red cells and give a more accurate count (*see* **Note 3**).
9. Add 5 mL of MACS buffer to wash (*see* **Note 4**).
10. Centrifuge at 400 × *g* for 5 min (with brake on). Resuspend pellet in 80 μL of MACS buffer per 10^7 cells.
11. For Anti-fibroblast Microbeads: Add 20 μL of Anti-fibroblast Microbeads per 10^7 cells. Mix well and incubate at room temperature for 30 min. For Anti-CD271 Microbeads: Add 20 μL of FCR block and 20 μL of Anti-CD271 Microbeads per 10^7 cells. Mix well and incubate at 4 °C for 20 min.
12. Add 5 mL of MACS buffer to wash. Centrifuge at 400 × *g* for 5 min (with brake on).
13. Resuspend in 500 μL of MACS buffer for up to 10^8 cells. For larger cell numbers scale up accordingly. For samples with large numbers of red cells resuspend in a larger volume.
14. Set aside 1×10^6 Microbead-labeled cells in a separate "Before separation" tube. Keep these cells on ice. They will be needed

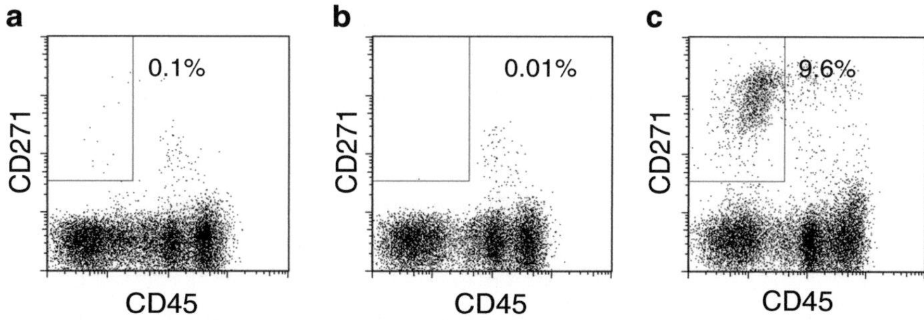

Fig. 1 Example profiles of CD45/CD271 staining (**a**) Before enrichment, (**b**) Negative fraction, and (**c**) Positive fraction, showing ~96-fold enrichment of the $CD45^{-/low}CD271^{+}$ cells in the positive fraction compared to Before enrichment

for flow cytometry (Fig. 1) or CFU-F assay [7, 17] to check efficiency of MSC isolation.

15. The choice of column is important and depends upon the number of cells. MS columns are recommended for up to 2×10^8 total cells or up to 10^7 labeled cells and LS columns for up to 2×10^9 total cells or 2×10^8 labeled cells. However, small MS columns provide better selection, but they are slow and may "clog" if too many cells are applied at once. Keep the number of cells low and do not put more than 5×10^7 total cells onto an MS column. Add cells in small aliquots and let them drip through before applying more. You can alternate adding the cells with drops of MACS buffer to allow for a better flow. LS columns are fast and hence MSC enrichment is normally not as good. Use LS columns if you have more than 5×10^7 initial cells. After eluting the cells from the LS column the "positive" fraction can be put through second (MS) column to achieve better MSC enrichment.
16. Prepare columns by washing in MACS buffer to remove the preservative; 0.5 mL for MS columns, 3 mL for LS columns, and discard the flow through.
17. Put MACS buffer containing cells through a 40 μm cell strainer and then apply to the column placed in MACS magnet (*see* **Note 5**). Collect negative fraction in "Negative fraction" tube (Fig. 1).
18. Wash column through with appropriate volume of MACS buffer; 3 × 500 μl for MS column, 3 × 3 mL for LS column. Add the buffer sequentially so that the reservoir empties of buffer before adding more (*see* **Note 6**).
19. Add appropriate volume of MACS buffer (1 mL for MS column, 5 mL for LS column) to release positive fraction. Remove from magnet and push the plunger into the column reservoir

to release the positive fraction from the column into a clean tube labeled "Positive fraction" (Fig. 1).

20. Count cells in "Before separation," "Negative," and "Positive" fractions. The "Negative" fraction may need to be concentrated by spinning ($400 \times g$ for 5 min) and resuspending the pellet in a smaller volume of MACS buffer.
21. Guide for efficiency of MSC selection: Ideally the positive fraction should represent 0.5–2 % of total recovered cells. If the positive fraction represents >10 % of total recovered cells, then a second column is needed. Use an MS column if you are putting cells through for a second time.
22. Use cells as required for flow cytometry analysis of MSC enrichment (Fig. 1) (*see* **Note 7**) or cell sorting (*see* **Note 9**). Set up at least two tubes for each fraction: "Before separation," "Negative." and "Positive," adding up to 1×10^6 cells into each 5 mL FACS tube (minimum 0.25×10^6 cells per tube).
23. Pellet the cells ($400 \times g$, 5 min).
24. Resuspend the cells in 100 μl of MACS buffer containing the appropriate amounts of antibody. Tube A: CD45 FITC, CD271 PE, 7AAD; Tube B: CD45 FITC, IgG1 PE, 7AAD (*see* **Note 8**).
25. Incubate on ice, in the dark for 20 min.
26. Wash cells with 500 μL of MACS buffer and pellet ($400 \times g$, 5 min).
27. Resuspend in 300 μL of MACS buffer and run samples on a FACS machine. Acquire as many events as possible (minimum 100,000). The MSC population can be identified as $CD45^{-/low}CD271^{+}$ (Fig. 1).

3.2 Phenotypic Analysis of Native MSCs Population in the MNC Fraction Without Pre-enrichment

1. Follow **steps 1–8** in Procedure 1 (Subheading 3.1).
2. Pellet MNCs ($400 \times g$, 5 min) and resuspend in MACS buffer at 2×10^7 cells/mL. Place 50 μL of cell suspension in each 5 mL FACS tube (*see* **Note 10**).
3. Add 5 μL of FCR block and incubate on ice for 10 min.
4. Add the appropriate amount of antibody to each tube and make the final volume up to 100 μL with MACS buffer (*see* **Note 11**).
5. Incubate cells on ice, in the dark for 20 min (*see* **Note 12**).
6. At the end of the incubation, wash the cells by adding ~1 mL MACS buffer and pellet the cells ($400 \times g$, 5 min).
7. Remove the supernatant without disturbing the cell pellet, and repeat the wash step (*see* **Note 13**).

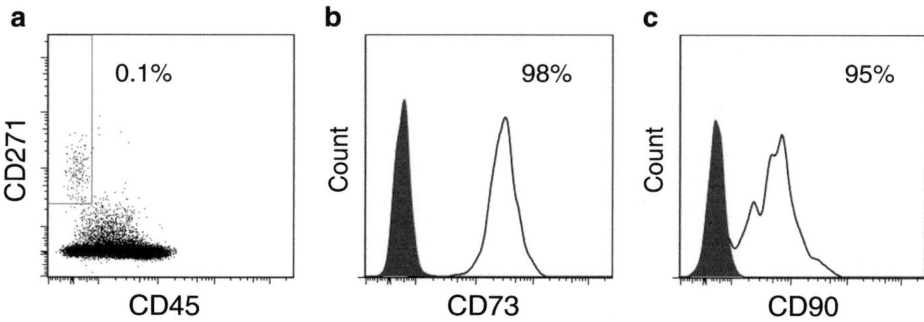

Fig. 2 Example profiles showing CD45$^{-/low}$CD271^{+} cells in the MNC fraction (**a**). Gating on these cells shows positive expression for (**b**) CD73 and (**c**) CD90 compared to isotype control (*grey peak* on histogram)

8. Resuspend the cells in 300–500 μL of MACS buffer. If there are cell "clumps" in the sample, filter through a 40 or 70 μm cell strainer before running on the FACS analyzer.
9. Use unstained and single antibody stained controls to set up the cytometer before each experiment to ensure accurate spectral compensation. Use isotype controls to confirm gating position for the cell populations of interest.
10. The MSC population can be identified as CD45$^{-/low}$CD271^{+} (Fig. 2). Aim to acquire at least 100 CD45$^{-/low}$CD271^{+} cells. Assuming the frequency of CD45$^{-/low}$CD271^{+} cells in the MNC fraction is 0.01–0.1 % this means collecting 1×10^5 to 1×10^6 total events.

3.3 Direct Native MSC Analysis Following Ammonium Chloride Lysis

1. Add 50 μL of whole bone marrow to a 5 mL FACS tube.
2. Add the appropriate amount of antibody to each tube (*see* Subheading 2.3).
3. Incubate at room temperature for 15 min.
4. Lyse red cells by the addition of 2 mL of ammonium chloride solution and incubate at room temperature for 10 min.
5. Pellet the cells ($400 \times g$, 5 min). Wash with 2 mL of MACS buffer ($400 \times g$, 5 min).
6. Resuspend the cell pellet in 300–500 μL of MACS buffer containing 0.5 μg/mL DAPI (*see* **Note 14**).
7. Collect a minimum of 250,000 events (*see* **Note 15**).

4 Notes

1. Good quality bone marrow is vital to cell recovery. Ensure that it is mixed with sufficient anticoagulant to prevent clotting. If micro-clotting occurs remove the clots by filtering through a 70 μm cell strainer before adding to a density gradient.

2. This is easier if done at an angle of approximately 45° and the sample is dripped down the side of the tube.
3. Frozen bone marrow MNCs can also be used for separation; if using a frozen sample thaw according to normal guidelines but ensure that there is 20 U/mL of DNase I in the thawing media and in all subsequent buffers.
4. Degassing the MACS buffer before use can remove microbubbles, which may help to prevent the column from blocking.
5. If applying more than 500 μL of sample wait for the first 500 μL to drip through before applying more sample; this is especially important for samples containing large amounts of red cells as these will settle quickly to the bottom of the reservoir and block the column.
6. Add the buffer gently to avoid creating air bubbles, which may block the column.
7. To calculate the efficiency of selection use the formula:

$$\text{Efficiency of selection} = \frac{\text{No. of cells in the positive fraction}}{\begin{pmatrix}\text{No. of cells in the positive fraction} \\ +\text{No. of cells in the negative fraction}\end{pmatrix}} \times 100.$$

A good selection is when the positive fraction represents between 0.5 and 2 % of total recovered cells. In our experience at Leeds (50+ separations) such a positive fraction should be ~100-fold enriched in CFU-Fs compared to BM MNCs (Before separation) [7]. In the positive fraction, CFU-Fs should represent ~1–5 % of total cells [7, 9]. The negative fraction should not contain any CFU-Fs. Occasional CFU-Fs could be found in the Negative fraction when LS columns are used. This is because LS columns are not as efficient as MS columns. Therefore, if there are few cells to start with, always use MS columns.

How to calculate % of cells recovered:

$$\%\,\text{Cell recovery} = \frac{\begin{pmatrix}\text{No. of cells in the positive fraction} \\ +\text{ No. of cells in the negative fraction}\end{pmatrix}}{\text{Initial No. of cells}} \times 100.$$

This should be between 85 and 95 %. To achieve good recovery be very careful at every stage, work quickly, and keep buffers cold to prevent antibody capping.

8. Using the CD271 PE clone C40-1457 (BD Biosciences) avoids any competitive binding issues. A dead cell marker is not necessarily required for fresh samples, but must be added

if frozen bone marrow was used for the enrichment. Two tubes per fraction is the minimum number required. Single, stained sample and isotype controls should also be stained, but when there are insufficient cell numbers it is sometimes necessary to use some of the un-enriched cells or compensation beads to set up the cytometer. Other fluorochromes can be used according to the cytometer set up, although PE and APC are the best fluorochromes for giving a distinct $CD271^{+}$ population. Red blood cells should have been removed by the selection process, but if there is any concern an anti-glycophorin (CD235a) antibody can be included for an exclusion gate.

9. For sorting of pre-enriched MSCs only the "positive" fraction should be processed. If using frozen samples, keep DNase 1 in the sample and filter the sample through a 70 μm cell strainer immediately before adding to the cell sorter. Set sorting gates on live cells + $CD45^{-/low}CD271^{+}$ cells. Collection tubes should be at least 30 % full with medium + 10 % FBS so that the cells fall directly into the medium and not onto the side of the tube. Vortex the collection tubes briefly before using to coat the side of the tube with medium. Collection tubes can be FACS tubes or any tube type compatible with the instrument collection holder. Sort into straight sided rather than V-shaped tubes whenever possible, to reduce the risk of the cells hitting the side of the tube. For very fragile cells, increase the concentration of FBS in the collection media to make it more viscous, providing a better "cushion" for the cells as they fall into the tube. The purity of the $CD45^{-/low}CD271^{+}$ fraction obtained by sorting should be 90–95 % and an average $CD45^{-/low}CD271^{+}$ cell yield from a 20 mL BM aspirate should be ~5,000 cells. For the data and images the reader is additionally referred to Churchman et al. [13].
10. For collection of larger numbers of cells scale up accordingly.
11. Use antibody combinations suitable for your instrument. For example, in a single tube we routinely use; CD90 FITC, CD105 PE, CD73 PcP-Cy5, CD45 PE-Cy7, CD271 APC, DAPI.
12. For frozen samples, dead cells must be excluded by the addition of 7AAD/DAPI or other dye appropriate to your cytometer.
13. For fresh samples containing a significant number of red blood cells, these can be lysed by adding 1 mL of ammonium chloride solution and incubating at room temperature for 10 min.
14. If volumetric enumeration is required, omit the wash step and add counting beads to the cell suspension immediately after the red cell lysis. Mix gently and analyze immediately on a flow cytometer.
15. The gating strategy and correlations between the number of $CD45^{-/low}CD271^{+}$ cells and CFU-Fs per milliliter of marrow

aspirate are described in Cuthbert et al. [16]. Note that the absolute number of aspirated MSCs depends on the aspiration technique and the volume of marrow drawn in a single draw as well as donor age [17].

References

1. Friedenstein AJ (1976) Precursor cells of mechanocytes. Int Rev Cytol 47:327–359. doi:10.1016/s0074-7696(08)60092-3
2. Friedenstein AJ (1995) Marrow stromal fibroblasts. Calcified Tissue Int 56:S17
3. Pittenger MF, Mackay AM, Beck SC et al (1999) Multilineage potential of adult human mesenchymal stem cells. Science 284: 143–147
4. Simmons PJ, Torok SB (1991) Identification of stromal cell precursors in human bone-marrow by a novel monoclonal-antibody, Stro-1. Blood 78:55–62
5. Stewart K, Monk P, Walsh S, Jefferiss CM et al (2003) STRO-1, HOP-26 (CD63), CD49a and SB-10 (CD166) as markers of primitive human marrow stromal cells and their more differentiated progeny: a comparative investigation in vitro. Cell Tissue Res 313:281–290
6. Quirici N, Soligo D, Bossolasco P et al (2002) Isolation of bone marrow mesenchymal stem cells by anti-nerve growth factor receptor antibodies. Exp Hematol 30:783–791
7. Jones EA, Kinsey SE, English A et al (2002) Isolation and characterization of bone marrow multipotential mesenchymal progenitor cells. Arthritis Rheum 46:3349–3360
8. Harichandan A, Buehring H-J (2011) Prospective isolation of human MSC. Best Pract Res Clin Haematol 24:25–36. doi:10.1016/j.beha.2011.01.001
9. Jones EA, English A, Kinsey SE et al (2006) Optimization of a flow cytometry-based protocol for detection and phenotypic characterization of multipotent mesenchymal stromal cells from human bone marrow. Cytometry B Clin Cytom 70:391–399
10. Tormin A, Li O, Brune JC et al (2011) CD146 expression on primary non-hematopoietic bone marrow stem cells correlates to in situ localization. Blood 117:5067–5077
11. Qian H, Le Blanc K, Sigvardsson M (2012) Primary mesenchymal stem and progenitor cells from bone marrow lack expression of CD44. J Biol Chem. doi:10.1074/jbc.M112.339622
12. Jones E, English A, Churchman SM (2010) Large-scale extraction and characterization of CD271+ multipotential stromal cells from trabecular bone in health and osteoarthritis: implications for bone regeneration strategies based on uncultured or minimally cultured multipotential stromal cells. Arthritis Rheum 62:1944–1954
13. Churchman SM, Ponchel F, Boxall SA (2012) Transcriptional profile of native CD271+ multipotential stromal cells: Evidence for multiple fates, with prominent osteogenic and Wnt pathway signaling activity. Arthritis Rheum 64:2632–2643. doi:10.1002/art.34434
14. Battula VL, Treml S, Bareiss PM et al (2009) Isolation of functionally distinct mesenchymal stem cell subsets using antibodies against CD56, CD271, and mesenchymal stem cell antigen-1. Haematologica 94:173–184. doi:10.3324/haematol.13740
15. Cox G, Boxall SA, Giannoudis PV et al (2012) High abundance of CD271(+) multipotential stromal cells (MSCs) in intramedullary cavities of long bones. Bone 50:510–517
16. Cuthbert R, Boxall SA, Tan HB et al (2012) Single-platform quality control assay to quantify multipotential stromal cells in bone marrow aspirates prior to bulk manufacture or direct therapeutic use. Cytotherapy 14:431–440. doi:10.3109/14653249.2011.651533
17. Stolzing A, Jones E, McGonagle D et al (2008) Age-related changes in human bone marrow-derived mesenchymal stem cells: consequences for cell therapies. Mech Ageing Dev 129:163–173. doi:10.1016/j.mad.2007.12.002

Chapter 12

High Yield Recovery of Equine Mesenchymal Stem Cells from Umbilical Cord Matrix/Wharton's Jelly Using a Semi-automated Process

Timo Z. Nazari-Shafti *, Ivone G. Bruno *, Rudy F. Martinez, Michael E. Coleman, Eckhard U. Alt, and Scott R. McClure

Abstract

Umbilical cord is an abundant source of perinatal, plastic adherent mesenchymal stem cells (UC-MSCs). UC-MSCs exhibit robust stemness and strong immunosuppressive and regenerative effects in vivo. This protocol describes enzymatic and mechanical dissociation of umbilical cord matrix (Wharton's jelly) that results in efficient isolation of large numbers of fresh nucleated umbilical cord regenerative cells (UC-RCs) that, when cultured on plastic, exhibit similar characteristics of UC-MSCs. This protocol potentially alleviates the need for culture expansion to obtain large numbers of cells required for clinical application. Dissociation is achieved with a blend of collagenase and neutral proteases with agitation at 37 °C in a semi-automatic system. Average expected yield is 1.65×10^6 cells/g tissue with 93 % viability. This protocol has been successfully used to isolate an uncultured nucleated regenerative cell population (also referred to as stromal vascular fraction or SVF) from surgically debrided skin and from human, equine, and canine adipose tissue. The procedure requires less than 30 min for tissue dissection and less than 100 min for cell extraction. Quickly obtaining a large number of UC-RCs that have pluripotent differentiation capacity without the complexity and risks of culture expansion could simplify and expand the use of UC-RCs in clinical as well as research applications.

Key words Equine, Wharton's jelly, Umbilical cord mesenchymal stem cells, Isolation

1 Introduction

Mesenchymal stem cells (MSCs) are a population of cells with the potential to differentiate into all three germ layers in vitro [1–5]. More importantly, in vivo, MSCs promote tissue regeneration and healing, and elicit potent immune-modulatory effects [6–9]. These characteristics of MSCs have generated significant interest in their use for cell-based regenerative therapies. MSCs reside in all tissues in the body and can be obtained from primary cell isolates of

*Author contributed equally with all other contributors.

Ivan N. Rich (ed.), *Stem Cell Protocols*, Methods in Molecular Biology, vol. 1235,
DOI 10.1007/978-1-4939-1785-3_12, © Springer Science+Business Media New York 2015

different sources in adult tissues and fetal tissues since they adhere to plastic in cell culture [10–16]. While isolation methods for adult tissues have been mainly established for adipose tissue and bone marrow, the main source for fetal tissue is the umbilical cord (UC). Due to their origin from fetal tissue, the perinatal MSCs in the UC exhibit higher stemness and immune-modulatory properties compared to MSCs originating from adult tissues [17, 18].

The UC of most mammals consists of two arteries and one or two veins that are embedded in a jelly-like ground substance of hyaluronic acid and chondroitin sulfate, named Wharton's jelly after Tomas Wharton, who first described it in 1656 [19]. Recent research has shown that the Wharton's jelly contains a high concentration of plastic adherent MSCs (UC-MSCs) that exhibit properties of embryologic stem cells [20]. In addition to their potential for differentiation to cells of all three germ layers [21–23], UC-MSCs have a higher rate of proliferation and may have more prolonged potential for self-renewal compared to adult MSCs. It has been shown that the presence of longer telomeres in UC-MSCs is responsible for this increased capacity for self-renewal prior to senescence [24]. Due to their close ontogenetic relationship to embryonic tissue, UC-MSCs possess broader plasticity and immune-privilege characteristics in vivo. Moreover, the low immunogenicity, ability to home to sites with ongoing tissue inflammation, and ability to promote neovascularization and tissue regeneration of UC-MSCs provide a compelling rationale for allogeneic transplantation [25]. Since Wharton's jelly provides a rich perinatal source of UC-MSCs, a method for efficiently isolating large numbers of cells quickly could advance the use of UC-MSCs in research and clinical applications [26].

The isolation of MSC from the Wharton's jelly was first described by McElreavey et al. [27]. His group cultured minced Wharton's jelly without prior enzymatic processing up to 2 weeks in order for the UC-MSCs to migrate out of the tissue and adhere to the culture dish. This method yielded very low initial cell numbers and required extensive culture expansion after the initial 2-week period. Currently, MSCs are isolated from the umbilical cord by the use of collagenase followed by a selection for plastic adherent subpopulation in cell culture [28–30]. As shown in Table 1 all published protocols for UC-MSCs isolation commonly result in a relatively low cell yield and require extensive cell culture expansion to obtain the high cell numbers that are recommended for preclinical studies and clinical use. Given that these protocols entail weeks to months of culture expansion [20, 22, 30, 31], differences in cell characteristics including regenerative potential, induction of chromosomal changes, and changes in cell surface antigens are not unlikely [32]. In addition, expansion in culture has potential for exposure to xenogenic proteins and contamination.

Table 1
Summary of previously published protocols for UC-MSCs isolation

Material	Duration	*n*	Cells/cord (10^6)	Viability	Reference
Wharton's jelly	≈18 h	30	–	–	[38]
Wharton's jelly	≈1 week	4	–	–	[39]
Wharton's jelly	≈1 day	15	1.22 ± 1.09	–	[37]
Umbilical cord	≈4.5 h	12	0.96	81 %	[40]
Wharton's jelly	≈2.5 h	5	2.28 ± 1.55	94.3 ± 2.2 %	[41]
Wharton's jelly	≈1–4 h	–	Explant: 0.042–0.0201 Enzyme: 0.003–0.027	–	[42]

As stated by Weiss: "The challenge for the future is to define industrial-grade procedures for isolation and cryopreservation of umbilical cord-derived MSCs and to generate Food and Drug Administration-approved standard operating procedures…" [20]. Accordingly, ex vivo expansion and differentiation of stem cell populations is considered to be substantial manipulation by the US Food and Drug Administration and the European Medicines Agency [33, 34]. Fresh preparation of nucleated regenerative cells from adipose tissue using a novel semi-automated system (generally referred to as the stromal vascular fraction or SVF) [35, 36], on the other hand, could bypass the need for cell culture expansion due to sufficient numbers of cells isolated. Therefore, this protocol aims at facilitating the ability to efficiently and more rapidly isolate fresh regenerative cells from Wharton's jelly (UC-RCs or umbilical cord regenerative cells) to enable potential therapeutic application without the need for culture expansion.

This protocol was developed based on a protocol for isolating the SVF from adipose tissue without ex vivo cell culture expansion. The SVF resulting from the aforementioned isolation process has been designated as a non-ATMP by the European Medicines Agency when used for regeneration, repair, or replacement of weakened or injured subcutaneous tissue (EMA/CAT/228/2013). This designation includes the determination that the SVF preparation has not been subjected to a substantial manipulation. Additionally, this protocol has been successfully used for the isolation of SVF from debrided skin and adipose tissue from burn victims as well as from equine lipoaspirate samples [35, 36].

Similar to the protocol for isolating the SVF from adipose tissue, the UC-RCs are recovered by an enzymatic and mechanical dissociation of the Wharton's jelly. In this protocol, tissue dissociation is achieved using a mammalian origin free, optimized blend of collagenase and neutral protease, and mechanical processing at elevated temperature in a novel semi-automatic system. This combination results in a high viability of the recovered cells and high yields of UC-RCs in shorter time with less operator involvement.

In this protocol, umbilical cords from Thoroughbred horses were obtained following parturition and dissected to obtain Wharton's jelly. The Wharton's jelly tissue was minced and subsequently processed to obtain UC-RCs. In order to compare the UC-RCs to previously published results, we characterized the plastic adherent fraction (UC-MSCs) of the UC-RCs. Initial cell yields were quantified and cultures of the primary UC-MCSs from passage 0 to passage 2 were characterized by assessing their colony-forming potential. Additionally, their capacity to differentiate along all three germ layers, including cell types of the mesoderm such as osteocytes, chondrocytes, adipocytes, cell types of the endoderm such as hepatocytes, and cell types of the ectoderm such as neurons, was assessed. Gene expression profiling by RT-PCR array for MSC-specific genes was also performed.

2 Materials

2.1 Collection of Umbilical Cords

Equine umbilical cords are obtained following parturition from normal full-term pregnancies with unassisted or assisted vaginal delivery. As soon as is feasible, umbilical tape is tied around the cord in two places; adjacent to where the cord breaks or is severed from the foal and approximately 40 cm toward the placenta. The ligations are placed to limit contamination into the lumen of the cord. The isolated portion of the cord between the ligations was placed on a clean surface and any visible gross contamination physically removed with sterile surgical instruments and gauze sponges. The cord is rinsed aggressively, for instance by shaking in a 1.5 L bottle with sterile 0.9 % saline solution three times, and then placed in cold (4 °C) saline solution until processed (*see* **Note 1**). The cord should be processed within 24 h after collection (*see* **Note 2**).

2.2 Dissection of Umbilical Cord

1. Hydrogen peroxide solution (H_2O_2, Sigma Aldrich Solution 3 wt% in H_2O).
2. Phosphate buffered saline (PBS, 1× diluted with MilliQ from OmniPur 10× concentrate).
3. MilliQ water.
4. 70 % Ethanol/diluted v/v with MilliQ water (Macron Fine Chemicals) (CAUTION: ethanol is flammable).
5. 5.5 in. Scissors (Supercut Mayo Scissors) and 4½ in. straight scissors.
6. Petri Dish (Falcon Petri Dish 150 × 15 mm).
7. Forceps (5½ in. thumb dressing and any smaller ones).

8. 50 mL Conical Tubes (Phenix Research).
9. Glass Beakers.

2.3 Cell Isolation Components and Equipment

1. Lactated Ringers for Injection USP (B Braun Medical).
2. Matrase™ Reagent (InGeneron, Inc.), store at −20 °C.
3. Sterile Water for Injection (APP Pharmaceuticals).
4. ARC™ processing unit (InGeneron Inc.).
5. ARC™ quad pack disposables (InGeneron Inc.).
   (a) Steriflips™ Vacuum filter (100 μm pore size).
   (b) Omniflix™ Irrigation syringe/vacuum syringe.
   (c) 60 mL luer lok syringe.
   (d) Centrifuge Tubes.
   (e) 10 mL syringe with 20G Needle.
6. Laminar Flow Hood (OPTIONAL).
7. Pipet Aid (Drummond).
8. Plastic Disposable Pipettes (10 mL) (VWR).
9. Test tube rack.
10. Micropipettes (Gilson and Eppendorf: 1 mL, 200 μL, 20 μL, and 10 μL).

2.4 Reagents

1. Penicillin.
2. Streptomycin.
3. Gentamicin.
4. Amphotericin B.
5. Phosphate buffered saline (PBS).
6. STYO 13 (Life Technologies).
7. Trypan blue (0.4 % w/v).
8. CryoStor 10 cryopreservation medium (BioLife Solutions).
9. Isopropanol.

2.5 Equipment

1. Fluorescence microscope.
2. Hemocytometer.
3. Mr. Frosty cryopreservation chamber.

3 Methods

3.1 Dissection of the Umbilical Cord

1. Place all equipment and reagents needed for dissection and processing of the UC under a biosafety hood (*see* **Note 3**).
2. Prepare wash solution by adding Penicillin and Streptomycin (10 IU and 10 ug/mL), gentamicin (2.5 μg/mL), and amphotericin B (250 ng/mL) to Phosphate Buffered Saline (PBS).

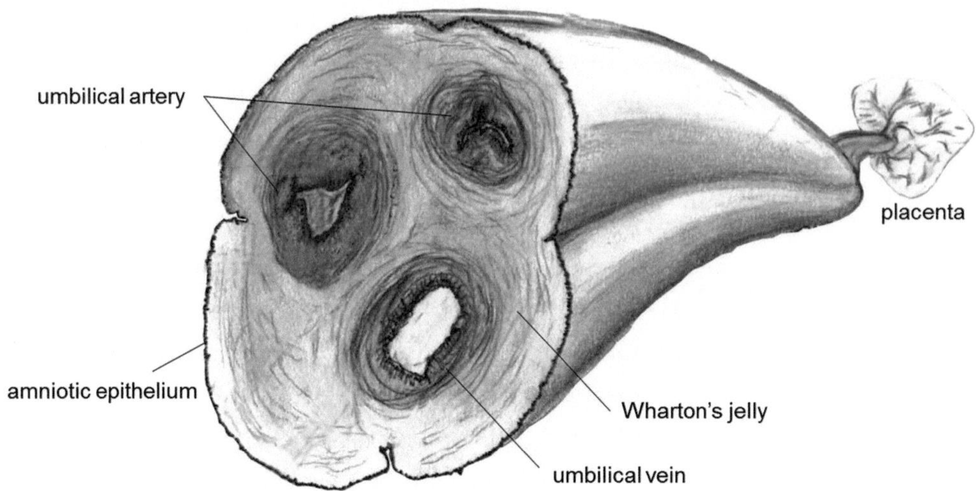

Fig. 1 Schematic representation of the equine umbilical cord diagram illustrates the umbilical cord vessels, surrounded by the tissue matrix known as Wharton's jelly. Wharton's jelly encompasses the perivascular, intervascular, and subamnionic regions. Wharton's jelly is walled by the amnion epithelium

3. Wash samples twice with 1 % (v/v) H_2O_2 in sterile water and three times with wash solution upon arrival.
4. The umbilical cord contains two arteries (firm, thick walled) and one vein (pliable, thin walled) surrounded by the Wharton's jelly, which insulates and protects the umbilical cord vessels (Figs. 1 and 2a). Place tissue samples on large sterile petri dishes for dissection (*see* **Note 4**). Remove the thin squamous epithelium with Metzenbaum scissors and forceps and discard this layer to prevent contamination of the internal tissue (Fig. 2b).
5. The remaining tissue is composed of the vessels and Wharton's jelly. Insert a large Debakey forceps into one of the vessels to provide stability for dissection. Using a small size Metzenbaum scissors blunt and sharp dissect to the fascial plane around the vessel to separate the surrounding Wharton's jelly from the vessel and then discard the vessel (*see* **Notes 5** and **6**) (Fig. 2c). Repeat for each vessel to isolate the Wharton's jelly (Fig. 2c–h).
6. Place the Wharton's jelly in a separate dissecting plate and mince into ~1–3 mm size pieces with a scalpel or scissors prior to processing the tissue (Fig. 2i) (*see* **Note 7**).

3.2 UC-RC Isolation

1. Pre-warmed tissue processing unit (ARC™ tissue processing unit) using the Preheat cycle (PHT). Preheat Lactated Ringers bag to an optimal temperature of 37 °C (*see* **Note 8**) (Fig. 3). The swing bucket must be placed in the upright position to assure mechanical processing of the sample.

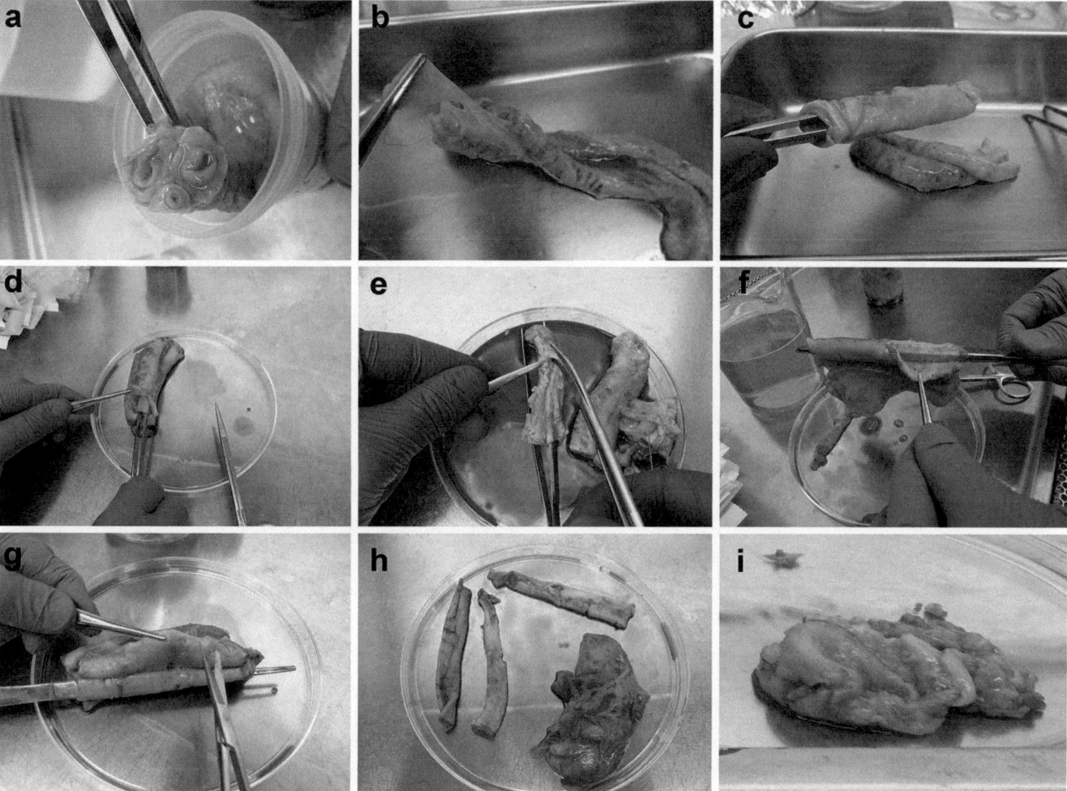

Fig. 2 Dissection of an equine umbilical cord (UC). (**a**) Photograph of the umbilical cord anatomy; (**b**) Removal of the UC amnion epithelium with a hemostat; (**c**, **d**) Stretching of the UC vessels using medium size scissors to facilitate the separation of the vessel from the Wharton's jelly intervascular tissue; (**e**–**g**) Stepwise dissection of the Wharton's jelly tissue from the vessel wall; (**h**) Photograph of UC vessel and Wharton's jelly tissue after dissection; (**i**) Photograph of resulting Wharton's jelly tissue

2. Approximately 10 g of tissue samples is placed in each sterile 50 mL conical tube.
3. Add 25 mL of pre-warmed Lactated Ringer's to each sample until they reach a final volume of 30 mL/sample.
4. Reconstitute enzyme (Matrase™ Reagent) vial by adding 10 mL of 4 °C sterile water with a 10 mL syringe with needle. The final concentration of the Matrase™ Reagent will be 10 units/mL. Invert until lyophilisate dissolves (Fig. 4).
5. Add 1 unit of reconstituted Matrase™ Reagent per gram of tissue (*see* **Note 9**).
6. Invert samples to mix. Place tubes containing samples in the processing unit. Lock tube holders in the upright position for processing. Process for 1 h in the ARC™ tissue processing unit. Include a balancing tube if necessary for counterbalance.
7. After 1 h of processing, place the samples on a rack and allow the samples to sediment for 2–3 min (Fig. 4a1).

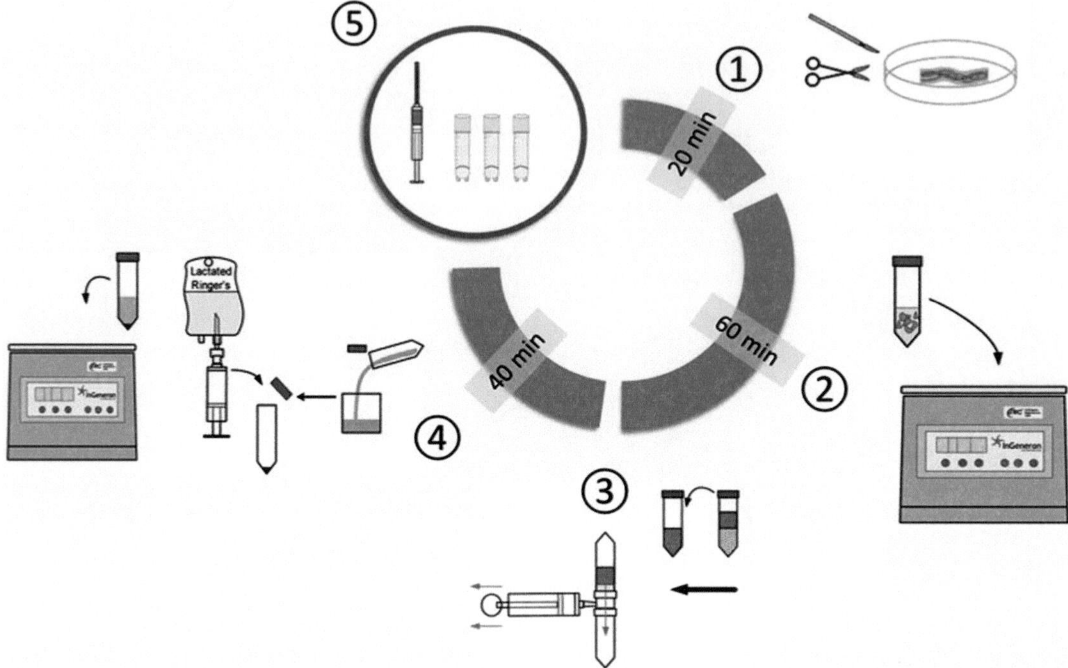

Fig. 3 Diagram of the processing of Wharton's tissue using a semi-automated system. *1.* Dissection of the umbilical cord matrix (UCM) (Wharton's jelly) and subsequent mincing of the UCM. *2.* Processing buffer and Matrase™ Reagent is added to the minced tissue and incubated for 1 h in the InGeneron processing unit at 40 °C under automated constant agitation. *3.* Supernatant of the processed tissue is passed through a 100 μm filter to remove debris and unprocessed tissue. *4.* After filtration the resulting cell suspension is spun down and the cell pellet is washed three times to remove any remaining enzyme. *5.* Final cell pellet can be resuspended in desired carrier solution and administered or cryopreserved

8. Collect supernatant and transfer to a fresh conical sterile 50 mL tube prior to filtration (*see* **Note 10**).
9. Tissue slurry is filtered through a Steriflip™ filter with 100 μm filter.
10. Concentrate cells into a pellet by centrifuging filtrate at 600 × *g* for 10 min. Cell pellet will look as in Fig. 4a2.
11. Wash cell pellet twice with Lactated Ringer's with centrifuging at 600 × *g* for 10 min between each wash step (Fig. 4a3).
12. Resuspend the final cell pellet, consisting of UC-RCs, in 5 mL of lactated Ringer's to assess cell viability and cell counts. The cell suspension can either be used immediately or cryopreserved for future applications. Representative examples for in vitro characterization of cultured UC-MSCs demonstrating potential for differentiation into cell types of all three germ layers, CFU-F assay, and gene expression profile are shown in Figs. 5 and 6, and Table 2.

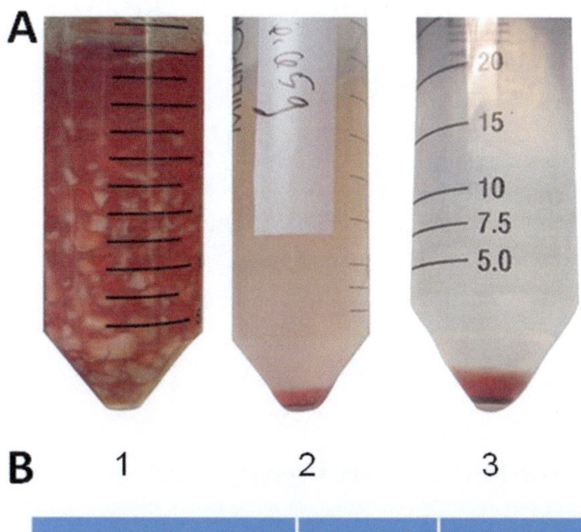

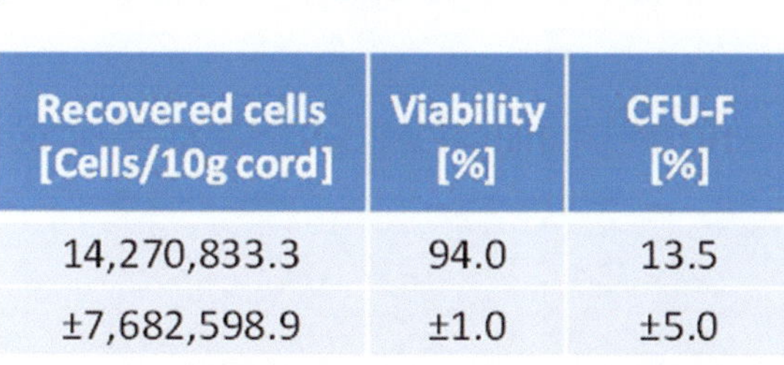

Recovered cells [Cells/10g cord]	Viability [%]	CFU-F [%]
14,270,833.3	94.0	13.5
±7,682,598.9	±1.0	±5.0

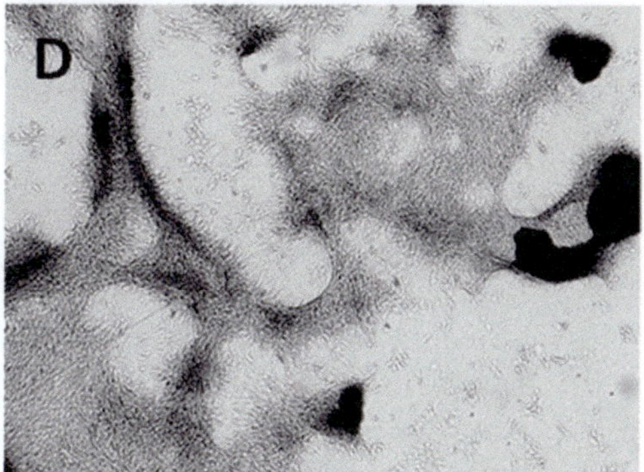

Fig. 4 Stepwise appearance of Wharton's jelly tissue and cells during processing and culture. (**a**) (*1*) Wharton's jelly after 1 h of processing with enzyme with automated mechanical agitation, (*2*) cell pellet appearance after the first filtration and concentration step, (*3*) cell pellet appearance after washes (UC-RC). (**b**) Total nucleated cell counts, percent viability, and percent CFU-F of freshly isolated cells from Wharton's Jelly tissue. (**c**) Light microscopy of passage 0 adherent cell fraction (UC-MSC) from Wharton's Jelly tissue after 4 days in culture; (**d**) Hematoxylin and Eosin (H&E) staining of umbilical cord stem cells plated at very low density and forming large colonies after 14 days in culture

3.3 Cell Viability and Nucleated Count Assessment

1. Obtain a uniform single cell suspension.
2. For nucleated cell counts, thaw an aliquot a SYTO® 13 Stock (5 mM) and make enough volume of working solution 2× dilution (10 μM) with Lactated Ringer's Solution or 1× PBS. Mix thoroughly and make a 1:1 dilution of 2× SYTO® 13 and cell suspension. Incubate at room temperature in the dark for at least 5 min.
3. For assessment of viability prepare a 1:1 dilution of Trypan Blue 0.4 % (w/v) in PBS and single cell suspension. Note! Be sure to perform the Trypan Blue counts within 15 min after staining.
4. For SYTO® 13 nucleated cell count view the cells under fluorescent microscopy using an Absorption 488 nm and Emission 509 nm. For Trypan Blue staining use light microscopy to visualize and count the dead cells that uptake the Trypan Blue.

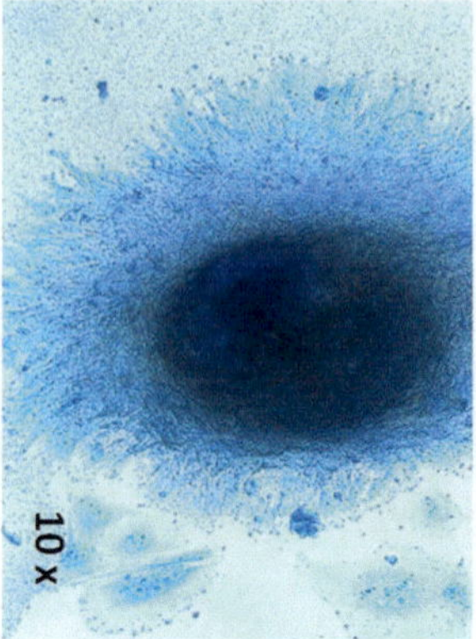

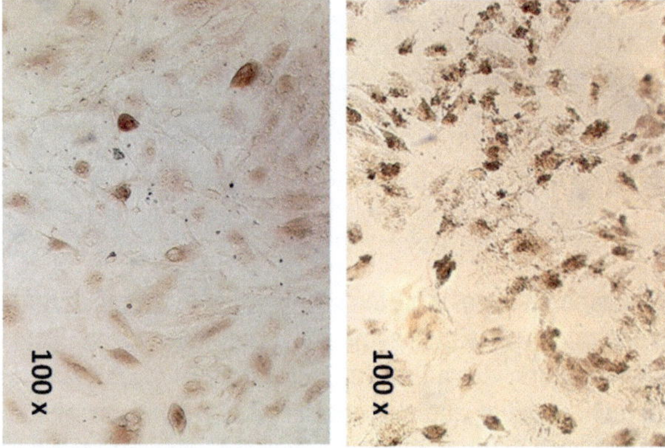

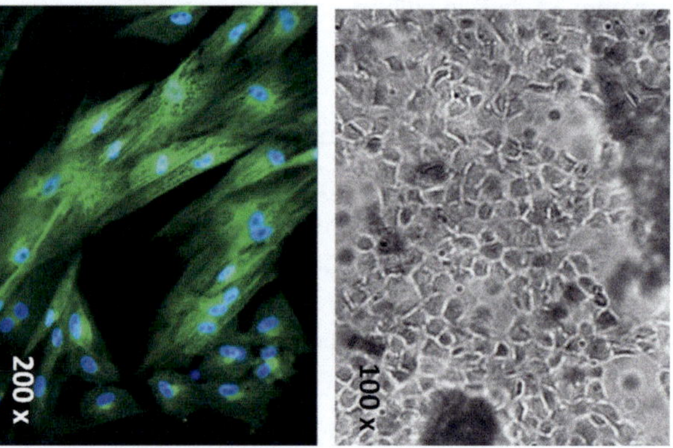

Fig. 5 Differentiation Potential of Wharton's Jelly Stem Cells in Culture. UC-MSCs obtained using the semi-automated protocol were cultured for 2–3 weeks under differentiation-inducing conditions. Light microscopy images of UC-MSC differentiated in adipogenic, osteogenic, chondrogenic, neurogenic, and hepatogenic media. Differentiation assays confirmed the ability of UC-MSCs to differentiate into cells of the endo-, ecto-, and mesodermal germ layers as previously reported. Adipogenic, osteogenic, chondrogenic, and neural-basal media were purchased from Life Technologies, Inc., Carlsbad, CA and used according to manufacturer's instructions. Hepatogenic media and differentiation were as described previously [37]

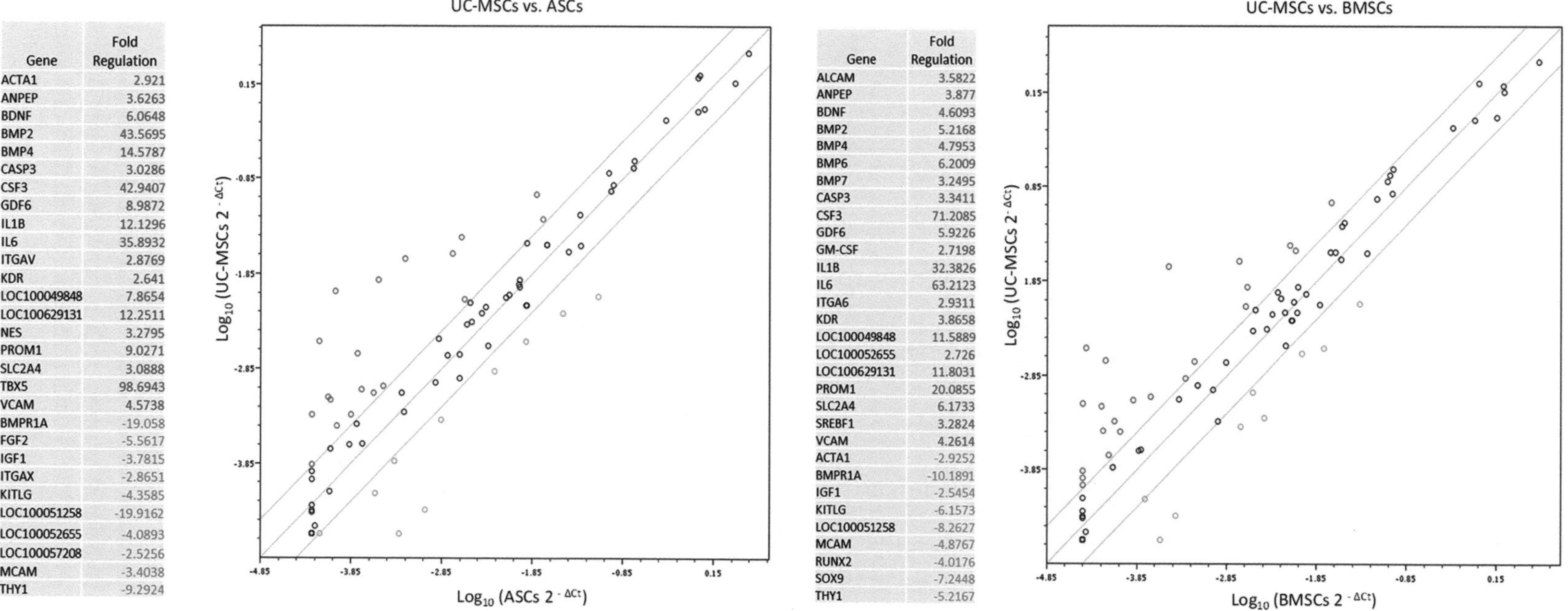

Gene	Fold Regulation
ACTA1	2.921
ANPEP	3.6263
BDNF	6.0648
BMP2	43.5695
BMP4	14.5787
CASP3	3.0286
CSF3	42.9407
GDF6	8.9872
IL1B	12.1296
IL6	35.8932
ITGAV	2.8769
KDR	2.641
LOC100049848	7.8654
LOC100629131	12.2511
NES	3.2795
PROM1	9.0271
SLC2A4	3.0888
TBX5	98.6943
VCAM	4.5738
BMPR1A	-19.058
FGF2	-5.5617
IGF1	-3.7815
ITGAX	-2.8651
KITLG	-4.3585
LOC100051258	-19.9162
LOC100052655	-4.0893
LOC100057208	-2.5256
MCAM	-3.4038
THY1	-9.2924

Gene	Fold Regulation
ALCAM	3.5822
ANPEP	3.877
BDNF	4.6093
BMP2	5.2168
BMP4	4.7953
BMP6	6.2009
BMP7	3.2495
CASP3	3.3411
CSF3	71.2085
GDF6	5.9226
GM-CSF	2.7198
IL1B	32.3826
IL6	63.2123
ITGA6	2.9311
KDR	3.8658
LOC100049848	11.5889
LOC100052655	2.726
LOC100629131	11.8031
PROM1	20.0855
SLC2A4	6.1733
SREBF1	3.2824
VCAM	4.2614
ACTA1	-2.9252
BMPR1A	-10.1891
IGF1	-2.5454
KITLG	-6.1573
LOC100051258	-8.2627
MCAM	-4.8767
RUNX2	-4.0176
SOX9	-7.2448
THY1	-5.2167

Fig. 6 Comparison of gene MSC expression profiling of equine UC-MSCs isolated with this protocol compared to equine bone marrow MSCs and equine adipose tissue MSCs. The scatterplot comparison of the normalized expression of genes associated with stemness and mesenchymal characteristics (equine MSC array, SABioscience, Qiagen, Inc., Valencia, CA). UC-MSC expression levels were plotted against adipose-derived stem cells (ASC), and against bone marrow-derived stem cells (BMSC) to quickly visualize large differences in gene expression. The *central line* indicates unchanged gene expression. Genes *above* the *central lines* indicated higher expression levels, whereas genes *below central lines* indicate lower expression levels for UC-MSC

Table 2
UC-RCS isolation and culture establishment efficiency obtained by this protocol from 4 umbilical cords

Mean tissue weight-size	25 g
Mean cell/gram of tissue	$1.65 \times 10^6 \pm 0.45$
Mean viability (%)	93 ± 4
Mean CFU formation (%)	13 ± 2
Mean subculture viability (%)	98 ± 2

Adjust cell density to 10–100 cells per counting field in a hemocytometer. Alternatively, a nuclear cell counter can be used in these steps.

3.4 Cell Cryopreservation

1. From the cell suspension, pellet cells by centrifugation at $400 \times g$ for 10 min.
2. Remove supernatant and resuspend cells in 1 mL of CryoStor 10 or equivalent cryopreservation media.
3. Transfer cells to a cryovial and place cryovial in a 100 % isopropanol containing cryopreservation chamber (Mr. Frosty) to avoid freezing cell damage or use a controlled rate freezing unit if available.
4. Place Mr. Frosty container in −80 °C freezer overnight.
5. For long-term storage transfer sample to the gas phase of liquid nitrogen.

4 Notes

1. One challenge of isolating equine UC-RCs cells is contamination of samples during the process of obtaining the umbilical cord in the field. Since foals are usually delivered in the open field or in barns, contamination of the umbilical cord is inevitable. Tissue collection during warm weather conditions can expedite bacterial and fungal growth on tissue samples. We, therefore, suggest following certain interventions to decrease the risk of contamination of cells during and downstream of the isolation process. The most important step is to minimize bacterial and fungal growth right after obtaining the tissue sample. If possible, a sterile field should be set up for working with the freshly obtained tissue sample. The umbilical cord should be rinsed with cold lactated Ringer's or saline solution. Additional washes with a povidone-iodine solution may also help deplete the bacterial and fungal flora on tissue samples. To prevent pathogen growth between tissue collection and tissue

processing the cord should then be kept chilled in Ringer's solution substituted with 1 % penicillin, 1 % streptomycin, 0.01 % gentamicin, and 0.2 % amphotericin B. This antibiotic regimen corresponds with the typical antibiotic combination in the regular growth media for primary MSC cultures. Identifying the bacterial and fungal flora of the tissue sample right after collection and testing for antibiotic resistance may help in adjusting the antibiotic regimen.

2. Isolation of UC-RCs is feasible in samples stored for 24 h post-collection. To maximize cell viability and minimize microbial growth the cord sample should be stored at 4 °C immediately after collection.

3. Performing the isolation process under a biosafety hood significantly decreases the risk of secondary contamination. Nevertheless, this process can also be performed on a sterile field or in an operation room setting.

4. Equine umbilical cords are about 80 cm in length. The described dissection technique works best with samples of 5–6 cm length. We therefore recommend sectioning the cord after the washing step.

5. It is crucial to dissect the vessels in the right plane between the outer smooth muscle cell layer of the vessel and the Wharton's jelly. Once in the right histological plane the vessels can be stripped out of the Wharton's jelly. This can reduce the dissection time significantly.

6. This protocol has been optimized to isolate the UC-RCs from equine umbilical cords. However, we have successfully isolated UC-RCs from human UC tissue samples, demonstrating that the protocol is not limited to equine samples. When processing of shorter 2–3 cm umbilical cord samples, the dissection step may be skipped and the cord may be minced prior to processing.

7. As seen in Fig. 4 even after 1 h of processing the tissue pieces will not be fully processed by the enzyme. To optimize the cell yield, it is recommended to mince the tissue as thoroughly as possible to increase the surface area and therefore the exposure area for the enzyme.

8. The lactated Ringer's bag may be placed inside the tissue processing unit drum with the swing buckets in the upright position and the bag in the bottom of the processing unit's drum during the preheat cycle. A 1 L bag of lactated Ringer's will need approximately 30 min to reach 37 °C.

9. Reconstituted enzyme that was not used immediately may be stored frozen at −20 °C in small aliquots for up to 3 months for later use.

10. Due to the high content of hyaluronic acid in the Wharton's jelly, the supernatant may have a gelatinous consistency that makes it more difficult to transfer the supernatant from the processing tube into the sterile filter without disturbing the debris pellet. If too much debris is transferred to the Steriflip filters it may clog the filter. In that case, the remaining supernatant can be transferred to a new filter.

References

1. Zhang YY, Yue J, Che H et al (2013) BKCa and hEag1 channels regulate cell proliferation and differentiation in human bone marrow-derived mesenchymal stem cells. J Cell Physiol. doi:10.1002/jcp.24435
2. Zhou J, Xu C, Wu G et al (2011) In vitro generation of osteochondral differentiation of human marrow mesenchymal stem cells in novel collagen-hydroxyapatite layered scaffolds. Acta Biomater 7:3999–4006
3. Young RG, Butler DL, Weber W et al (1998) Use of mesenchymal stem cells in collagen matrix for Achilles tendon repair. J Orthop Res 16:406–413
4. Hung MJ, Wen MC, Huang YT et al (2013) Fascia tissue engineering with human adipose-derived stem cells in a murine model: Implications for pelvic floor reconstruction. J Formos Med Assoc. 167–168. doi:pii: S0929-6646
5. Caballero M, Skancke MD, Halevi AE et al (2013) Effects of connective tissue growth factor on the regulation of elastogenesis in human umbilical cord-derived mesenchymal stem cells. Ann Plast Surg 70:568–573
6. Anzalone R, Lo IM, Loria T et al (2011) Wharton's jelly mesenchymal stem cells as candidates for beta cells regeneration: extending the differentiative and immunomodulatory benefits of adult mesenchymal stem cells for the treatment of type 1 diabetes. Stem Cell Rev 7:342–363
7. Chen QQ, Yan L, Wang CZ et al (2013) Mesenchymal stem cells alleviate TNBS-induced colitis by modulating inflammatory and autoimmune responses. World J Gastroenterol 19:4702–4717
8. Wada N, Gronthos S, Bartold PM (2013) Immunomodulatory effects of stem cells. Periodontol 63:198–216
9. Takehara N (2013) Cell therapy for cardiovascular regeneration. Ann Vasc Dis 6:137–144
10. Ruzzini L, Longo UG, Rizzello G et al (2012) Stem cells and tendinopathy: state of the art from the basic science to clinic application. Muscles Ligaments Tendons J 2:235–238
11. Stewart AA, Barrett JG, Byron CR et al (2009) Comparison of equine tendon-, muscle- and bone marrow-derived cells cultured on tendon matrix. Am J Vet Res 70:750–757
12. Izadpanah R, Trygg C, Patel B et al (2006) Biologic properties of mesenchymal stem cells derived from bone marrow and adipose tissue. J Cell Biochem 99:1285–1297
13. Yoshimura H, Muneta T, Nimura A et al (2007) Comparison of rat mesenchymal stem cells derived from bone marrow, synovioum, periosteum, adipose tissue and muscle. Cell Tissue Res 27:449–462
14. Harris DT (2013) Umbilical cord tissue mesenchymal stem cells: characterization and clinical applications. Curr Stem Cell Res Ther 8:394–399
15. Ryu YJ, Seol HS, Cho TJ et al (2013) Comparison of the ultrastructural and immunophenotypic characteristics of human umbilical cord-derived mesenchymal stromal cells and in situ cells in Wharton's jelly. Ultrastruct Pathol 37:196–203
16. Kadar K, Kiraly M, Porcsalmy B et al (2009) Differentiation potential of stem cells from human dental origin—promise for tissue engineering. J Physiol Pharmacol 60(Suppl): 167–175
17. Hsieh JY, Fu YS, Chang SJ et al (2010) Functional module analysis reveals differential osteogenic and stemness potentials in human mesenchymal stem cells from bone marrow and Wharton's jelly of umbilical cord. Stem Cells Dev 19:1895–1910

18. Lovati AB, Corradetti B, Lange CA et al (2011) Comparison of equine bone marrow-, umbilical cord matrix and amniotic fluid-derived progenitor cells. Vet Res Commun 35:103–121
19. Warton T (1656) Adenographia: sive glandularum totius corporis descriptio. London: Wharton. pp 243–244
20. Weiss ML, Troyer DL (2006) Stem cells in the umbilical cord. Stem Cell Rev 2:155–162
21. Karahuseyinoglu S, Kocaefe C, Balci D et al (2008) Functional structure of adipocytes differentiated from human umbilical cord stroma-derived stem cells. Stem Cells 26:682–691
22. Sarugaser R, Lickorish D, Baksh D et al (2005) Human umbilical cord perivascular (HUCPV) cells: a source of mesenchymal progenitors. Stem Cells 23:220–229
23. Kita K, Gauglitz GG, Phan TT et al (2010) Isolation and characterization of mesenchymal stem cells from the sub-amniotic human umbilical cord lining membrane. Stem Cells Dev 19:491–502
24. Gotherstrom C, West A, Liden J et al (2005) Difference in gene expression between human fetal liver and adult bone marrow mesenchymal stem cells. Haematologica 90: 1017–1026
25. Prasanna SJ, Jahnavi VS (2011) Wharton's Jelly Mesenchymal stem cells as off-the-shelf cellular therapeutics: a closer look into their regenerative and immunomodulatory properties. Open Tissue Eng Regen Med J 4:28–38
26. Can A, Karahuseyinoglu S (2007) Concise review: human umbilical cord stroma with regard to the source of fetus-derived stem cells. Stem Cells 11:2886–2895
27. McElreavey KD, Irvine AI, Ennis KT et al (1991) Isolation, culture and characterisation of fibroblast-like cells derived from the Wharton's jelly portion of human umbilical cord. Biochem Soc Trans 19:29S
28. Baudin B, Bruneel A, Bosselut N et al (2007) A protocol for isolation and culture of human umbilical vein endothelial cells. Nat Protoc 2:481–485
29. Iacono E, Brunori L, Pirrone A et al (2012) Isolation, characterization and differentiation of mesenchymal stem cells from amniotic fluid, umbilical cord blood and Wharton's jelly in the horse. Reproduction 143:455–468
30. Cardoso TC, Ferrari HF, Garcia AF et al (2012) Isolation and characterization of Wharton's jelly-derived multipotent mesenchymal stromal cells obtained from bovine umbilical cord and maintained in a defined serum-free three-dimensional system. BMC Biotechnol 12:18
31. Weiss ML, Medicetty S, Bledsoe AR et al (2006) Human umbilical cord matrix stem cells: preliminary characterization and effect of transplantation in a rodent model of Parkinson's disease. Stem Cells 24: 781–792
32. Wagner W, Horn P, Castoldi M et al (2008) Replicative senescence of mesenchymal stem cells: a continuous and organized process. PLoS One 3(5):e2213
33. European Medicines Agency. Reflection paper on stem cell-based medicinal products. 14 January 2011
34. Liras A (2010) Future research and therapeutic applications of human stem cells: general, regulatory, and bioethical aspects. J Transl Med 8:131
35. Chan RK, Zamora DO, Wrice NL et al (2012) Development of a vascularized skin construct using adipose-derived stem cells from debrided burned skin. Stem Cells Int. Article ID 841203. http://dx.doi.org/10.1155/2012/841203
36. Bruno I, Nazari-Shafti T, Martinez R, et al (2013) Rapid and efficient preparation of equine regenerative cells from umbilical cord matrix using a semi-automated process. 4th N. American Vet Regen Med Assoc, Atlanta, GA (Abstract)
37. Campard D, Lysy PA, Najimi M et al (2008) Native umbilical cord matrix stem cells express hepatic markers and differentiate into hepatocyte-like cells. Gastroenterology 134:833–848
38. Wang HS, Hung SC, Peng ST et al (2004) Mesenchymal stem cells in Wharton's jelly of the human umbilical cord. Stem Cells 22:1330–1337
39. Bailey MM, Wang L, Bode CJ et al (2007) A comparison of human umbilical cord matrix stem cells and temporomandibular joint condylar chondrocytes for tissue engineering temporomandibular joint condylar cartilage. Tissue Eng 13:2003–2010
40. Tsagias N, Koliakos I, Karagiannis V et al (2011) Isolation of mesenchymal stem cells using the total length of umbilical cord for transplantation purposes. Transfus Med 21:253–261

41. Christodoulou I, Kolisis FN, Papaevangeliou D, et al (2013) Comparative evaluation of human mesenchymal stem cells of fetal (Whartons's Jelly) and adult (Adipose Tissue) origin during prolonged in vitro expansion: considerations for cytotherapy. Stem Cells Int. Article ID 246134

42. Hua J, Gong J, Meng M et al (2014) Comparison of different methods for the isolation of mesenchymal stem cells from umbilical cord matrix: proliferation and multilineage differentiation as compared to mesenchymal stem cells from umbilical cord blood and bone marrow. Cell Biol Int 38:198–210

Chapter 13

Isolation and Functional Assessment of Cutaneous Stem Cells

Yanne S. Doucet and David M. Owens

Abstract

The epidermis and associated appendages of the skin represent a multi-lineage tissue that is maintained by perpetual rounds of renewal. During homeostasis, turnover of epidermal lineages is achieved by input from regionalized keratinocytes stem or progenitor populations with little overlap from neighboring niches. Over the last decade, molecular markers selectively expressed by a number of these stem or progenitor pools have been identified, allowing for the isolation and functional assessment of stem cells and genetic lineage tracing analysis within intact skin. These advancements have led to many fundamental observations about epidermal stem cell function such as the identification of their progeny, their role in maintenance of skin homeostasis, or their contribution to wound healing. In this chapter, we provide a methodology to identify and isolate epidermal stem cells and to assess their functional role in their respective niche. Furthermore, recent evidence has shown that the microenvironment also plays a crucial role in stem cell function. Indeed, epidermal cells are under the influence of surrounding fibroblasts, adipocytes, and sensory neurons that provide extrinsic signals and mechanical cues to the niche and contribute to skin morphogenesis and homeostasis. A better understanding of these microenvironmental cues will help engineer in vitro experimental models with more relevance to in vivo skin biology. New approaches to address and study these environmental cues in vitro will also be addressed.

Key words Skin, Stem cell, Epidermis, Keratinocytes, Cell culture

1 Introduction

Mammalian skin epithelium undergoes perpetual rounds of renewal during homeostasis. The turnover of the epithelium is ensured by compartmentalized, phenotypically distinct niches harboring epidermal stem cells and their differentiated progeny [1, 2]. Cutaneous epithelial lineages form a multilayered stratified interfollicular epidermis (IFE), which is continuous with skin appendages including as pilosebaceous units composed of hair follicles and associated sebaceous glands, as well as sweat glands that are composed of a coiled secretory gland found deep in the dermis and extends up to the epidermis via the sweat duct [3] (Fig. 1). In many of these epithelial compartments, a pool of stem or progenitor cells has been

Ivan N. Rich (ed.), *Stem Cell Protocols*, Methods in Molecular Biology, vol. 1235,
DOI 10.1007/978-1-4939-1785-3_13, © Springer Science+Business Media New York 2015

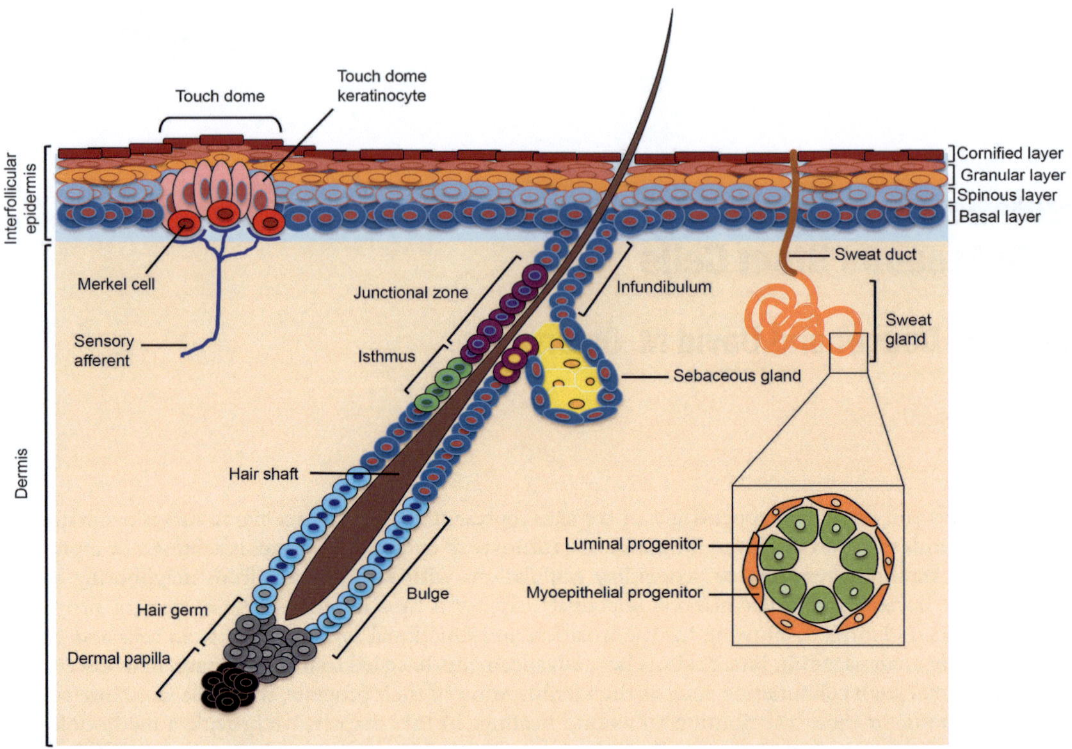

Fig. 1 Schematic illustration of adult skin stem cell niches. Each stem cell or progenitor population is color-coded and corresponds to a different set of markers (*see* Table 1 for details). The hair follicle represented is in telogen (resting) phase

identified and shown to give rise to a number of terminal differentiation products including: hair shafts (hair follicles) [4–6], sebum (sebaceous glands) [7, 8], stratum corneum (IFE) [9–12], and Merkel cells (touch domes) [13, 14]. These differentiated products facilitate a variety of skin functions including thermoregulation, barrier protection against biological, chemical, and physical pathogens, hydration, and perception of mechanical stimuli. As these terminal differentiation products become exhausted or damaged, they must be replenished to ensure proper skin function. Therefore, to maintain skin homeostasis, epidermal stem cells are charged with the task of sustaining these various niches along with the turnover of their differentiated progeny.

Skin homeostasis requires epithelial stem cells to strike a balance between epidermal proliferation and differentiation. To accomplish this, epidermal stem cells self-renew by symmetrical cell division in order to maintain pools of undifferentiated cells in the IFE or in pilosebaceous units. To sustain differentiated lineages, stem cells may also undergo asymmetric division, giving rise to cells committed to terminal differentiation [15, 16]. Though self-renewal is a criterion, the definition of stem cell potential is more inclusive.

Table 1
Markers of progenitor cells in adult mouse and human skin

Progenitor compartment	Progenitor markers	References
Dermal papilla	Sox2 CD133	[52] [53]
Hair germ	P cadherin Lgr5	[54] [55]
Bulge	CD34 Krt15 Lgr5 Sox9 NFATc1 Tcf3/4 Lhx2 Krt19 CD200	[5] [6] [55] [56, 57] [58] [59] [60] [61, 62] [63]
Isthmus	Lgr6 $\alpha6^{low}$CD34^{-}Sca-1^{-}	[64] [46]
Sebaceous gland	MTS24—(Plet-1) Blimp1	[8] [7]
Junctional zone	MTS24—(Plet-1) Lrig1	[8] [12]
IFE and infundibulum	Lrig1 $\alpha6^{+}$CD34^{-}Sca-1^{+}	[12] [46]
Touch dome	CD200 Krt17	[13] [14]
Sweat gland	Krt18 (gland luminal cells) Krt15 (gland myoepithelial cells) $\alpha6^{high}$B1^{high}Sca-1^{-}CD24^{+} (gland) $\alpha6^{high}$B1^{med}Sca-1^{+}CD24^{+} (duct)	[65] [45] [45]

Skin homeostasis results from epidermal stem cells fate decisions and the stem cell potential of a given population cannot be addressed by a single assessment. To address these questions, a number of assays have been developed. Stem cells are historically defined as being multipotent, meaning that they give rise to multiple differentiated epidermal cell lineages. A highly effective tool to measure epidermal stem cell potency is the skin reconstitution assay [17, 18]. Initially, putative progenitor keratinocytes are isolated from donor mouse skin, purified by fluorescent activated cell sorting (FACS), then grafted onto the skin of a recipient host mouse. In this manner, the capacity of grafted cells to give rise to a stratified epidermis or fully formed pilosebaceous unit can be tested. High clonogenic capacity is also a hallmark of keratinocyte stem cells. This property can be addressed by using the colony forming assay [19], in which freshly harvested epidermal cells are seeded at clonal density and their capacity to form large, proliferative colonies in vitro is measured.

Defining stem cell properties necessitates proper isolation of the cell population of interest. Two general methods have been used to gain physical access to stem cells: tissue microdissection or FACS. Microdissection is a suitable approach when stem cell markers are unknown, as is the case with sebaceous gland stem cells in human skin, or if isolation by enzymatic dissociation is hardly possible as for dermal papilla cells in the hair follicle [20, 21]. However, if cell surface markers have been identified for a stem cell population or if cells are genetically labeled with fluorescent proteins (Table 1), epidermal cells can be enzymatically dissociated into a single cell suspension, labeled with antibodies against surface proteins if necessary, and separated using FACS analysis [6, 22, 23]. These cells can subsequently be plated for clonogenic assays [19] or used for engraftment in skin reconstitution assay [17, 18].

It is important to note that the removal of cells from their natural microenvironment may complicate the assessment of skin stem cell properties. Indeed, stem cells deprived of their niche's cues may show significant variations in their gene expression profile [24], morphology, or exit the stem cell compartment and undergo terminal differentiation. However, cell culture media can be supplemented with known stem cell factors that partially restore the molecular cues normally supplied by the natural microenvironment [25–28]. Therefore, it may be necessary to modify these ex vivo assays to account for the fact that the microenvironment is composed of different cell types transmitting multiple signals to epidermal stem cells. Indeed, cell–cell interaction and communication via soluble factors [29] as well as mechanical cues provide a suitable environment that contributes to stem cell homeostasis. For instance, fibroblasts or adipocytes are known to secrete factors (TGFβs and PDGFs) that control hair follicle stem cell homeostasis [30, 31]. The same observation has been made in other niches. In hairy skin, Merkel cells are found in the basal layer of the epidermis, juxtaposed with columnar-shaped keratinocyte progenitors in specialized structures termed touch domes. Recent evidence has shown that touch dome keratinocyte progenitors may regulate proper innervation of the structure by sensory neurons [14]. Moreover, other elements of epithelial stem cell niche in the skin, including extracellular matrix integrins, maintain epidermal stem cell homeostasis by anchoring these cells and regulating keratinocyte differentiation [32].

Given these observations, recreating such a complex environment in vitro is technically challenging and pursuit in the upcoming years will be to engineer and reconstruct intact skin in a petri dish using organotypic models.

Modeling skin structure in vitro broadens our understanding of epidermal cell behavior, and importantly expands the possibilities for improving of skin therapies for burn victims and patients suffering from chronic wounds [33]. Current clinical applications to treat skin wounding utilize skin grafts propagated from the patient's cultured keratinocytes, but do not include any skin appendages such as hair follicles or sweat glands. More than an aesthetic issue, areas of grafted skin lacking the pilosebaceous unit, sweat glands or innervations, will not be able to regulate hydration, barrier protection, thermoregulation, or restore sensitivity. For this reason, regenerative medicine has explored different methods attempting to induce de novo hair follicle growth. As hair morphogenesis and cycling largely depends on extrinsic signaling from a group of mesenchymal cells found in dermal condensates called the dermal papilla (DP) [34, 35], researchers have aimed at cultivating these specialized cells to recreate a hair organ in vitro. It has been shown that human DP cells grown in 2-dimensional (2D) culture outside of their natural environment lose their molecular, transcriptional signatures and hair growth inductive properties due to the lack of epithelial cues [26, 36]. However, recent advancements have shown that a 3-dimensional (3D) spheroid culture could be a way to restore de novo hair induction. Indeed, DP cells cultured using the hanging drop method rearrange into condensates, recapitulating their in vivo structure. Importantly, this correlates with the partial restoration of the molecular and transcriptional signature observed in intact DP [37]. In a similar fashion, keratinocyte 3D organotypic cultures recreate the structural organization (i.e. stratification) and the identity (i.e. expression of differentiation markers) of all the layers of the epidermis [38, 39]. Overall, these studies demonstrate that 3D organotypic cultures may significantly improve the study of skin biology in vitro.

Both supplementing culture media by known crucial soluble factors and inducing mechanical cues by recreating a 3D environment, might not be enough to fully mimic skin behavior in vitro. Other non-epithelial cell types surrounding a stem cell niche are known to regulate skin homeostasis, as in the bulge region neighbored by blood vessels, adipocytes, mesenchymal cells, nerves and the arrector pili muscle [40]. As we have not fully elucidated the exact contribution or mechanisms by which these different cell types impact skin homeostasis, one promising solution is to directly incorporate these cell types into the 3D culture. Therefore, by integrating the skin with appendages vasculature [41], melanocytes [42], immune cells, and nerves, a more comprehensive study of wound healing, skin barrier, inflammation, auto-immune process, and mechanosensation will be made possible.

This book chapter will therefore focus on the identification of the different stem cell and epidermal niches, then present the different methods used to characterize these cells in culture. In particular, the clonogenic and skin reconstitution assays will be highlighted. Recent advancements in vitro skin models will also be discussed.

2 Materials

2.1 Skin Dissociation Reagents and Equipment

1. 0.25 % Trypsin/1 mM EDTA in Hank's Balanced Salt Solution (HBSS, without calcium) (Invitrogen) (*see* **Note 1**).
2. 1× PBS, pH = 7.6 (Invitrogen), sterilized.
3. Fibroblast growth medium: DMEM (Invitrogen) supplemented with 10 % Donor Bovine Serum (Invitrogen) and 2 % penicillin-streptomycin (Invitrogen).
4. Collagenase Type I (Worthington Biochemical), 10 mg/mL stock solution in PBS.
5. Humidified incubator at 37 °C with 5 % CO_2.
6. DNAse I (Worthington Biochemical), 20,000 U/mL stock solution in PBS.
7. 40 and 70 μm strainers.
8. Betadine 1 % solution in H_2O.
9. 70 % ethyl alcohol solution.
10. Isolation medium: 10 % Fetal Bovine Serum (HyClone) in DMEM (Invitrogen) with 1× Antibiotic-Antimycotic (Life Technologies).

2.2 Skin Cell Culture Solutions

2.2.1 Clonogenic Assay

1. If using serum-containing medium is required, cells must be grown on a feeder layer. For instance, complete FAD growth medium can be used: Three parts DMEM (Invitrogen), one part Ham's F12 Supplement (Invitrogen), 10 % Defined Fetal Bovine Serum (HyClone), 10 ng/mL EGF (Peprotech), 0.5 mg/mL hydrocortisone (Sigma-Aldrich), 10^{-10} M cholera enterotoxin (Sigma-Aldrich), 5 mg/mL insulin (Sigma-Aldrich), 1.8×10^{-4} M adenine (Sigma-Aldrich), 100 U/mL penicillin (Invitrogen), and 100 mg/mL streptomycin (Invitrogen).
2. If serum-free medium such as Cnt-07 (CELLnTEC) is preferred; cells can be plated without feeders.
3. 3T3 fibroblasts (ATCC) mitotically arrested with either mitomycin c (Sigma) or γ-radiation.
4. 0.25 % trypsin/1 mM EDTA stock solution (Invitrogen).
5. Nunclon 6-well dishes (Fisher Scientific).
6. Versene: 0.48 mM EDTA in PBS (Fisher Scientific).
7. Rhodamine B, 1 % solution in H_2O (Sigma).

2.2.2 Skin Reconstitution Assay

1. Silicon culture chambers—Upper F2U #30-268; Lower F2L #30-269 (Renner GmbH).
2. Surgical instruments including forceps, curved scissor, stapler, and staple remover (all from Temin); sterile drapes; alcohol swabs; and anesthetics.

3. Immunodeficient mice Nude mice (NCR nude), male, 7–9 weeks old, supplied by Taconic or preferably NSWNU-M (homozygote females) (*see* **Note 2**).
4. Small heating pad.

2.2.3 Antibodies

1. α6 Integrin (CD49f, BD Biosciences).
2. CD34 (RAM 34, BD Biosciences).
3. Sca-1 (Ly6G, BD Biosciences).
4. CD200 (OX-2, BD Biosciences).

3 Methods

3.1 Isolation of Epidermal Stem Cell Populations

The skin epithelium is regionalized in phenotypically distinct niches (Fig. 1). Thanks to numerous molecular markers (Table 1), it is now possible to localize and isolate these stem cell or progenitor populations for downstream characterization and ex vivo assays. The listed markers can be used for mouse and human epidermal cell progenitor populations.

3.2 Isolation of Mouse Keratinocytes by FACS

1. Euthanize the mouse and shave the dorsal skin using electric clippers (*see* **Note 3**).
2. For colony forming and skin reconstitution assay, it is recommended to start the protocol in a sterile environment. Sterile harvests are not necessary for biochemical end points (microarray analysis, etc.). In a biological cabinet, submerge euthanized mice in 1 % betadine for 2 min and wash in sterile H_2O. Submerge mice in 70 % EtOH for 1 min and wash in sterile H_2O.
3. Remove as much of the shaved dorsal skin as possible using forceps and surgical scissors (*see* **Note 4**).
4. Float the skin epidermis-side down in PBS in a 10 cm Petri dish and gently scrape the subcutaneous fat tissue off the dermis using a sterile scalpel (*see* **Note 5**). Replace with fresh PBS as necessary.
5. Float the skin, dermis-side down, in a 10 cm Petri dish with 10 mL 0.25 % Trypsin/1 mM EDTA. Cut the skin in three pieces (anteroposterior axis) and incubate for 1.5–2 h at 37 °C (*see* **Note 6**).
6. Aspirate the trypsin, wash twice with PBS. Remove PBS and add 10 mL of cold isolation medium containing at least 10 % serum. Gently scrape the epidermis off and discard the dermis. Dissociate the epidermis in small pieces using a scalpel and forceps.
7. Pipet the dissociated epidermis into a 100 mL autoclaved bottle with a stirring bar (*see* **Note 7**). Wash the plate twice with

10 mL of isolation medium and pipet into the bottle. Stir the epidermal cell suspension gently at RT for 30 min.

8. Filter cells through a 40 μm strainer into a 50 mL Falcon tube and wash the strainer with 5 mL of cold isolation media. Centrifuge cells at 200 × *g* at 4 °C for 10 min.
9. Resuspend the pellet at 20×10^6 cell/mL in cold isolation medium (*see* **Note 8**). Usually, stem cells represent a small percentage of the total skin population (3–8 %). Thus, pooling several dorsal skins depending on how many isolated cells are required can be considered. Between 10 and 20×10^6 epidermal cells can typically be harvested from one dorsal skin.
10. Aliquot the cells at desired cell number for antibody labeling (*see* **Note 9**) and centrifuge at 200 × *g* at 4 °C. Resuspend in 300–500 μL of cold isolation medium and stain with the appropriate antibodies for the selection of the stem cell population of interest (Fig. 1, Table 1) for 30 min on ice.
11. If the use of unconjugated antibodies is preferred, block the cell suspension with species-specific serum after the incubation with the primary antibody.
12. Centrifuge cells at 200 × *g* at 4 °C and wash once with 10 mL of cold PBS after each antibody incubation. For the final step, resuspend in isolation medium (without serum if used previously) supplemented with a nuclear stain to select for live cells (*see* **Note 10**).
13. Proceed to FACS analysis or sort (Fig. 2). To maintain high viability, it is recommended to sort the cells into cold glucose-containing medium supplemented with 20 % DBS.
14. For further use of sorted cells (culture, engraftment, RNA extraction), add a washing step with 5–10 mL of cold PBS.

3.3 FACS Isolated Epidermal Cells Characterization

3.3.1 BrdU Quantification

In addition to the selection by surface markers, the proliferative or quiescent status of a cell population can be addressed by quantifying the uptake of thymidine analogs such as BrdU or EdU by FACS analysis [23, 43].

3.4 Culture of Sorted Skin Stem Cells

The culture of sorted cells allows testing for the proliferative capacity or the lineage commitment of a population. It may also be used to screen pathways implicated in the regulation of the stem cell population [26]. Furthermore, if the sorted population is small, expanding them in culture will optimize the engraftment process (skin reconstitution assay) [17, 43]. Genetic modification can also be performed while cells are in culture to induce expression or to knock-out a gene of interest before grafting [26]. However, it is important to consider that 2D culture may alter the stem cell phenotype or induce terminal differentiation.

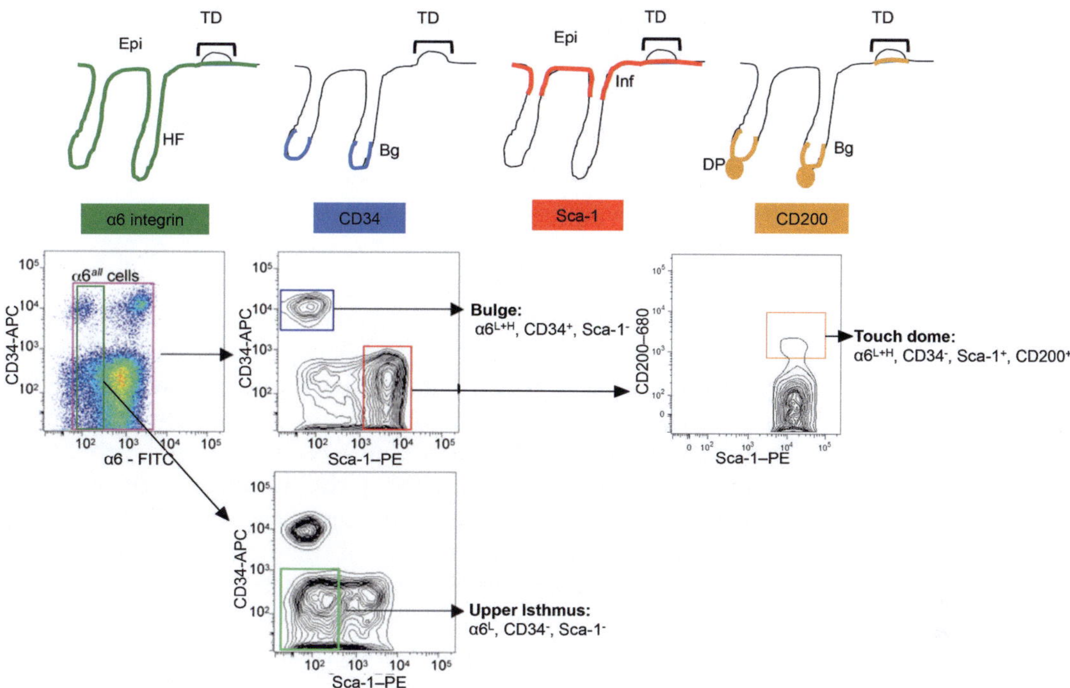

Fig. 2 Expression of surface markers in the epidermis. *Top panel* is a schematic of the localization of four surface markers α6, CD34, Sca-1, and CD200. *Bottom panel* shows representative FACS plots after labeling with these markers. Each FACS profile allows the isolation a specific stem cell population of the skin. Abbreviations: *Epi* epidermis, *TD* touch dome, *HF* hair follicle, *Bg* bulge, *DP* dermal papillae

3.4.1 Preparation of Highly Inductive Dermal Fibroblasts

1. Surgically excise the dorsal skin from euthanized postnatal day 1–2 mice using sterile forceps and scissors (*see* **Note 11**).
2. In a dry 10 cm Petri dish, lay skins flat dermis side down with no folded edges. Slowly pour in ice-cold 0.25 % trypsin/1 mM EDTA and avoid getting the tops of the skins wet. Incubate skins overnight at 4 °C.
3. Remove skins, one at a time, from trypsin and place on dry 10 cm plate, dermis side up. Flatten it again and use fine forceps to separate epidermis from dermis, starting at one edge of skin and flipping the dermis up and off the epidermis, which should stay on the plate.
4. Transfer each dermis, one at a time, to a plate containing 10 mL of media on ice.
5. For eight dermises, use 0.5 mL collagenase Type I stock solution (10 mg/mL in H_2O) plus 12 mL HBSS in a 50–100 mL sterile beaker. Transfer dermises into a beaker and mince into small pieces using sharp scissors.
6. Transfer to a 250 mL flask with a magnetic stir bar. Stir at 37 °C for 30 min (Optional: For the last 5 min add 20 μL of DNAse I (stock at 20,000 U/ml in PBS)).

7. Dilute three- to fourfold with media and filter through sterile gauze or 70 μm filter. Rinse the flask with 5 mL media and pass through the filter.
8. Centrifuge the cells at $450 \times g$ for 5 min at 4 °C.
9. Resuspend and wash the pellet once in HBSS.
10. Count the cells and use at 1×10^6 cells per 10 cm dish for later use.

3.4.2 Colony Forming Assay

The formation of colonies in culture is indicative of both the clonogenic and proliferative capacity of a cell population. Typically sorted cells are grown on a feeder layer of mitotically arrested 3T3 fibroblasts in complete FAD growth medium [19] to provide nutrients to the epithelial cells. Defined serum-free medium in the absence of fibroblasts can also be used to assay the number of adherent cells shortly after plating. Keratinocyte stem cells typically form large round colonies with smooth edges of small cells (holoclones) that can be passaged several times [44]. This technique has been extensively used to help identifying epidermal stem cells in the bulge [6] and more recently in the sweat glands [45].

1. Preparing the 6-well plates 1 day in advance will allow the fibroblasts to fully attach, spread, and condition the growth medium. Plate 1×10^6 mitotically arrested 3T3 fibroblasts in 3 mL complete FAD growth medium per well.
2. The next day, harvest mouse keratinocytes from a single mouse dorsal skin and FACS sort desired keratinocyte subpopulations as described above. The FACS instrument can be optimized towards purity and accuracy in counting since cell numbers will be in excess. Propidium iodide or DAPI should be used to exclude dead cells. Many of the dead cells are post-mitotic suprabasal cells that are sensitive to the 70 % ethanol washes during cell harvesting. Use either the FACSAria Automated Cell Deposition Unit (ACDU) function or manually seed at a clonal density of 1×10^3 sorted keratinocytes per well (a range of $0.5–2 \times 10^3$ cells can be used).
3. Incubate cells for 2 weeks at 32 °C in a humidified incubator with 5 % CO_2 and change the medium every 48 h.
4. For serum-containing approaches, after 2 weeks, aspirate off the culture medium and replace with 3 mL versene per well. After 1–2 min at room temperature, detach the feeder layer by repeated pipetting of versene over the plate (keratinocytes will not detach). Gently wash plates two times with PBS (take care not to detach colonies).
5. Stain wells with Rhodamine B for 1 h (*see* **Note 12**) at room temperature (use just enough Rhodamine solution to cover the cells). Aspirate off the Rhodamine solution and gently wash wells three times with PBS (take care not to detach colonies).

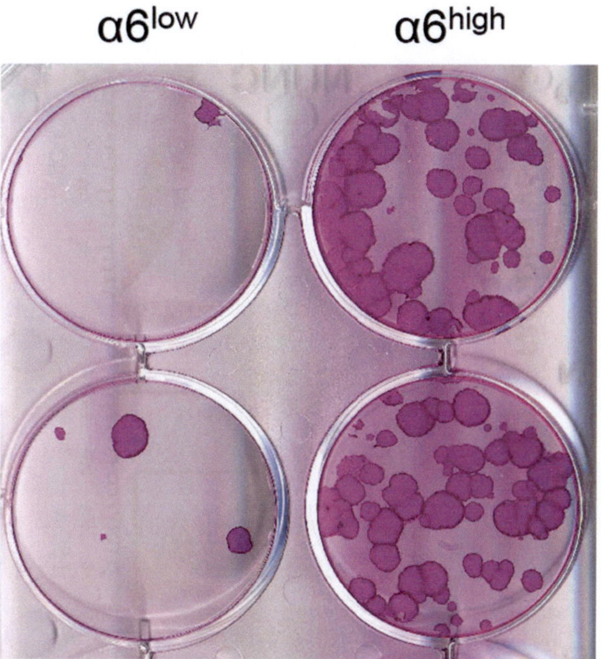

Fig. 3 Example of a clonogenic assay in duplicate. $\alpha6^{low}$ and $\alpha6^{high}$ FACS-sorted cell populations were plated at equal density and cultured for 2 weeks. Cultures were stained with Rhodamine to visualize the colonies in order to quantify them

6. Aspirate off the final wash and allow wells to completely dry by turning plates upside down. Thereafter, plates can be imaged and colonies may be manually counted (Fig. 3). Typically the total number of colonies as well as the number of colonies greater than 4 mm in diameter are counted and compared between keratinocyte subpopulations.

3.4.3 Skin Reconstitution Chamber Assay

To address whether a cell population is committed to a specific epidermal lineage, FACS-sorted cells can be engrafted into a host immunodeficient mouse. This not only addresses their pluripotency and regenerative capacity but also their interaction with the microenvironment (recruitment of other cells or response to inductive signal) [6]. To be engrafted in a recipient mouse, sorted keratinocytes are mixed with neonatal dermal fibroblasts, which provide inductive cues for grafted epidermal cells and give rise to a living dermis.

1. Clip hair with electric clippers if necessary, clean the dorsal skin with 1 % betadine and place anesthetized mice on heating pad.
2. Use scissors to make a small incision on the back of the mouse (approximately 1 cm in diameter). Better areas for chamber placement are interscapular or suprapelvic. Do not make incisions directly on the spinal protrusion.

3. Assemble the upper and lower grafting chambers together and insert through the incision so that the rims of the chamber are under the skin (*see* **Note 13**).
4. Secure the chamber to the skin with surgical stapler clips (two staples are usually enough).
5. Allow the chamber to adhere to the dorsal surface overnight prior to implanting cells.
6. Mix the desired number of epidermal cells and 2×10^6 neonatal dermal cells together as a slurry in HBSS [13, 43, 46]. Between 1×10^5 and 6×10^6 epidermal cells per graft can be successfully implanted. Spin cells at $200 \times g$ for 5 min, resuspend the pellet in 100 μL HBSS, and store on ice until used.
7. Gently mix cell suspensions before pipetting entire aliquot into chamber of the hat through the hole on top.
8. Replace each mouse in individual cages (on belly and away from the spout of the water bottle).
9. After 1 week (*see* **Note 14**), anesthetize mice and remove staples and gently tug on chamber to release it from mouse's back. Use tweezers to loosen skin around edge of chamber. The grafted area may be moist with puss. Just replace mouse in cage, as in **step 8**.
10. Chambers are retained, cleaned (soak overnight in soapy water), and autoclaved for reuse.
11. Grafts are usually biopsied at 5–10 weeks post-grafting. Hair usually appears after approximately 2–3 weeks (Fig. 4).

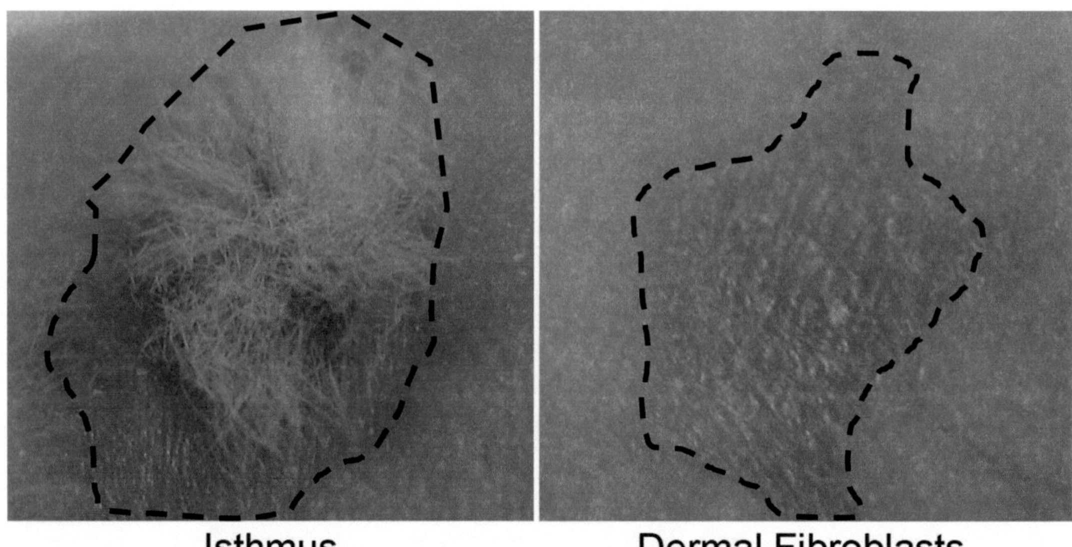

Fig. 4 Hair reconstitution assay showing the multipotency of isthmus stem cells compared to the control (dermal fibroblasts). Images were taken 5 weeks after removal of the chamber and 6 weeks post-grafting. *Dashed lines* circle the graft perimeter

3.4.4 Intradermal Injections

When the sorted cell population is relatively small, cells can be mixed with dermal fibroblasts in a 1:1 ratio and injected subcutaneously [6]. Basement membrane matrix gels (BD Matrigel™) are also available to replace dermal fibroblasts [45].

3.5 Identification and Isolation by Microdissection

The use of specific markers to localize or isolate skin stem cells has been crucial to elucidate some of their characteristics. However, surface markers are not always known or the structure where the cells of interest reside might prevent from FACS isolation. This is the case for both the human sebaceous gland where no markers are available and for the dermal papilla (mouse and human), which is surrounded by extracellular matrix that cannot be enzyme digested (collagenase). For those two niches, microdissection followed by expansion in culture remains the preferred option [20, 21].

3.6 Organotypic Models of Skin

3.6.1 3D Organotypic Keratinocyte Culture

Addressing the stemness of a population requires testing their differentiation potential. Therefore, to fully mimic the stratified layers of the epidermis, 3D organotypic cultures have been established using both human [38] and mouse [39] keratinocytes. It has been shown that cell proliferation and differentiation of keratinocytes in culture depends on the composition of the media [47], but might also rely on the mechanical cues from their environment [48, 49]. Thus, compared to 2D tissue culture plastic, the organotypic culture gives a more adequate and realistic in vitro model with which we can assess behavior of epidermal stem cell populations and homeostasis to ultimately recapitulate the skin architecture. Furthermore, beyond constituting a new tool for homeostasis, wounding studies and small molecule screening, 3D reconstructed skins are promising regenerative therapies for burn victims and skin disease patients [33].

3.6.2 Dermal Spheroids

Dermal spheroids (DS) are a unique cell population located at the base of the hair follicle, capable of inducing de novo hair formation in a variety of recipient epithelia [50]. Within the hair regeneration field, researcher efforts are focused on exploiting human DP cells for their hair-inducing properties. However, a major obstacle has been trying to overcome the loss of inductive potential of the DP cells that ensues following a brief period in culture [26, 36]. Therefore, recent studies have focused on maintaining or restoring hair-inductive properties of human DP cells. In the context of regenerative medicine, inductive restoration would allow for cells isolated and expanded from several dermal papilla to give rise to hundreds of new hair follicles. Several groups have utilized keratinocyte-conditioned media, and added Wnt and Bmp proteins to the dermal papilla cultures, in an effort to reintroduce the crucial epithelial influence to the cells [25–27]. Most recently, Higgins et al. demonstrated that growth of dermal papilla cells in a hanging drop culture enabled them to cluster as dermal spheroids. When papilla cells are grown as spheroids, they bear a close morphological resemblance to intact papillae in skin [37]. Moreover, growth as a spheroid enables a partial restoration of the intact papilla transcriptional signature, where genes that are not usually expressed in culture become re-expressed. The authors established a human-to-human grafting assay to assess the inductive potential of dermal spheroids, demonstrating that the partial restoration of transcriptional signature was sufficient to restore the inductive capacity within the cells, and induce a de novo human follicle [37]. This methodology demonstrates that the microenvironment of papilla cells dramatically affects their inductive potential.

3.6.3 Bioengineering a Hair Follicle Niche

The identification of epidermal stem cells has led to a better understanding of the various skin niches and interactions with their microenvironment. Recently, Toyoshima et al. successfully regenerated a functional hair organ by intracutaneous transplantation into a Nude mouse of bioengineered pelage and vibrissae follicles [51]. They used mouse embryonic skin-derived epithelial and mesenchymal cells for pelage hairs or adult vibrissa stem cells from the bulge region with cultured DP cells. The intracutaneous transplantation process was optimized using an inter-epithelial tissue-connecting device with a nylon thread that could serve as a guide for the infundibulum, avoiding cyst formation. Surprisingly, bioengineered hair follicles showed complete re-establishment of their in vivo properties, with appropriate hair cycles and capacity of piloerection demonstrating successful development of proper connections with their microenvironment (arrector pili muscle, nerve fibers). Histological analysis confirmed the presence of all pelage hair types derived from the pelage hair germ, and transmission electron microscopy revealed that all hair follicle layers (medulla, cortex, cuticule) and even melanin granules were present. Importantly, epithelial stem cell markers were used to identify the bulge and the DP region. Overall, this demonstrates that these bioengineered hair follicles can reconstitute and maintain all epidermal stem cell niches as well as the DP, and can re-establish a favorable microenvironment for normal hair follicle development.

4 Notes

1. To optimize cell viability use HBSS or suitable high glucose containing medium.
2. Hairless immunodeficient mouse strains such as Nude are more amenable to skin grafting procedure.
3. Hair follicle cycle needs to be considered. Scraping the subcutaneous fat during anagen (growing phase of the hair follicle) can trigger a loss of bulge cells. Also, the hair cycle might have a detrimental impact on the scatter profile of stem cell populations

by FACS analysis. For instance, the bulge has a different FACS profile during anagen versus telogen (resting phase of the hair follicle) stages of the hair cycle.

4. To remove the dorsal skin, make a small incision in the lower back area. Delicately insert scissors between the skin and the underlying tissue to dissociate the skin from the fascia.
5. Skin requires being handled with care to optimize the isolation of live cells. In addition, female mice tend to have a thinner and more fragile epidermis. While scrapping, hold the skin gently with forceps.
6. After 1.5 h, use a scalpel to check if the epidermis can be gently scraped off the dermis. If the digestion is complete, the epidermis should come off easily. If not, place the Petri dish back at 37 °C and check every 15 min.
7. To transfer the epidermis into the bottle, use a 5 mL pipette with the tip broken off. Pipette up and down several times to facilitate a single cell suspension.
8. If there is a lot of debris, it will be necessary to add an additional wash step with 10 mL of isolation medium.
9. It is recommended to use the minimum amount of cells for controls (unstained and single-stained) and save the remainder for the sample with all the markers. Usually 1×10^6 cells are sufficient for controls.
10. For sorting, the usual cell concentration to allow a time efficient sort is 20×10^6 cells/mL. It is also recommended to set aside medium in case of a clog during the procedure. To optimize viability, sheath pressure should be decreased to 11 psi.
11. Euthanized pups are washed in sterile water once, followed by two washes in 70 % ethanol. Pups are then placed in a sterile petri dish in the hood. When processing several pups, float skins in PBS until all are harvested.
12. Cells can be stained from 1 h to 1 week in Rhodamine B.
13. Prior to implanting the chambers, make sure there is a hole in the top half of the hat. A small punch can be used.
14. If necessary, the grafting caps can be left on the skin for 2 weeks.

References

1. Fuchs E, Tumbar T, Guasch G (2004) Socializing with the neighbors: stem cells and their niche. Cell 116:769–778
2. Yan X, Owens DM (2008) The skin: a home to multiple classes of epithelial progenitor cells. Stem Cell Rev 4:113–118
3. Lobitz C, Dobson L (1961) Dermatology: the eccrine sweat glands. Annu Rev Med 12;289–298
4. Cotsarelis G, Sun TT, Lavker RM (1990) Label-retaining cells reside in the bulge area of pilosebaceous unit: implications for follicular stem cells, hair cycle, and skin carcinogenesis. Cell 61:1329–1337
5. Trempus CS, Morris RJ, Bortner CD et al (2003) Enrichment for living murine keratinocytes from the hair follicle bulge with the cell

surface marker CD34. J Invest Dermatol 120: 501–511

6. Morris RJ, Liu Y, Marles L et al (2004) Capturing and profiling adult hair follicle stem cells. Nat Biotechnol 22:411–417
7. Horsley V, O'Carroll D, Tooze R et al (2006) Blimp1 defines a progenitor population that governs cellular input to the sebaceous gland. Cell 126:597–609
8. Nijhof JGW, Braun KM, Giangreco A et al (2006) The cell-surface marker MTS24 identifies a novel population of follicular keratinocytes with characteristics of progenitor cells. Development 133:3027–3037
9. Ghazizadeh S, Taichman LB (2001) Multiple classes of stem cells in cutaneous epithelium: a lineage analysis of adult mouse skin. EMBO J 20:1215–1222
10. Levy V, Lindon C, Harfe BD et al (2005) Distinct stem cell populations regenerate the follicle and interfollicular epidermis. Dev Cell 9:855–861
11. Clayton E, Doupé DP, Klein AM et al (2007) A single type of progenitor cell maintains normal epidermis. Nature 446:185–189
12. Jensen KB, Collins CA, Nascimento E et al (2009) Lrig1 expression defines a distinct multipotent stem cell population in mammalian epidermis. Cell Stem Cell 4:427–439
13. Woo S-H, Stumpfova M, Jensen UB et al (2010) Identification of epidermal progenitors for the Merkel cell lineage. Development 139:622
14. Doucet YS, Woo S-H, Ruiz ME et al (2013) The touch dome defines an epidermal niche specialized for mechanosensory signaling. Cell Rep 3:1759–1765
15. Morrison SJ, Kimble J (2006) Asymmetric and symmetric stem-cell divisions in development and cancer. Nature 441:1068–1074
16. Lechler T, Fuchs E (2005) Asymmetric cell divisions promote stratification and differentiation of mammalian skin. Nature 437:275–280
17. Weinberg W, Goodman L (1993) Reconstitution of hair follicle development in vivo: determination of follicle formation, hair growth and hair quality by dermal cells. J Investig Dermatol 100:229–236
18. Kamimura J, Lee D, Baden H (1997) Primary mouse keratinocyte cultures contain hair follicle progenitor cells with multiple differentiation potential. J Investig Dermatol 109:534–540
19. Rheinwald JG, Green H (1975) Serial cultivation of strains of human epidermal keratinocytes: the formation of keratinizing colonies from single cells. Cell 6:331–343
20. McNairn AJ, Doucet Y, Demaude J et al (2013) TGFβ signaling regulates lipogenesis in human sebaceous glands cells. BMC Dermatol 13:2
21. Gledhill K, Gardner A, Jahoda C (2013) Isolation and establishment of hair follicle dermal papilla cell cultures. In: Turksen K (ed) Skin Stem Cells: Methods and Protocols, Methods in Molecular Biology, vol 989. Springer, New York, pp 285–292
22. Tani H, Morris RJ, Kaur P (2000) Enrichment for murine keratinocyte stem cells based on cell surface phenotype. Proc Natl Acad Sci U S A 97:10960–10965
23. Rendl M, Lewis L, Fuchs E (2005) Molecular dissection of mesenchymal-epithelial interactions in the hair follicle. PLoS Biol 3:e331
24. Birgersdotter A, Sandberg R, Ernberg I (2005) Gene expression perturbation in vitro–a growing case for three-dimensional (3D) culture systems. Semin Cancer Biol 15:405–412
25. Kishimoto J, Burgeson RE, Morgan BA (2000) Wnt signaling maintains the hair-inducing activity of the dermal papilla Wnt signaling maintains the hair-inducing activity of the dermal papilla. Genes Dev 14:1181–1185
26. Rendl M, Polak L, Fuchs E (2008) BMP signaling in dermal papilla cells is required for their hair follicle-inductive properties. Genes Dev 22:543–557
27. Ouji Y, Nakamura-Uchiyama F, Yoshikawa M (2013) Canonical Wnts, specifically Wnt-10b, show ability to maintain dermal papilla cells. Biochem Biophys Res Commun 438:493–499
28. Ohyama M, Kobayashi T, Sasaki T et al (2012) Restoration of the intrinsic properties of human dermal papilla in vitro. J Cell Sci 125:4114–4125
29. Kandyba E, Leung Y, Chen Y-B et al (2013) Competitive balance of intrabulge BMP/Wnt signaling reveals a robust gene network ruling stem cell homeostasis and cyclic activation. Proc Natl Acad Sci U S A 110:1351–1356
30. Nolte SV, Xu W, Rennekampff H-O et al (2008) Diversity of fibroblasts—a review on implications for skin tissue engineering. Cells Tissues Organs 187:165–176
31. Festa E, Fretz J, Berry R et al (2011) Adipocyte lineage cells contribute to the skin stem cell niche to drive hair cycling. Cell 146:761–771
32. Watt FM, Hogan BL (2000) Out of Eden: stem cells and their niches. Science 287:1427–1430
33. Itoh M, Umegaki-Arao N, Guo Z et al (2013) Generation of 3D skin equivalents fully reconstituted from human induced pluripotent stem cells (iPSCs). PLoS One 8:e77673
34. Millar SE (2002) Molecular mechanisms regulating hair follicle development. J Invest Dermatol 118:216–225

35. Rompolas P, Deschene ER, Zito G et al (2012) Live imaging of stem cell and progeny behaviour in physiological hair-follicle regeneration. Nature 487:496–499
36. Kishimoto J, Ehama R, Wu L et al (1999) Selective activation of the versican promoter by epithelial- mesenchymal interactions during hair follicle development. Proc Natl Acad Sci U S A 96:7336–7341
37. Higgins CA, Chen JC, Cerise JE, et al (2013) Microenvironmental reprogramming by three-dimensional culture enables dermal papilla cells to induce de novo human hair-follicle growth. Proc Natl Acad Sci U S A [Epub ahead of print] PMID: 24145441
38. Gangatirkar P, Paquet-Fifield S, Li A et al (2007) Establishment of 3D organotypic cultures using human neonatal epidermal cells. Nat Protoc 2:178–186
39. Carroll JM, Molès JP (2000) A three-dimensional skin culture model for mouse keratinocytes: application to transgenic mouse keratinocytes. Exp Dermatol 9:20–24
40. Goldstein J, Horsley V (2012) Home sweet home: skin stem cell niches. Cell Mol Life Sci 69:2573–2582
41. Chan RK, Zamora DO, Wrice NL et al (2012) Development of a vascularized skin construct using adipose-derived stem cells from debrided burned skin. Stem Cells Int 2012:841203
42. Liu Y, Suwa F, Wang X et al (2007) Reconstruction of a tissue-engineered skin containing melanocytes. Cell Biol Int 31: 985–990
43. Blanpain C, Lowry W, Geoghegan A (2004) Self-renewal, multipotency and the existence of two cell populations within an epithelial stem cell niche. Cell 118:635–648
44. Barrandon Y, Green H (1987) Three clonal types of keratinocyte with different capacities for multiplication. Proc Natl Acad Sci U S A 84:2302–2306
45. Lu CP, Polak L, Rocha AS et al (2012) Identification of stem cell populations in sweat glands and ducts reveals roles in homeostasis and wound repair. Cell 150:136–150
46. Jensen U, Yan X, Triel C et al (2008) A distinct population of clonogenic and multipotent murine follicular keratinocytes residing in the upper isthmus. J Cell Sci 121:609–617
47. Yuspa SH, Kilkenny AE, Steinert PM et al (1989) Expression of murine epidermal differentiation markers is tightly regulated by restricted extracellular calcium concentrations in vitro. J Cell Biol 109:1207–1217
48. Yano S, Komine M, Fujimoto M et al (2004) Mechanical stretching in vitro regulates signal transduction pathways and cellular proliferation in human epidermal keratinocytes. J Invest Dermatol 122:783–790
49. Reichelt J (2007) Mechanotransduction of keratinocytes in culture and in the epidermis. Eur J Cell Biol 86:807–816
50. Fliniaux I, Viallet JP, Dhouailly D et al (2004) Transformation of amnion epithelium into skin and hair follicles. Differentiation 72:558–565
51. Toyoshima K, Asakawa K, Ishibashi N, Toki H, Ogawa M, Hasegawa T, Irié T, Tachikawa T, Sato A, Takeda A, Tsuji T (2012) Fully functional hair follicle regeneration through the rearrangement of stem cells and their niches. Nat Commun 3:784
52. Clavel C, Grisanti L, Zemla R et al (2012) Sox2 in the dermal papilla niche controls hair growth by fine-tuning BMP signaling in differentiating hair shaft progenitors. Dev Cell 23:981–994
53. Ito Y, Hamazaki TS, Ohnuma K et al (2007) Isolation of murine hair-inducing cells using the cell surface marker prominin-1/CD133. J Invest Dermatol 127:1052–1060
54. Greco V, Chen T, Rendl M et al (2009) A two-step mechanism for stem cell activation during hair regeneration. Cell Stem Cell 4:155–169
55. Jaks V, Barker N, Kasper M et al (2008) Lgr5 marks cycling, yet long-lived, hair follicle stem cells. Nat Genet 40:1291–1299
56. Vidal VPI, Chaboissier M-C, Lützkendorf S et al (2005) Sox9 is essential for outer root sheath differentiation and the formation of the hair stem cell compartment. Curr Biol 15: 1340–1351
57. Nowak JA, Polak L, Pasolli HA et al (2008) Hair follicle stem cells are specified and function in early skin morphogenesis. Cell Stem Cell 3:33–43
58. Horsley V, Aliprantis AO, Polak L et al (2008) NFATc1 balances quiescence and proliferation of skin stem cells. Cell 132:299–310
59. Nguyen H, Merrill BJ, Polak L et al (2009) Tcf3 and Tcf4 are essential for long-term homeostasis of skin epithelia. Nat Genet 41:1068–1075
60. Rhee H, Polak L, Fuchs E (2006) Lhx2 maintains stem cell character in hair follicles. Science 312:1946–1949
61. Youssef KK, Van Keymeulen A, Lapouge G et al (2010) Identification of the cell lineage at the origin of basal cell carcinoma. Nat Cell Biol 12:299–305
62. Lapouge G, Youssef KK, Vokaer B et al (2011) Identifying the cellular origin of squamous skin tumors. Proc Natl Acad Sci U S A 108: 7431–7436

63. Ohyama M, Terunuma A, Tock CL et al (2006) Characterization and isolation of stem cell—enriched human hair follicle bulge cells. J Clin Invest 116:249–260
64. Snippert HJ, Haegebarth A, Kasper M et al (2010) Lgr6 marks stem cells in the hair follicle that generate all cell lineages of the skin. Science 327:1385–1389
65. Leung Y, Kandyba E, Chen Y-B et al (2013) Label retaining cells (LRCs) with myoepithelial characteristic from the proximal acinar region define stem cells in the sweat gland. PLoS One 8:e74174

Chapter 14

Isolation of Adult Stem Cell Populations from the Human Cornea

Matthew J. Branch, Wing-Yan Yu, Carl Sheridan, and Andrew Hopkinson

Abstract

Corneal blindness is a leading cause of vision loss globally. From a tissue engineering perspective, the cornea represents specific challenges in respect to isolating, stably expanding, banking, and effectively manipulating the various cell types required for effective corneal regeneration. The current research trend in this area focuses on a combined stem cell component with a biological or synthetic carrier or engineering scaffold. Corneal derived stem cells play an important role in such strategies as they represent an available supply of cells with specific abilities to further generate corneal cells in the long term. This chapter describes the isolation protocols of the epithelial stromal and endothelial stem cell populations.

Key words Keratocytes, Mesenchymal stem cells, Limbal stem cells, CD34, Cornea, Epithelium, Stroma, PLCSC, Endothelium

1 Introduction

The cornea is the window of the eye and its clarity is vital for vision. It is an avascular structure comprised of six layers (including the recently discovered Dua's layer [2]). Three of these layers, the epithelium, stroma, and endothelium, are cellularized with distinct cell types and different properties. Under homeostasis, the uppermost layer of stratified epithelium is continuously shed and renewed. Quiescent stromal cells, known as keratocytes, sparsely populate the corneal stroma, which accounts for around 90 % of the corneal thickness. The endothelium actively maintains corneal hydration, which is vital for maintaining its shape and clarity, but does not actively proliferate. The limbus which lies at the corneal-conjunctival and scleral interfaces represents a region of stemness for all three layers, and is known to contain epithelial, stromal, and endothelial progenitor populations. The organization and function of all of these layers are essential for corneal transparency and vision, and insult to any one of these layers, particularly at the limbus, can affect the others and compromise transparency. Second

Ivan N. Rich (ed.), *Stem Cell Protocols*, Methods in Molecular Biology, vol. 1235,
DOI 10.1007/978-1-4939-1785-3_14, © Springer Science+Business Media New York 2015

only to cataracts, corneal opacities caused by injury or disease are the most significant causes of vision loss worldwide [1], creating an important drive to develop innovative engineered cellular therapies to regenerate the cornea, restore transparency, and regain sight. Emerging regenerative strategies ultimately strive to restore transparency to the damaged cornea and often involve a stem cell component. Replacement stem cells may be sourced from a wide variety of different cell types including limbal, buccal mucosal, induced pluripotent, embryonic, and mesenchymal stem cells.

Clinically the most widely utilized cell type for corneal regeneration is corneal epithelial stem cells, also known as limbal stem cells (LSC) because of their anatomical localization. LSC transplants can be taken from autologous or allogeneic (either living related or cadaveric) donors. Both autologous and allogeneic LSC sources can be transplanted directly onto the eye or ex vivo expanded LSC are also used for transplantation. An outline of conventional surgical decision making on this area can be found in Dua et al. 2010 [1]. Although allogeneic grafts offer the patient short-term improvements in corneal transparency, the long-term outcomes are still poor (around 25 % failure at 5 years). Thus, alternative sources of cells are sought which may improve this. LSC, and the corneal epithelial lineage, have been extensively characterized with comprehensive marker profiles representing LSC and the various stages of differentiation. LSC in particular can be defined by their characteristic expression of cytokeratins 14, 15, and 19, ABCG2, ΔNp63α, C/EBPΔ, vimentin, SOD2, OCT4, nestin, NOTCH1, HES1, FRZ1, CDH1 and by their lack of desmoglein 3, desmocollin 3, and connexin 43 [2–14]. Terminally differentiated corneal epithelium can be clearly recognized by their expression of cytokeratins 3, 12, and 24, desmocollin 2, and desmoglein 2 [3, 8, 10, 15, 16]. Additional markers of stratification (E cadherin), extracellular matrix (tenascin C), and cell-basement membrane interactions (integrins α6 β4) are also used [2, 4, 17]. Many of these markers can be and are used to isolate and refine LSC populations; however true LSC are present in low numbers, and possess limited ability to be banked, thus creating a therapeutic challenge.

The characteristics of corneal stromal stem cells however are less defined. This is because until recently they were thought to be simply scar-forming fibroblasts when active. Recently, we identified cultured corneal stromal cells (termed peripheral and limbal corneal stromal cells or PLCSC) as mesenchymal stem cells (MSC) [18], research subsequently built upon by others [19], and therefore opening up their therapeutic potential. However disparities between culture conditions (including, for example, varying methods of isolation, substrate, and media) and differing (stem) cell requirements mean that MSC characterization is not always appropriate and instead, a number of other criteria are used by different groups. These include different markers including CD34 which is a marker of the keratocyte phenotype [20] or SSEA-4, PAX6, Oct4, Sox2, and Nanog all of which are markers of early development in the embryo and eye [21, 22]. Of these only the CD34 and SSEA-4 markers are cell surface antigens and thus are the only current markers that provide the opportunity for viable cell sorting. The predominant therapeutic property exploited by corneal stromal stem cells appears to be either epithelial [21, 23] or endothelial [24] regeneration. Ironically these stromal cells are seemingly less often investigated for stromal regeneration, although this is also researched [22]. Furthermore, non-corneal MSC have been differentiated into a keratocyte phenotype [25, 26], which suggests that corneal derived MSC could also be used for stromal regeneration.

With the advances in surgical technique such as Descemet stripping endothelial keratoplasty (DSEK), the use of corneal endothelial stem cells to regenerate the endothelium represents a promising therapeutic strategy. Previous studies have provided convincing evidence for the presence of a stem cell niche in the transition zone between the peripheral corneal endothelium and the trabecular meshwork. Whikehart and colleagues detected telomerase activity at the peripheral corneal endothelium and showed bromodeoxyuridine (BrdU) labeling in the transition zone and trabecular meshwork [27]. The BrdU staining was more intense and extended further into the corneal endothelium following experimental mechanical injuries. These findings suggest that stem cells at the transition zone may help renew the corneal endothelium, especially after trauma. McGowan et al. 2007 [28] later identified stem cell markers including nestin, alkaline phosphatase, and telomerase in some cells at the posterior limbus. Additional markers (Oct3/4, Pax6, Wntl, and Sox2) were detected in wounded corneas. He et al. found that the expression of nestin and telomerase on flat-mount corneas was largely restricted in the extreme periphery of the CE, whilst ABCG2 staining was found both at the centre and extreme periphery [29]. Purification of the CE-specific stem or progenitor cells remains a difficult task to date. There has been a lack of specific markers for both precursor and differentiated corneal endothelial cells (reviewed in [30]). The shortage of specific cell-surface markers hinders the prospective isolation of stem cells using fluorescence-activated cell sorting or magnetic immunosorting. However, attempts have been made to isolate and expand corneal endothelial stem cells using sphere culture. Yokoo and coworkers successfully applied this cultivation technique to isolate human corneal endothelial progenitor cells [31]. The peripheral corneal endothelium was found to generate more spheres compared to the central area [32, 33]. The corneal endothelial progenitor cells showed longer telomere length and higher telomerase activity than primary human CE cells [34]. Furthermore, the corneal endothelial spheres were shown to be effective for the treatment of CE deficiency and bullous keratopathy in rabbit models [35, 36]. Some other studies explored the feasibility of using bone marrow or umbilical cord blood MSC to reconstitute the damaged corneal endothelium [37–39]. Nevertheless, their differentiation and functional properties need to be rigorously characterized before clinical trials could be contemplated.

Increasingly, research is focused on the exploitation of corneal-derived stem cells (particularly from the limbus), and their regenerative potential, with the view that these cells not only possess the ability to generate the requisite corneal cell types but do so readily and with minimal undesired deviations into non-corneal cell types. The use of corneal derived cells for corneal regenerative therapies is attractive and will likely facilitate regulatory approval for clinical use and also inform our fundamental understanding of the cornea. As corneal regenerative medicine progresses, there are many diverse methodologies for isolation, characterization, and expansion of corneal stem cells for both clinical and research settings and a standardized protocol for the isolation of these cell types is key. This protocol aims to provide a standardized method of isolation and thus help reduce variation.

2 Materials

This protocol concentrates on the isolation of stem cells from corneoscleral rims, containing the limbus, for research purposes. Amongst other reasons, this trend of isolating corneal stromal stem cells from the limbus may be due to the limbal stroma yielding stem cell populations with properties that do not seem to appear in central corneal stroma. Postoperative eye bank rims from penetrating grafts are the most prevalent source of tissue available to researchers and because tissue is precious, it is logical that an individual rim is used to isolate epithelial, stromal, and endothelial cell types. The rim, which has had the central 7.5 or 8 mm section removed, constitutes the peripheral one third of the cornea and the limbus with some scleral tissue. This protocol mainly utilizes two-dimensional cultures currently more compatible for translational manufacturing procedures. However three-dimensional cultures including sphere [21, 40], hydrogel [41, 42], and hanging drop [43] and feeder layer systems [40, 44] are also used for corneal stromal stem cells and co-cultures between corneal stromal and corneal epithelial cells are common [40, 41]. Amniotic membrane is widely used in ophthalmology as a culture substrate and carrier for both stromal [45] and epithelial cell populations [46–48].

Unless otherwise stated, the preparation of all materials and all methods are performed in a class II laminar air flow hood under aseptic conditions, all cell suspensions are pelleted in a centrifuge at 200 × *g* for 5 min, all media is warmed to 37 °C, all cells are cultured at 37 °C, 5 % v/v CO_2 and 95 % humidity, and the culture media is changed every 2 days.

2.1 Reagents

1. Dispase (CellnTec).
2. CNT-20 Medium (CellnTec).
3. Calcium chloride.
4. Collagenase (Sigma Aldrich).
5. Dulbecco's phosphate buffered saline (DPBS).
6. Ethylenediaminetetraacetic acid (EDTA).
7. Bovine serum albumin (BSA, Affymetrix).
8. Anti-human biotin-conjugated CD34 (Life Technologies).
9. Anti-biotin microbeads (Miltenyi Biotech).
10. Fibronectin (Sigma Aldrich).
11. M199 Medium (Sigma Aldrich).
12. Plasmocin (Autogen Bioclear).
13. Gentamicin/Amphotericin B (combination, Life Technologies).
14. l-Glutamine (Sigma Aldrich).
15. DMEM/F12 Medium (Life Technologies).
16. Penicillin /streptomycin (combination, Life Technologies).
17. Fungizone (Life Technologies).
18. Basic fibroblast growth factor (bFGF, Peprotech).
19. B27 supplement minus vitamin A (Life Technologies).
20. Epidermal growth factor (EGF, Peprotech).
21. Heparin (Sigma Aldrich).

2.2 Supplies

1. Sterile filters (Sartorius).
2. Watchmakers forceps.
3. Vortex mixer.
4. 35 mm sterile Petri dishes.
5. Hemocytometer.
6. Trypan blue.
7. Accutase® Cell Detachment Solution.
8. Sterilized filter holder and nylon net filter (Swinnex, Millipore).
9. TripLE™ Express.

2.3 Isolation Components

2.3.1 Dispase Digestion Solution Preparation

Dilute dispase stock solution to 2.4 U/mL using complete CNT-20 medium adding 2 mM calcium chloride and filter-sterilize using 0.20 μm filter. Prepare 3 mL aliquots in polypropylene tubes and store at −20 °C. One corneoscleral rim will require one 3 mL aliquot and thus repeated freeze/thaw is prevented.

2.3.2 Collagenase Solution Preparation

Dissolve powdered collagenase (stored at −20 °C), to 0.1 % w/v with serum-free CM-M (Subheading 2.3) and filter-sterilize using 0.20 μm filter. Prepare collagenase solution fresh each time to avoid loss of activity.

2.3.3 MACS Buffer Preparation

Supplement Dulbecco's phosphate buffered saline (DPBS) with 2 mM EDTA and 0.5 % w/v bovine serum albumin. For a single rim isolation make up approximately 200 mL, and filter-sterilize using a 0.2 μm vacuum filter. In a sterile glass bottle, degas the buffer using a vacuum chamber and pump. Seal the bottle and only disturb when needed.

2.3.4 MACS Materials

This protocol requires anti-human CD34 (or cell surface antigen of choice) conjugated with biotin, MACS Anti-Biotin Microbeads, MS Columns, and MiniMACS magnetic separator and board.

2.4 Culture and Reagent Preparation

2.4.1 Limbal Epithelial Culture Preparation

Fibronectin Coating: Coat tissue culture plastic (wells and flasks), glass slides, or coverslips intended for epithelial cell culture with fibronectin. Prepare coating solution by reconstituting 1 mg fibronectin in 1 mL distilled H_2O for 30 min at 37 °C and then further diluting the solution 1:30 with CnT-20 and coat using minimal volume to surface area (e.g., 1 mL for one well of a six well plate). Air-dry, at ambient temperature, in a sterile environment for 45 min.

2.4.2 Corneal Stromal Cell Culture Preparation

Corneal Stromal Cell Culture Medium: Culture extracted corneal stromal cells in M199 supplemented with 20 % v/v fetal bovine serum, 2.5 μg/mL Plasmocin, 0.02 μg/mL gentamicin, 0.5 ng/mL amphotericin B, and 1.59 mM l-glutamine. This media is subsequently abbreviated to "CM-M."

2.4.3 Corneal Endothelial Cell Culture Preparation

Corneal Endothelial Cell Culture Medium: Supplement DMEM/F12 media (1:1 ratio) with 10 % v/v fetal bovine serum, 100 U/mL penicillin, 100 μg/mL streptomycin, 2.5 μg/mL Fungizone, 2.5 mM l-glutamine, and 2 ng/mL basic fibroblast growth factor. This media is subsequently abbreviated to "CM-E."

Corneal Endothelial Sphere Culture Medium: Freshly prepare DMEM/F12 media (1:1 ratio) supplemented with 100 U/mL penicillin, 100 μg/mL streptomycin, 2.5 μg/mL Fungizone, B27 supplement minus vitamin A, 20 ng/mL epidermal growth factor, 20 ng/mL basic fibroblast growth factor, 5 μg/mL heparin,

0.1 mg/mL bovine serum albumin, and 2.5 mM l-glutamine. The media expires within a week. This media is subsequently abbreviated to "CM-S."

Serum-Free Endothelial Media: DMEM/F12 media supplemented with 100 U/mL penicillin, 100 μg/mL streptomycin, 2.5 μg/mL Fungizone, and 2.5 mM l-glutamine. This media is subsequently abbreviated to "CM-F."

3 Methods

3.1 Isolation and Culture of Limbal Epithelial Cells

1. Remove the corneoscleral rim from the organ culture bottle and remove excess sclera (leaving a 1–2 mm scleral border to preserve the limbus).
2. At this point the endothelial layer can be mechanically peeled away at the descemets, using watchmaker's forceps, and independently cultured.
3. Using a No. 22 blade quarter the rim and place each quarter segment, epithelial side up in a 35 mm petri dish and add a 3 mL dispase aliquot to the tissue.
4. Incubate for 16 h at 4 °C (*see* **Note 1**).
5. Whilst still submerged in the dispase digestion solution, gently remove released epithelial cells from the rim segments using a cell scraper (Corning Incorporated).
6. Pipette the cell suspension into a centrifuge tube and wash the segments in 3 mL CnT-20 to collect any remaining epithelium. Pellet the limbal epithelial cells by centrifugation. The remaining rim issue is then processed for isolation of stromal cells (*see* Subheading 3.2).
7. Suspend the pelleted epithelial cells in 1 mL of CnT-20. If cell clumps persist, separate using a vortex mixer or aspirate using a 25G hypodermic needle and syringe. Once thoroughly disaggregated, count the cells using a standard hemocytometer method with or without Trypan blue to assess viability.
8. When the cells have been counted adjust the cell concentration with the CnT-20 CM to achieve optimal seeding density. Culture using CnT-20 onto fibronectin coated chamber slides, wells, or in small flasks (we recommend no more than one 25 cm^2 flask per rim). Change CnT-20 every other day.
9. Passage confluent limbal epithelium by discarding media and washing the cells in DPBS before incubating with Accutase at 37 °C for 10 min. If the cells are not fully dissociated incubate for a further 5 min.
10. When the cells are detached, pipette the suspension into a centrifuge tube. Wash off any remaining cells with CnT-20, which

should also be added to the same centrifuge tube. Pellet the cells by centrifugation.

11. Discard the supernatant and seed the cells according to requirements for experimental work or continued culture for expansion of cell numbers (*see* **Note 2**).

3.2 Isolation and Culture of Corneal Mesenchymal Stem Cells

1. Finely chop the remaining stroma of each rim segment and add to the collagenase solution and incubate for 7 h.
2. Once incubated, gently agitate and pour the suspension into a 10 mL syringe. Place the syringe onto a sterilized filter holder containing a 41 μm nylon net filter. Pass the solution through the filter in order to remove any undigested debris whilst allowing the released cells to pass through and be collected into a centrifuge tube and pelleted.
3. Suspend the corneal stromal cells (not yet MSC) in 7 mL CM-M and decant into a 25 cm^2 tissue culture treated flask (one rim per flask).
4. Seeding densities are not usually calculated for primary culture as there are very few cells. However, once these cells reach confluence in their initial culture flask they are passaged (*see* **step 5**) and counted. For each subsequent passage, our typical seeding density for continuous culture is 3,000 cells per cm^2 and 10,000 cells per cm^2 for differentiation assays (*see* **Note 3**).
5. To passage corneal stromal cells cultured in CM-M wait until the flask is 80–90 % confluent and then remove culture medium and wash the cells with DPBS. Incubate the cells with 3 mL TripLE, or other enzymatic dissociation reagent, for 5 min at 37 °C. If cells are not fully suspended return flask to incubator for another 5 min, checking every minute.
6. When the cells are detached, pipette the suspension into a centrifuge tube. Wash the flask with CM-M, which should also be added to the same centrifuge tube. Pellet the cells by centrifugation.
7. Discard the supernatant and seed the cells according to requirements for experimental work, stem cell isolation (Subheading 3.3), or continued culture for expansion of cell numbers. Old CM-M should be exchanged for fresh every other day.

3.3 Isolation of Corneal Stromal Stem Cell Populations

1. For MACS antibody binding, prepare cultured cells as described in **steps 5** and **6** of Subheading 3.2.
2. Gently suspend pelleted cells in 80 μL MACS buffer and count the cell number.
3. Add 5 μL of biotinylated antibody for every 1×10^6 cells, and incubate for 30 min.

4. Wash cells in 5 mL of MACS buffer, pellet by centrifugation, and suspend the cells in 80 μL MACS buffer per 1×10^7 cells. Add 20 μL MACS anti-biotin microbeads per 1×10^7 cells and incubate for 15 min at 4 °C, with gentle agitation.
5. Pellet and wash the cells in 5 mL MACS buffer. Re-pellet the cells and, ensuring all supernatant has been removed, suspend cells in 500 μL MACS buffer, for up to 1×10^8 cells.
6. For MACS sorting attach the magnet unit to the separation stand and place the MS column in the magnet unit with a collection tube.
7. Fill the column by pipetting 500 μL MACS buffer on top of column and let the buffer flow through.
8. Discarding the flow through and replacing the collection tube, pipette the magnetically labeled cell suspension through the MS column.
9. Collect the flow through as the CD34 negative fraction.
10. Remove MS column from the magnet unit and place it on a fresh collection tube. Apply 1 mL buffer to the MS column and firmly flush out positive fraction using plunger. The cell fraction of interest may be passed through the column again to reduce the potential of cell contamination (*see* **Note 4**).

3.4 Isolation and Expansion of Corneal Endothelial Stem Cell Populations

1. Wash the isolated endothelial layer (**step 2**, Subheading 3.1) gently with 3 mL CM-F.
2. Place the endothelial side up on a 35 mm petri dish and digest it with 3 mL Accutase at 37 °C for 15 min.
3. After the incubation, carefully scrape off the endothelial cells with a cell scraper.
4. Transfer the cell suspension into a centrifuge tube and wash to collect the remaining endothelial cells with 3 mL CM-E. Pellet the cells by centrifugation.
5. Resuspend the cells in corneal endothelial cell CM-E. Seed the cells in a 35 mm petri dish pre-coated with fibronectin.
6. At the initial passage, primary cells may continue to be cultured as a monolayer (**step 8**) or as spheres (**steps 9–13**). To passage the primary corneal endothelial cells wash in DPBS and incubate with 1 mL Accutase at 37 °C for 5 min.
7. When the cells are detached, pipette the cell suspension into a centrifuge tube. Wash off any remaining cells with CM-F also adding this to the centrifuge tube. Pellet the cells by centrifugation.
8. For monolayer culture, suspend the cell pellet in CM-E and plate. Endothelial cells may be passaged at a 1:2 to 1:4 split ratio with Accutase when they are 90 % confluent. Endothelial

monolayers are cultured using CM-E. Change media every other day.

9. For sphere cultures, suspend the cells in the CM-F and triturate through a 1 mL pipette tip approximately 30 times until dissociated into a single cell suspension. Finally, count the cells using a hemocytometer and seed at 2,500 viable cells/cm^2 (trypan blue exclusion test) on suspension culture dishes (Corning) and culture in CM-S. Add fresh media every other day (*see* **Note 5**).
10. Spheres can be collected for subsequent studies (e.g., self-renewal property on subculture, differentiation assay, and immunocytochemical analysis) on day 7.
11. To subculture corneal endothelial spheres, transfer the sphere suspension from the culture dishes to a centrifuge tube. Allow the tube to stand for 5–10 min for the spheres to settle at the bottom (this method produces less damage to the spheres than centrifugation).
12. Carefully aspirate the supernatant and gently wash the spheres once in DPBS. Let the spheres settle for 5 min, then aspirate the DPBS. Resuspend the spheres in 1 mL Accutase and incubate at 37 °C in a water bath for 10 min.
13. Dissociate the spheres into a single cell suspension by triturating through a 1 mL pipette tip approximately 30 times. Count and replate as described in **step 9**.

4 Notes

1. Limbal epithelial cells may be grown out of *explant* cultures from rims that have been split to remove the majority of the stromal layer. Stromal cell isolation may be performed on the remnants. However, when isolation of epithelial and stromal cell types is required, explant culture is a less efficient use of the rim than the protocol above, given that the amount of stroma removed for explant culture is likely to be thicker than any removed by the dispase digestion step. Furthermore, if the stromal stem cells are most prevalent around the LSC niche, splitting the rim may remove a key population of stem cells. Finally the more immature epithelial cells residing in the limbus may not migrate from the explant but may be released by dispase digestion.
2. Limbal epithelial cells are difficult to passage and, unlike corneal stromal cells, do not seem to have the same extensive proliferative capacities. If possible experiments should be planned for primary cultures although one or two passages may be achieved with these cells.

3. MSC characterization protocols are widely available and often combine a profile of cell surface markers analyzed by flow cytometry, with mesenchymal differentiation analyzed by histological staining [18, 50]. Perhaps the most widely known criterion for this is the ISCT's position statement on "*Minimal criteria for defining multipotent mesenchymal stromal cells*" [51]. Others also argue for analysis of colony forming units fibroblasts (CFU-F) to demonstrate purity of populations, for which protocols are also available [52].
4. The corneal stromal stem cell culture method will differ depending on the type of stem cell required and that this will also influence the expression of markers. Culture on two-dimensional tissue culture plastics does not promote the retention of stem cell or keratocyte properties or markers and thus the current trend leans towards sphere cultures [21, 40].
5. Ensure that the CM-E is adequately aspirated off the cell pellet as remnants of serum in the media promote undesirable sphere attachment onto the dish.

Acknowledgements

Matthew Branch is funded by Fight for Sight. Corneal tissue for this work is kindly supplied by Manchester Eye Bank and Rakesh Jayaswal at the Eye Unit, Queen Alexandra Hospital, Portsmouth. The authors also wish to thank Laura Sidney for providing her help in reviewing the MACS sections. Wing-Yan Yu was funded by the Postgraduate Scholarship awarded from the University of Hong Kong.

References

1. Dua HS, Miri A, Said DG (2010) Contemporary limbal stem cell transplantation—a review. Clin Exp Ophthalmol 38:104–117
2. Chen Z, de Paiva CS, Luo L et al (2004) Characterization of putative stem cell phenotype in human limbal epithelia. Stem Cells 22:355–366
3. Schlotzer-Schrehardt U, Dietrich T, Saito K et al (2007) Characterization of extracellular matrix components in the limbal epithelial stem cell compartment. Exp Eye Res 85: 845–860
4. Schlotzer-Schrehardt U, Kruse FE (2005) Identification and characterization of limbal stem cells. Exp Eye Res 81:247–264
5. Dua HS, Azuara-Blanco A (2000) Limbal stem cells of the corneal epithelium. Surv Ophthalmol 44:415–425
6. Nieto-Miguel T, Calonge M, de la Mata A et al (2011) A comparison of stem cell-related gene expression in the progenitor-rich limbal epithelium and the differentiating central corneal epithelium. Mol Vis 17:2102–2117
7. Lyngholm M, Hoyer PE, Vorum H et al (2008) Immunohistochemical markers for corneal stem cells in the early developing human eye. Exp Eye Res 87:115–121
8. Pauklin M, Thomasen H, Pester A et al (2011) Expression of pluripotency and multipotency factors in human ocular surface tissues. Curr Eye Res 36:1086–1097
9. Zhou S, Schuetz JD, Bunting KD et al (2001) The ABC transporter Bcrp1/ABCG2 is expressed in a wide variety of stem cells and is a molecular determinant of the side-population phenotype. Nat Med 7:1028–1034

10. Joseph A, Powell-Richards AO, Shanmuganathan VA et al (2004) Epithelial cell characteristics of cultured human limbal explants. Br J Ophthalmol 88:393–398
11. Kulkarni BB, Tighe PJ, Mohammed I et al (2010) Comparative transcriptional profiling of the limbal epithelial crypt demonstrates its putative stem cell niche characteristics. BMC Genomics 11:526
12. Barbaro V, Testa A, Di Iorio E et al (2007) C/EBPdelta regulates cell cycle and self-renewal of human limbal stem cells. J Cell Biol 177: 1037–1049
13. Di Iorio E, Barbaro V, Ruzza A et al (2005) Isoforms of DeltaNp63 and the migration of ocular limbal cells in human corneal regeneration. Proc Natl Acad Sci U S A 102: 9523–9528
14. Grueterich M, Espana E, Tseng SC (2002) Connexin 43 expression and proliferation of human limbal epithelium on intact and denuded amniotic membrane. Invest Ophthalmol Vis Sci 43:63–71
15. Schermer A, Galvin S, Sun TT (1986) Differentiation-related expression of a major 64K corneal keratin in vivo and in culture suggests limbal location of corneal epithelial stem cells. J Cell Biol 103:49–62
16. Messent AJ, Blissett MJ, Smith GL et al (2000) Expression of a single pair of desmosomal glycoproteins renders the corneal epithelium unique amongst stratified epithelia. Invest Ophthalmol Vis Sci 41:8–15
17. Yeung AM, Schlotzer-Schrehardt U, Kulkarni B et al (2008) Limbal epithelial crypt: a model for corneal epithelial maintenance and novel limbal regional variations. Arch Ophthalmol 126:665–669
18. Branch MJ, Hashmani K, Dhillon P et al (2012) Mesenchymal stem cells in the human corneal limbal stroma. Invest Ophthalmol Vis Sci 53:5109–5116
19. Bray LJ, Heazlewood CF, Munster DJ et al (2013) Immunosuppressive properties of mesenchymal stromal cell cultures derived from the limbus of human and rabbit corneas. Cytotherapy. doi:pii: S1465-3249(13) 00635-X
20. Joseph A, Hossain P, Jham S et al (2003) Expression of CD34 and L-selectin on human corneal keratocytes. Invest Ophthalmol Vis Sci 44:4689–4692
21. Katikireddy KR, Dana R, Jurkunas UV (2013) Differentiation potential of limbal fibroblasts and bone marrow mesenchymal stem cells to corneal epithelial cells. Stem Cells. doi:10.1002/stem.1541
22. Funderburgh ML, Du Y, Mann MM et al (2005) PAX6 expression identifies progenitor cells for corneal keratocytes. FASEB J 19: 1371–1373
23. Hashmani K, Branch MJ, Sidney LE et al (2013) Characterisation of corneal stromal stem cells with the potential for epithelial transdifferentiation. Stem cell Res Ther 4:75
24. Hatou S, Yoshida S, Higa K et al (2013) Functional corneal endothelium derived from corneal stroma stem cells of neural crest origin by retinoic acid and Wnt/beta-catenin signaling. Stem cells Dev 22:828–839
25. Park SH, Kim KW, Chun YS et al (2012) Human mesenchymal stem cells differentiate into Keratocyte-like cells in Keratocyte-Conditioned medium. Exp Eye Res 101: 16–26
26. Liu H, Zhang J, Liu CY et al (2012) Bone marrow mesenchymal stem cells can differentiate and assume corneal keratocyte phenotype. J Cell Mol Med 16:1114–1124
27. Whikehart DR, Parikh CH, Vaughn AV et al (2005) Evidence suggesting the existence of stem cells for the human corneal endothelium. Mol Vis 11:816–824
28. McGowan SL, Edelhauser HF, Pfister RR et al (2007) Stem cell markers in the human posterior limbus and corneal endothelium of unwounded and wounded corneas. Mol Vis 13:1984–2000
29. He Z, Campolmi N, Gain P et al (2012) Revisited microanatomy of the corneal endothelial periphery: new evidence for continuous centripetal migration of endothelial cells in humans. Stem Cells 30:2523–2534
30. Yu WY, Sheridan C, Grierson I et al (2011) Progenitors for the corneal endothelium and trabecular meshwork: a potential source for personalized stem cell therapy in corneal endothelial diseases and glaucoma. J Biomed Biotechnol. 412743. doi:10.1155/2011/412743
31. Yokoo S, Yamagami S, Yanagi Y et al (2005) Human corneal endothelial cell precursors isolated by sphere-forming assay. Invest Ophthalmol Vis Sci 46:1626–1631
32. Mimura T, Yamagami S, Yokoo S et al (2005) Comparison of rabbit corneal endothelial cell precursors in the central and peripheral cornea. Invest Ophthalmol Vis Sci 46:3645–3648
33. Yamagami S, Yokoo S, Mimura T et al (2007) Distribution of precursors in human corneal stromal cells and endothelial cells. Ophthalmology 114:433–439
34. Mimura T, Yamagami S, Yokoo S et al (2010) Selective isolation of young cells from human

corneal endothelium by the sphere-forming assay. Tis Eng Part C Methods 16:803–812

35. Mimura T, Yamagami S, Yokoo S et al (2005) Sphere therapy for corneal endothelium deficiency in a rabbit model. Invest Ophthalmol Vis Sci 46:3128–3135
36. Mimura T, Yokoo S, Araie M et al (2005) Treatment of rabbit bullous keratopathy with precursors derived from cultured human corneal endothelium. Invest Ophthalmol Vis Sci 46:3637–3644
37. Joyce NC, Harris DL, Markov V et al (2012) Potential of human umbilical cord blood mesenchymal stem cells to heal damaged corneal endothelium. Mol Vis 18:547–564
38. Shao C, Fu Y, Lu W et al (2011) Bone marrow-derived endothelial progenitor cells: a promising therapeutic alternative for corneal endothelial dysfunction. Cells Tissues Organs 193:253–263
39. Liu XW, Zhao JL (2007) Transplantation of autologous bone marrow mesenchymal stem cells for the treatment of corneal endothelium damages in rabbits. Zhonghua Yan Ke Za Zhi 43:540–545
40. Xie HT, Chen SY, Li GG et al (2012) Isolation and expansion of human limbal stromal niche cells. Invest Ophthalmol Vis Sci 53:279–286
41. Wilson SL, Yang Y, El Haj AJ (2013) Corneal stromal cell plasticity: in vitro regulation of cell phenotype through cell-cell interactions in a three-dimensional model. Tissue Eng A 20(1–2):225–238
42. Chien Y, Liao YW, Liu DM et al (2012) Corneal repair by human corneal keratocyte-reprogrammed iPSCs and amphiphatic carboxymethyl-hexanoyl chitosan hydrogel. Biomaterials 33:8003–8016
43. Huang L, Harkenrider M, Thompson M et al (2010) A hierarchy of endothelial colony-forming cell activity displayed by bovine corneal endothelial cells. Invest Ophthalmol Vis Sci 51:3943–3949
44. Pellegrini G, Traverso CE, Franzi AT (1997) Long-term restoration of damaged corneal surfaces with autologous cultivated corneal epithelium. Lancet 349:990–993
45. Zhang W, Ge W, Li C et al (2004) Effects of mesenchymal stem cells on differentiation, maturation, and function of human monocyte-derived dendritic cells. Stem Cells Dev 13: 263–271
46. Koizumi N, Inatomi T, Suzuki T et al (2001) Cultivated corneal epithelial stem cell transplantation in ocular surface disorders. Ophthalmology 108:1569–1574
47. Nakamura T, Koizumi N, Tsuzuki M et al (2003) Successful regrafting of cultivated corneal epithelium using amniotic membrane as a carrier in severe ocular surface disease. Cornea 22:70–71
48. Pauklin M, Steuhl KP, Meller D (2009) Characterization of the corneal surface in limbal stem cell deficiency and after transplantation of cultivated limbal epithelium. Ophthalmology 116:1048–1056
49. Branch MJ, Hashmani K, Dhillon P et al (2012) Mesenchymal stem cells in the human corneal limbal stroma. Invest Ophthalmol Vis Sci 53:5109–5116
50. Reger RL, Tucker AH, Wolfe MR (2008) Differentiation and characterization of human MSCs. Methods Mol Biol 449:93–107
51. Dominici M, Le Blanc K, Mueller I et al (2006) Minimal criteria for defining multipotent mesenchymal stromal cells. The International Society for Cellular Therapy position statement. Cytotherapy 8:315–317
52. Pochampally R (2008) Colony forming unit assays for MSCs. Methods Mol Biol 449: 83–91

Chapter 15

Advanced Imaging and Tissue Engineering of the Human Limbal Epithelial Stem Cell Niche

Isobel Massie, Marc Dziasko, Alvena Kureshi, Hannah J. Levis, Louise Morgan, Michael Neale, Radhika Sheth, Victoria E. Tovell, Amanda J. Vernon, James L. Funderburgh, and Julie T. Daniels

Abstract

The limbal epithelial stem cell niche provides a unique, physically protective environment in which limbal epithelial stem cells reside in close proximity with accessory cell types and their secreted factors. The use of advanced imaging techniques is described to visualize the niche in three dimensions in native human corneal tissue. In addition, a protocol is provided for the isolation and culture of three different cell types, including human limbal epithelial stem cells from the limbal niche of human donor tissue. Finally, the process of incorporating these cells within plastic compressed collagen constructs to form a tissue-engineered corneal limbus is described and how immunohistochemical techniques may be applied to characterize cell phenotype therein.

Key words Limbal epithelial stem cell niche, 3View imaging, Limbal epithelial stem cells, Limbal fibroblasts, Corneal stromal stem cells, Immunohistochemistry, Tissue engineering

1 Introduction

Limbal epithelial stem cells (LESC) are responsible for maintenance of the corneal epithelium during both normal homeostasis and in response to injury [1]. They reside within the limbal epithelial stem cell niche, a unique environment that ensures their normal function [2–4].

Using advanced imaging techniques (serial block face scanning electron microscopy (SBFSEM)), imaging of the limbal epithelial stem cell niche within native human cornea in three dimensions (3-D) is now possible. In this technique, ultrathin sections (100 nm) are serially cut from the embedded tissue and the newly exposed tissue surface is subsequently imaged. This process can be repeated many times enabling a complete 3-D reconstruction on a micrometer scale such that cellular interactions and organization can be observed [5].

Ivan N. Rich (ed.), *Stem Cell Protocols*, Methods in Molecular Biology, vol. 1235,
DOI 10.1007/978-1-4939-1785-3_15,

The limbal epithelial stem cells themselves along with accessory cell types (limbal fibroblasts and corneal stromal stem cells found in the underlying stroma) may be isolated from donor tissue following specific combinations of enzymatic digestion and/or mechanical disruption of tissue architecture to selectively isolate the different cell populations. Culture conditions for each cell type differ and are designed to promote proliferation/preserve phenotype of each [4, 6].

Following culture of these different cell types, a tissue-engineered limbus may be formed from a collagen hydrogel. Collagen hydrogels can be produced by neutralizing the pH of acid-solubilized type I collagen with heating, allowing fibrillogenesis to occur. Then, by applying a hydrophilic absorber to the top surface of the hydrogel, water can be wicked upwards to produce a thin, mechanically strong, transparent construct (~150 μm thick) that may be utilized as an experimental model of the native limbus. This technique was first proposed by Brown et al. [7] but we have since iteratively improved the method to increase reproducibility and reliability [8, 9]. Collagen constructs produced using this newer methodology are referred to as Real Architecture For 3D Tissue, or RAFT. Most recently, in collaboration with an industrial partner, we have developed a method for producing absorbers with a capability to print different topographies onto the surface of RAFT constructs whilst simultaneously wicking water away [10]. These surface topographies recreate the physical aspects of the LESC niche. LESC can be seeded onto the surface of the constructs and air-lifted to induce stratification to produce a multilayered epithelium, similar to that of native cornea. Furthermore, limbal fibroblasts may be incorporated within the collagen constructs if desired without a loss of cell viability to further support overlying LESC [8].

Once formed, immunohistochemistry (either whole-/flat-mount or sections) may be performed to characterize cell phenotype within and/or on collagen constructs. The transcription factor, p63α, along with markers of corneal differentiation such as cytokeratin (CK) 3 [11] is typically used for expression of putative LESC markers.

2 Materials

Prepare all solutions using deionized water. Prepare and store all reagents at room temperature (unless indicated otherwise). Dispose of all waste materials according to waste disposal regulations. Read all appropriate COSHH and Risk Assessment forms prior to beginning work.

1. Human donor corneal tissue: either whole cornea or corneoscleral rims (*see* **Note 1**).
2. Dulbecco's phosphate buffered saline (DPBS) 1×, without calcium or magnesium ions (Life Technologies).

3. DMEM with 4.5 g/L glucose and GlutaMAX (Life Technologies). Store at 4 °C.
4. DMEM/F12 (Life Technologies). Store at 4 °C.
5. Fetal bovine serum (FBS). Store at −20 °C.
6. 100× antibiotic, anti-mycotic (anti-anti) (Life Technologies). Store at −20 °C.
7. 0.5 % Trypsin-EDTA without phenol red (10×) (Life Technologies). Store at −20 °C.
8. 0.05 % Trypsin-EDTA without phenol red (1×): Dilute 5 mL of 0.5 % trypsin-EDTA without phenol red in 45 mL DPBS. Store at 4 °C.
9. Sodium hydroxide (NaOH).
10. Hydrochloric acid (HCl).
11. TT solution: Add 1.36 mg T3 (Sigma-Aldrich) to 2 mL of 0.02 N NaOH. Add 8 mL of DPBS. Store 100 μL T3 stock aliquots at −20 °C. Add 50 mg of transferrin (Sigma-Aldrich) to 6 mL of DPBS. Add 100 μL of T3 stock. Make up to 100 mL with DPBS. Filter sterilize using a 0.22 μm filter. Store at −20 °C.
12. Adenine solution: Add 250 mg adenine powder (Sigma-Aldrich) to 10 mL of 1 M HCl. Make up to 100 mL with water. Filter sterilize using a 0.22 μm filter. Store at −20 °C.
13. Hydrocortisone solution: Dissolve 50 mg of hydrocortisone powder (Sigma-Aldrich) in 10 mL of ethanol. Filter sterilize using a 0.22 μm filter. Add 1 mL of this to 11.5 mL of DPBS and store at −20 °C.
14. Cholera toxin solution: Add 1 mg cholera toxin (Sigma-Aldrich) to 1.18 mL of water. Add 100 μL of this to 10 mL of DMEM/F12 containing 10 % FBS. Filter sterilize using a 0.22 μm filter. Store at −20 °C.
15. Corneal epithelial cell culture medium (CECM): Mix together 250 mL DMEM/F12 (Life Technologies) with 250 mL DMEM. Remove 71.25 mL of this mixture. Add 50 mL of FBS, 5 mL of anti-anti, 5 mL of TT solution, 5 mL of 18.5 mM adenine solution, 0.5 mL of 5 mg/mL hydrocortisone solution, 0.5 mL of 10 μM cholera toxin solution, 250 μL of 10 mg/mL insulin (Sigma-Aldrich). Store at 4 °C. Add 50 μL of 10 μg/mL epidermal growth factor (EGF) (Life Technologies) in DPBS (store at −20 °C) to 50 mL CECM immediately prior to use.
16. hLF medium: Remove 55 mL from a 500 mL bottle of DMEM. Add 50 mL of FBS and 5 mL of anti-anti to give final concentrations of 10 % FBS and 1× anti-anti. Store at 4 °C.
17. 4 % Paraformaldehyde (PFA).
18. Goat serum (Sigma-Aldrich). Store at −20 °C.

19. Triton X-100.
20. Isotype controls. Store at 4 °C.
21. Alexa Fluor secondary antibodies (Life Technologies) (store at 4 °C for short-term storage or at −20 °C for long-term storage).
22. Phalloidin (FITC or TRITC conjugates, Sigma-Aldrich). Store at −20 °C.
23. Vectashield mounting medium (with DAPI or PI, Vector Laboratories). Store at 4 °C.
24. Capture antibody solution: 2 % goat serum in DPBS.
25. Ethanol.
26. Trypan blue.
27. Cell culture inserts with PTFE membrane, 300 mm diameter, 0.4 μm pore (Millipore).
28. 0.22 μm sterile filters (Fisher Scientific).
29. Tissue culture plates and flasks.
30. Glass microscope slides.
31. Coverslips.
32. Hemocytometer.
33. Nail polish.
34. Forceps, small scissors, scalpels.
35. Hot plate.
36. Water bath.
37. Dissecting microscope.
38. Bright-field light microscope.
39. (Confocal) fluorescence microscope.

2.1 Specific Reagents Required for 3View Imaging

1. *0.3 M sodium cacodylate buffer stock*: Dissolve 64 g of sodium cacodylate in 900 mL of water. Adjust the pH to 7.4 and make up to 1 L with distilled water.
2. *10 % aqueous paraformaldehyde*: Add 10 g of paraformaldehyde to 100 mL of distilled water and heat to 65 °C in a fume cupboard. Add 2–5 drops of 1 M NaOH to clarify the solution. Filter through a 0.22 μm filter and allow to cool before use.
3. *Karnovsky's fixative*: Add 100 mL of freshly prepared 10 % paraformaldehyde to 120 mL of 25 % glutaraldehyde (Agar Scientific) and combine with 400 mL of 0.2 M sodium cacodylate buffer. Adjust the pH to 7.4 and make up to 1 L with distilled water. Store at −20 °C.
4. *3 % potassium ferricyanide solution*: Add 3 g of potassium ferricyanide (TAAB Laboratories Equipment Ltd.) to 100 mL of 0.3 M sodium cacodylate buffered to pH 7.4. Store at 4 °C.

5. *2 % aqueous osmium tetroxide solution*: (*see* **Note 2**).
6. *1 % thiocarbohydrazide solution*: Add 0.1 g of thiocarbohydrazide (Sigma-Aldrich) to 10 mL of distilled water and place in a 60 °C oven for 1 h. Agitate every 15 min to facilitate dissolving. Filter through a 0.22 μm filter and leave to cool.
7. *1 % uranyl acetate*: Prepare solution just before use. Dissolve 0.1 g of uranyl acetate (Agar Scientific) in 10 mL of water. Store the solution at 4 °C (*see* **Note 3**).
8. *Conductive silver epoxy kit*: Prepare a fresh mix of an equal volume of the two components (Agar Scientific) just before attachment of the resin block onto the cryopin.
9. *1 % toluidine blue*: Dissolve 1 g of sodium borate (Sigma-Aldrich) in 100 mL of water. Add 1 g of toluidine blue (Agar Scientific). Gently mix until complete dissolved. Filter through a 0.22 μm filter.
10. *Walton's lead aspartate solution*: Dissolve 0.066 g of lead nitrate in 10 mL of 0.03 M aspartic solution and adjust the pH to 5.5 using 1 N potassium hydroxide. Place the solution in a 60 °C oven for 30 min prior to use.
11. *Durcupan ACM resins*: (Sigma-Aldrich).
    (a) Single component A, M epoxy resin.
    (b) Single component B, hardener 964.
    (c) Single component C, accelerator 960.
    (d) Single component D.
12. *Durcupan ACM resin*: Combine 11.4 g of component A, 10 g of component B, 0.3 g of component C, and 0.1 g of component D.
13. Acetone.
14. DePeX mounting media (VWR).
15. Flat embedding molds.
16. Glass knives (Agar Scientific).
17. Diamond knife (dEYEmond, 45 Histo).
18. Single edge stainless steel razor blades (Agar Scientific).
19. Cryopin (Agar Scientific).
20. Sputter coater (Emitech).
21. Scanning electron microscope.
22. 3View ultramicrotome (Gatan).
23. Wacom workstation Cintiq 22HD touch and interactive pen.
24. AMIRA software (Visualization Science Group, FEI).

2.2 Specific Reagents Required for Isolation and Culture of Human Limbal Epithelial Cells (hLEC)

1. *1.2 IU/mL dispase*: Dissolve 10 IU dispase II (Roche Diagnostics) in 1 mL of DPBS. Store at −20 °C. Immediately prior to use, combine 0.6 mL of 10 IU/mL dispase II with 4.4 mL of CECM without EGF.
2. *3T3-J2 medium*: Remove 55 mL from a bottle of DMEM and add 50 mL of adult bovine serum (Labtech) and 5 mL of anti-anti. Store at 4 °C.
3. *Growth-arrested 3T3-J2 cells*: Maintain 3T3-J2 cells in 3T3-J2 medium. To expand, passage cells one in ten twice per week using 0.05 % trypsin-EDTA with mechanical agitation. Growth arrest 3T3-J2 cells at 70–80 % confluency using 4ug/mL mitomycin-C in 3T3-J2 medium for 2 h at 37 °C in a humidified incubator with 5 % CO_2.

2.3 Specific Reagents Required for Isolation and Culture of Human Limbal Fibroblasts (hLF)

1. *2 mg/mL collagenase*: Dissolve 10 mg collagenase type 1 powder in 5 mL of hLF media. Filter sterilize through a 0.22 μm filter prior to use. Make fresh each time.

2.4 Specific Reagents Required for Isolation and Culture of Human Corneal Stromal Stem Cells (hCSSC)

1. *MCDB-201 stem cell medium*: Add 17.7 g of MCDB-201 powdered stem cell medium (Sigma-Aldrich) to 900 mL of water and stir gently. Adjust the pH to 7.1–7.3 using 1 N NaOH or HCl as needed. Filter sterilize through a 0.22 μm filter prior to use. Store at 4 °C.
2. *hCSSC medium*: Mix 300 mL of DMEM-low glucose (Life Technologies), 200 mL of MCDB-201 medium, 10 mL of FBS, 5 mL of AlbuMaxI (Life Technologies), 5 mL of 100× penicillin/streptomycin (CellGro), 1 mL of l-ascorbic acid-2-phosphate sesquimagnesium salt, 0.5 mL of insulin, transferrin, selenium solution (Life Technologies), 0.5 mL of platelet-derived growth factor (R&D Biosystems) together. Add EGF to achieve a final concentration of 10 mg/mL, dexamethasone (Sigma-Aldrich) to 50 mM, cholera toxin to 1 mg/mL, and gentamicin (Life Technologies) to 50 mg/mL. hCSSC medium should be stored protected from light at 4 °C.
3. Fibronectin-collagen (FNC) coating (Athena Enzyme Systems). Store at 4 °C.
4. TrypLE Express (1×) (Life Technologies).
5. Collagenase-L.

2.5 Specific Reagents Required for Production of RAFT Constructs

1. RAFT reagent kit (TAP Biosystems). Store at 4 °C.
   - (a) Type I Collagen.
   - (b) 10× MEM.
   - (c) Neutralizing solution.

2. Mixing vessel (TAP Biosystems).
3. 24-well plates (Greiner Bio-One).
4. 24-well plate absorbers (TAP Biosystems).
5. Guide plate (TAP Biosystems).
6. Plate heater (TAP Biosystems).

2.6 Specific Reagents Required for Wholemount Immunohistochemistry

1. *Blocking buffer*: 5 % goat serum in DPBS, with 0.25 % Triton-X 100.

2.7 Specific Reagents Required for Immunohistochemistry on Paraffin Wax Embedded Sections

1. Specimen wrapping paper.
2. Tissue cassettes.
3. Industrial methylated spirit (IMS).
4. Xylene.
5. Paraffin.
6. Plastic molds.
7. Embedding machine.
8. Microtome.
9. Superfrost ++ slides (VWR).
10. Coplin jars.
11. Sodium citrate buffer (pH 6.0): 10 mM sodium citrate containing 0.05 % v/v Tween 20, pH using 1 M hydrochloric acid (this is stable for 3 months at RT).
12. Blocking buffer: 10 % goat serum in DPBS.

3 Methods

3.1 3View Imaging

3.1.1 Tissue Preparation and Resin Embedding

1. Use a scalpel to dissect small limbal biopsies of approximately 2–3 mm under a dissecting microscope in Karnovsky's fixative (*see* **Note 4**).
2. Place the tissues in fresh Karnovsky's fixative and fix overnight at 4 °C.
3. Wash the tissue 3 × 5 min with ice-cold 0.15 M sodium cacodylate buffer in a fume cupboard.
4. Incubate the tissue in equal volumes of 2 % osmium tetroxide solution and 3 % potassium ferricyanide solution for 1 h on ice in a fume cupboard (*see* **Note 5**).
5. Wash 3 × 5 min in distilled water in a fume cupboard.
6. Transfer the tissues into 1 % thiocarbohydrazide solution for 20 min in a fume cupboard.

7. Wash 3 × 5 min in distilled water in a fume cupboard.
8. Transfer the tissue into 2 % aqueous osmium tetroxide for 30 min in a fume cupboard.
9. Wash 3 × 5 min in distilled water in a fume cupboard.
10. Place the tissues in 1 % aqueous uranyl acetate overnight at 4 °C.
11. The next day, wash the tissues for 3 × 5 min in distilled water.
12. Transfer the tissues into the freshly prepared Walton's lead aspartate solution and place in a 60 °C oven for 30 min.
13. Wash the tissues for 3 × 5 min in distilled water.
14. Dehydrate the tissues through increasing concentration of ethanol:
    (a) 5 min in 20 % (×2).
    (b) 5 min in 50 % (×2).
    (c) 5 min in 70 % (×3).
    (d) 5 min in 90 % (×2).
    (e) 10 min in 100 % (×3).
15. After dehydration, transfer the tissues into acetone (15 min × 2) prior to infiltration.
16. Infiltrate the tissues with a mixture of Durcupan ACM resin:acetone: 2 h in 25 % resin, 2 h in 50 % resin, 2 h in 75 % resin.
17. Place the tissues into 100 % resin overnight and transfer the tissues the next day into new molds containing 100 % of freshly prepared resin.
18. Place the molds into a 60 °C oven for 48 h.

3.1.2 Specimen Trimming and Mounting

1. Remove the samples from the embedding molds.
2. Trim away excess resin and attach the sample onto a cryopin using the conductive silver epoxy kit.
3. Leave the glue to set overnight.
4. The next day, trim away excess conductive glue and use a glass knife to obtain a flat surface on the resin block.
5. Use a razor blade to trim around the area of interest (this should be approximately 1 × 1 mm).
6. Use a clean diamond knife to cut semi-thin sections of approximately 0.7 μm, collect them onto microscope slides, and dry them on a hotplate.
7. Stain the sections by adding 1 drop of 1 % toluidine blue and return the slides to the hot plate for a further 30 s.
8. Once the edge of the staining drop is dry, gently rinse the slides with distilled water.
9. Mount the slides with DePeX, coverslip and observe using a bright-field light microscope.

10. Once the area of interest has been identified, use a razor blade to trim the resin block to obtain a square surface measuring approximately 0.5 × 0.5 mm (*see* Fig. 1a) (*see* **Note 6**).

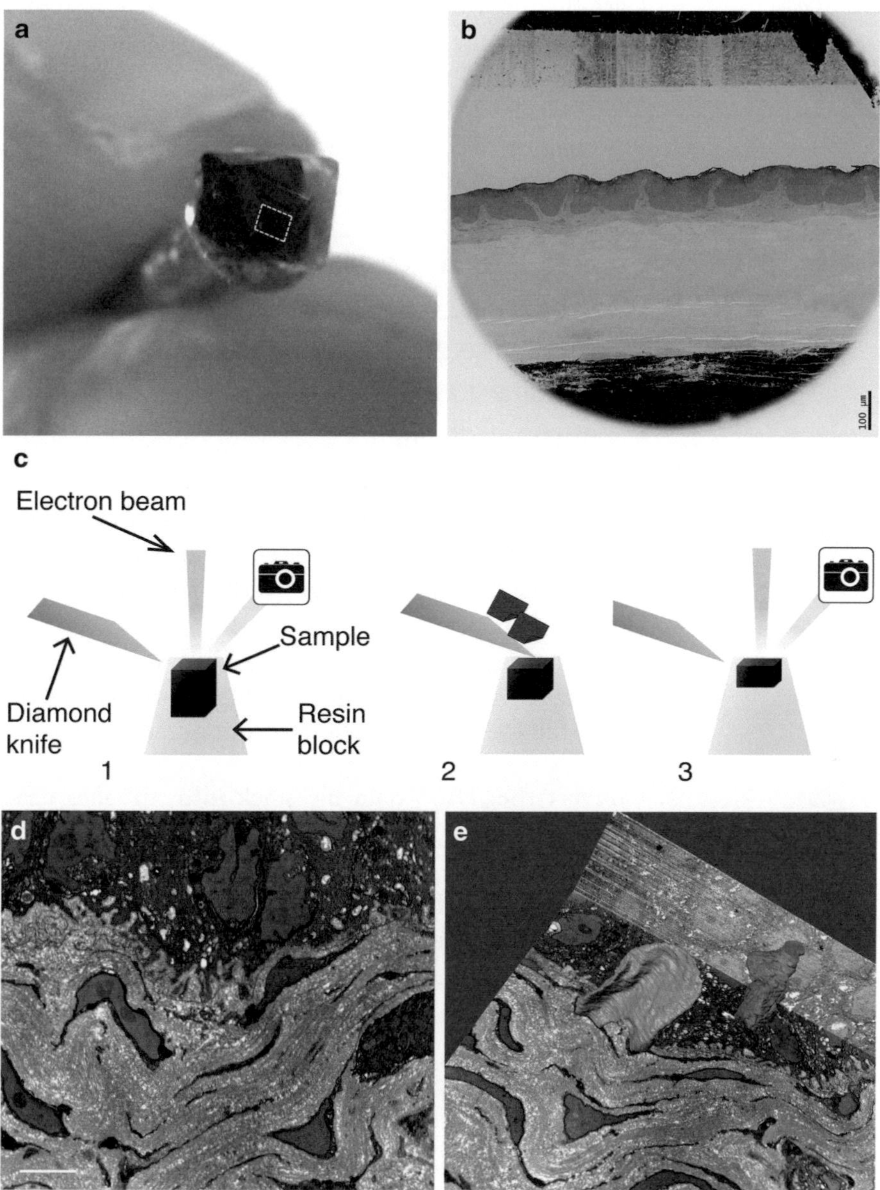

Fig. 1 Serial block face scanning electron microscopy. (**a**) Resin block attached to the cryopin, ready to be loaded into the SEM. The *dashed white line* outlines the surface of the resin block measuring approximately 0.5 × 0.5 mm. (**b**) Low magnification scanning electron micrograph of the resin block in **a**. (**c**) Schematic showing the principle of the SBFSEM (3View). *1*: Imaging of the surface of the resin block. *2*: A diamond knife inside microscope chamber cuts an ultrathin section away from the specimen. *3*: The freshly exposed edge is imaged. (**d**) Manual segmentation of the area of interest. The nuclei of a stromal cell and epithelial cell are outlined. (**e**) 3D reconstruction of the areas manually segmented in **d**. Scale bars: **b**: 100 μm and **d**: 5 μm

11. Sputter coat the surface of the resin block with a thin layer of gold palladium in order to get a conductive surface. This prevents the accumulation of negative charges at the surface of the specimen softening the resin block.

3.1.3 Sample Loading and Acquisition

1. Carefully load the specimen into the 3View system associated to the scanning electron microscope (*see* **Note 7**). Figure 1b shows the surface of the resin block once loaded inside the microscope chamber.
2. For imaging of the basal limbal epithelial cell layer of the cornea, the following setting were used:
   - Focus: 5.
   - Magnification: ×6,000.
   - Accelerating voltage: 4 kV.
   - Dwell time: 2 μs.
   - Pressure: 20 Pa.
   - Aperture: 60 μm.
   - Resolution: 4k × 4k.
3. Serially cut ultrathin sections (100 nm) from the resin block to expose a fresh surface to the electron beam. This allows the generation of a new image of the surface of the specimen (*see* Fig. 1c).
4. Repeat this automated process 999 times to generate a large data stack of 999 serial images. Serial images are collected as .Dm3 file format in Digital Micrograph.

3.1.4 Data Analysis, Manual Segmentation, and 3D Reconstruction

1. Convert the .Dm3 data file stack into .tiff files using Digital Micrograph.
2. Copy the data into the workstation and load the .tiff file data stack as a complete volume into AMIRA for conversion into voxels (volumetric picture elements).
3. Use the noise reduction median filter and manually segment the area of interest on every single slice using the interactive pen (*see* Fig. 1d).
4. Generate the 3D structure of the segmented surface (*see* Fig. 1e).

3.2 Isolation and Culture of Human Limbal Epithelial Cells (hLEC) Using Dispase Dissociation

3.2.1 Dissection of Corneal Tissue

1. Sterilize instruments in 70 % ethanol for 10 min prior to use.
2. Cut the human donor corneal rim into quarters.
3. Incubate the pieces in 5 mL of 1.2 IU/mL dispase II for 2 h at 37 °C or overnight at 4 °C.
4. Transfer the pieces to a fresh petri dish containing 1 mL of fresh CECM without EGF.

5. Release hLEC from the limbus by scraping the anterior limbal surface with the tips of fine forceps.
6. Triturate to get a single cell suspension and place into a T75 flask containing 1.8×10^6 growth-arrested 3T3-J2 feeder cells.
7. Incubate at 37 °C in a humidified incubator with 5 % CO_2 overnight.

3.2.2 hLEC Culture and Expansion

1. After 1 day, replace the medium with CECM and change the medium three times per week (*see* **Note 8**).
2. Colonies of hLEC will start to appear between the feeder cells (*see* Fig. 2), easily seen by eye as circles of dense cells. Once confluent (10–14 days), hLEC may be passaged.
3. Wash cells twice with DPBS.
4. Incubate for 3 min in 0.05 % in trypsin-EDTA to selectively detach 3T3-J2 feeder cells (this can be observed using a light microscope).
5. Aspirate and discard the feeder cells.
6. Add 2 mL of 0.5 % trypsin-EDTA, return to the incubator for no more than 5 min until hLEC detach when the side of the flask is hit.
7. Add 10 mL of CECM without EGF and centrifuge at $1{,}000 \times g$ for 5 min.
8. Resuspend the hLEC pellet in fresh CECM without EGF and continue culture either on RAFT constructs (*see* Subheading 3.5) or on tissue culture plastic.

3.3 Isolation and Culture of Human Limbal Fibroblasts (hLF)

Air-dry method for hLF isolation and culture.

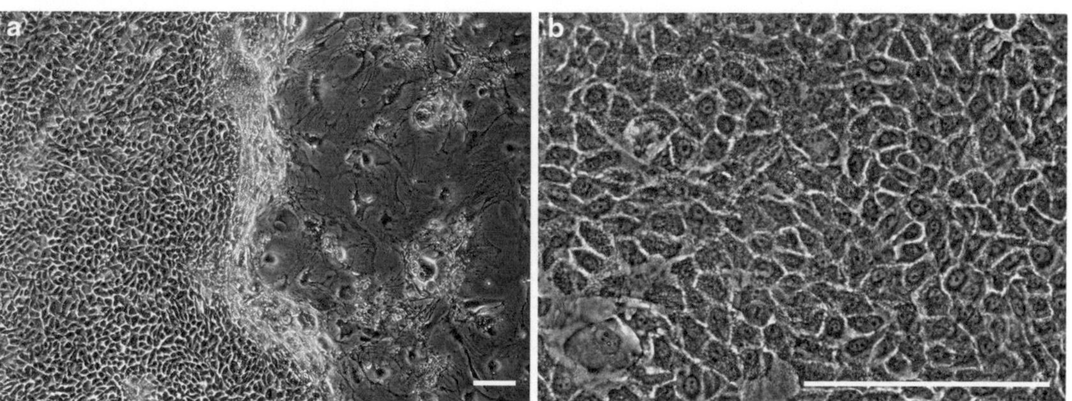

Fig. 2 hLEC cultures. (**a**) hLEC form colonies (*left*) between the feeder cells (*right*). (**b**) hLEC display typical cobblestone morphology, with scant cytoplasm. Scale bars: 200 μm

3.3.1 Dissection of Corneal Tissue

1. Sterilize instruments in 70 % ethanol for 10 min prior to use.
2. Following hLEC isolation using the dispase-dissociation method, dissect the corneal rim into 2–3 mm sections using a scalpel.

3.3.2 Attachment of Tissue to T25 Culture Flask

1. Place 5–6 sections into a T25 flask.
2. Allow to air-dry for 15 min in a biological safety cabinet by placing the flask on its side without the lid.
3. Gently add 6 mL of hLF media.
4. Allow to grow undisturbed in a 37 °C humidified incubator with 5 % CO_2 for 3 weeks.

3.3.3 hLF Culture and Expansion

1. Check for growth of hLF at the explant edge and gently change media without dislodging the tissue from the flask.
2. After a sufficient number of fibroblasts have grown out from the edge of the explant, passage cells into a new T25 flask by rinsing with 2 mL of DPBS and adding 1 mL of 0.05 % trypsin-EDTA for 5 min at 37 °C.
3. Mechanically detach cells by tapping the flask and add 4 mL of hLF media to inactivate trypsin.
4. Centrifuge at 1,000 × *g* for 5 min.
5. Resuspend the cells in hLF cell pellet in 1 mL of hLF media and reseed into a new T25 flask with 5 mL of hLF media.
6. Change media three times per week.
7. Amplify cell stock to use hLF for experimental use at passages 1–6. Figure 3 shows typical hLF morphology.

3.3.4 Collagenase-Dissociation Method for hLF Isolation and Culture

1. Sterilize instruments in 70 % ethanol for 10 min prior to use.
2. Following hLEC isolation using the method described in Subheading 3.2, dissect the corneal rim into 2–3 mm sections using a scalpel.

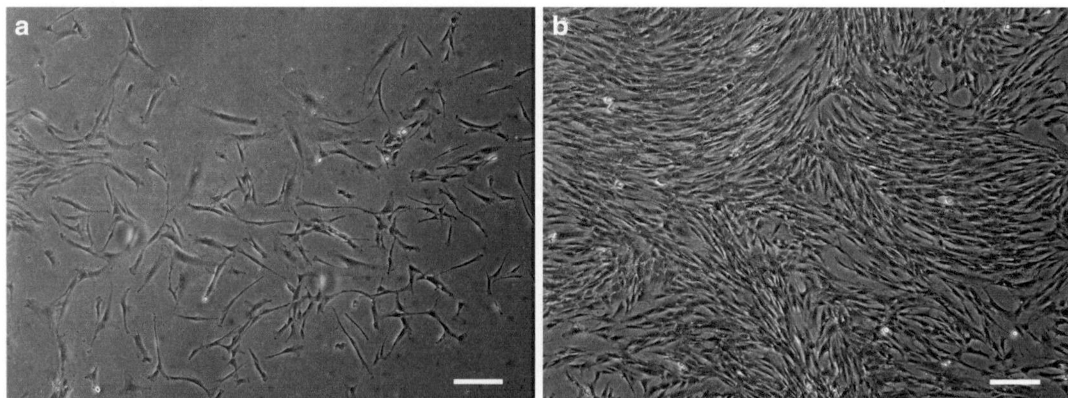

Fig. 3 hLF cultures. (**a**) hLF in culture at sub-confluency. (**b**) A confluent hLF layer, ready for trypsinization. hLF appear dendritic. Scale bars: 200 μm

3.3.5 Enzymatic Dissociation of Corneal Tissue

1. Add dissected sections to 5 mL of collagenase solution in a 35 mm dish.
2. Incubate overnight in a humidified incubator at 37 °C with 5 % CO_2.
3. The next day, transfer the collagenase solution to a 15 mL centrifuge tube and centrifuge at 1,000 × *g* for 5 min.
4. Aspirate and discard the supernatant.
5. Resuspend the cell pellet in 1 mL of hLF media.

3.3.6 hLF Culture and Expansion

1. Add a further 5 mL of hLF media and transfer to a T25 flask.
2. Change hLF media three times per week until 70–80 % confluent.
3. At this confluency, the hLF may be trypsinized.
4. Wash hLF with DPBS twice.
5. Add 1 mL of 0.05 % trypsin-EDTA and incubate for 5 min at 37 °C in a humidified incubator with 5 % CO_2.
6. Mechanically detach cells by tapping the flask and add 4 mL of hLF media to inactivate trypsin.
7. Centrifuge at 1,000 × *g* for 5 min and resuspend the pellet in 1 mL of hLF medium before reseeding into a new T25 flask with 5 mL of hLF media.
8. Change media three times per week.
9. Amplify cell stock to use hLF for experimental use at passages 1–6.

3.4 Isolation and Culture of Human Corneal Stromal Stem Cells (hCSSC)

This method has been previously described by Du et al. [6].

3.4.1 Dissection of Corneal Tissue

1. Sterilize instruments in 70 % ethanol for 10 min prior to use.
2. Place the whole cornea or corneoscleral rim in a single well of a 12-well plate containing DMEM/F12 (supplemented with Gentamicin and pen/strep only).
3. Wash the tissue three times for 10 min each using this media.
4. If there is any trace of the Tenon's capsule (fibrous tissue) (*see* **Note 9**), remove this using a pair of forceps and scissors.
5. Dissect the superficial limbus 360° (*see* Fig. 4a, b): cut part of the sclera and cornea with the limbus sandwiched between (*see* **Note 10**).
6. Cut cornea into two halves to ease dissection, ensuring that the tissue is moistened with hCSSC media whilst dissecting.
7. Cut approximately 100 μm deep to obtain the superficial stroma only.

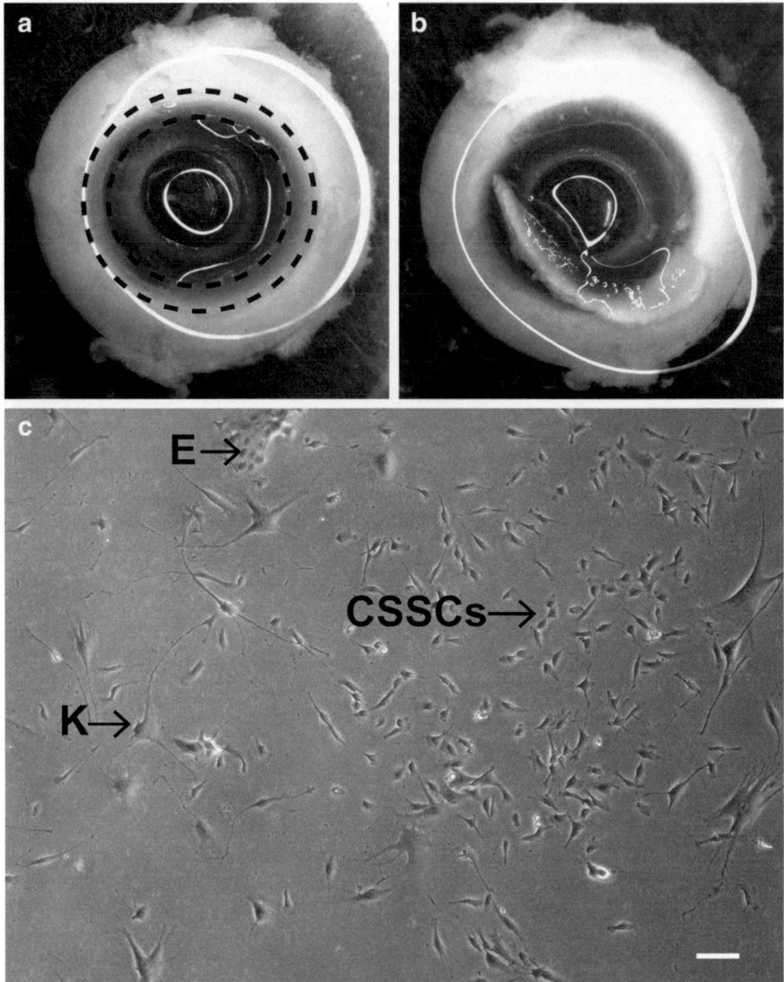

Fig. 4 hCSSC isolation and culture using enzymatic dissociation. (**a**) *Black dotted lines* illustrate the limbal region before dissection. (**b**) Upper half of superficial limbal rim is dissected. (**c**) hCSSC in culture (passage 1). CSSCs appear small and square in sparsely arranged colonies. Some epithelial colonies (E) and keratocyte-like cells (K) remain visible in early passages. Scale bar: 100 μm

3.4.2 Enzymatic Digestion of Limbal Stroma

1. Cut the limbal stroma into smaller fragments using scissors and place into collagenase solution in a well of a 6-well plate.
2. Incubate overnight at 37 °C in a 5 % CO_2 humidified incubator. The next day, pipette the mixture up and down to break up any remnants of fibrous tissue that might be remaining in media.
3. Centrifuge mixture in 10 mL of DMEM/F12 (containing Gentamicin and pen/strep) at 1,500 rpm* for 5 min at 4 °C.
4. Remove supernatant leaving approximately 0.5 mL at bottom of tube (*see* **Note 11**).
5. Add 1 mL of hCSSC media to the tube containing cells and mix gently before adding a further 4 mL of hCSSC media.

3.4.3 hCSSC Culture

1. Coat T25 flask with FNC. Add approximately 1 mL of FNC to cover the surface of the flask, leave for 30 s, then aspirate the excess.
2. Add the hCSSC suspension to the FNC-coated flask.
3. Incubate at 37 °C in a 5 % CO_2 humidified incubator (*see* **Note 12**) overnight. hCSSC will adhere to the plastic surface of the flask by the next day.
4. Wash cells with DPBS and replace with fresh hCSSC medium (*see* **Note 13**).
5. Small hCSSC colonies should appear by day 2 or 3 (*see* Fig. 4c). Cells should be passaged as soon as small colonies are visible using 1 mL of TrypLE Express followed by 2 mL of hCSSC media to neutralize.
6. Culture hCSSC at 37 °C in a 5 % CO_2 humidified incubator (*see* **Note 14**).

3.5 Production of RAFT Constructs

3.5.1 hLF Cell Solution Preparation

1. Trypsinize hLF cells as in Subheading 3.3.
2. Resuspend hLF in 1 mL of hLF medium and perform a cell count.
3. Resuspend hLF at a cell density of 100,000 cells per mL of collagen in an appropriate volume of culture medium, c, depending on the final number of constructs (*see* **Note 15**) (*see* Table 1).
4. Keep cells on ice until required.

3.5.2 Collagen Solution Preparation

1. Determine the number of collagen constructs required and refer to Table 1 for reagent volumes (*see* **Note 16**).
2. Add volume × of 10 × MEM to the mixing vessel.
3. Slowly add volume *y* of the collagen solution to the mixing vessel.
4. Swirl the solution gently to mix until a homogenous color is achieved.
5. Add volume *z* of neutralizing solution evenly distributed across the surface of the mixture, swirl the solution again gently to mix.
6. Add volume *c* of hLF cell suspension (from Subheading 3.5.1) (*see* **Note 17**), evenly distributed across the surface of the mixture and swirl again gently to mix.
7. Leave the collagen solution on ice for 30 min to allow any bubbles that might have formed to disperse (*see* **Note 18**).

3.5.3 Hydrogel Formation

1. Allow the plate heater to reach 37 °C.
2. Using a 5 mL pipette, transfer 2.4 mL of collagen/hLF mix to each well of a 24-well plate (*see* **Note 19**).

Table 1
Reagent volume guide for RAFT production in 24-well Greiner plates (all volumes are in mL)

Number of constructs	10× MEM *x*	Collagen *y*	Neutralizing solution *z*	Cells or medium *c*	Total volume	Excess
1	0.4	3.5	0.261	0.180	4.4	2
2	0.7	5.4	0.404	0.279	6.8	2
3	0.9	7.4	0.546	0.377	9.2	2
4	1.2	9.3	0.689	0.476	11.6	2
5	1.4	11.2	0.832	0.574	14.0	2
6	1.6	13.1	0.974	0.672	16.4	2
7	1.9	15.0	1.117	0.771	18.8	2
8	2.1	17.0	1.259	0.869	21.2	2
9	2.4	18.9	1.402	0.968	23.6	2
10	2.6	20.8	1.544	1.066	26.0	2
11	2.8	22.7	1.687	1.164	28.4	2
12	3.1	24.6	1.830	1.263	30.8	2
13	3.3	26.6	1.972	1.361	33.2	2
14	3.6	28.5	2.115	1.460	35.6	2
15	3.8	30.4	2.257	1.558	38.0	2
16	4.0	32.3	2.400	1.656	40.4	2
17	4.3	34.2	2.542	1.755	42.8	2
18	4.5	36.2	2.685	1.853	45.2	2
19	4.8	38.1	2.827	1.952	47.6	2
20	5.0	40.0	2.970	2.050	50.0	2
21	5.2	41.9	3.113	2.148	52.4	2
22	5.5	43.8	3.255	2.247	54.8	2
23	5.7	45.8	3.398	2.345	57.2	2
24	6.0	47.7	3.540	2.444	59.6	2

3. Place the 24-well plate on the plate heater and close the lid (*see* **Note 20**).
4. Allow 30 min for fibrillogenesis to occur and hydrogels to form.
5. Whilst gelling is occurring, UV sterilize the desired number of absorbers.

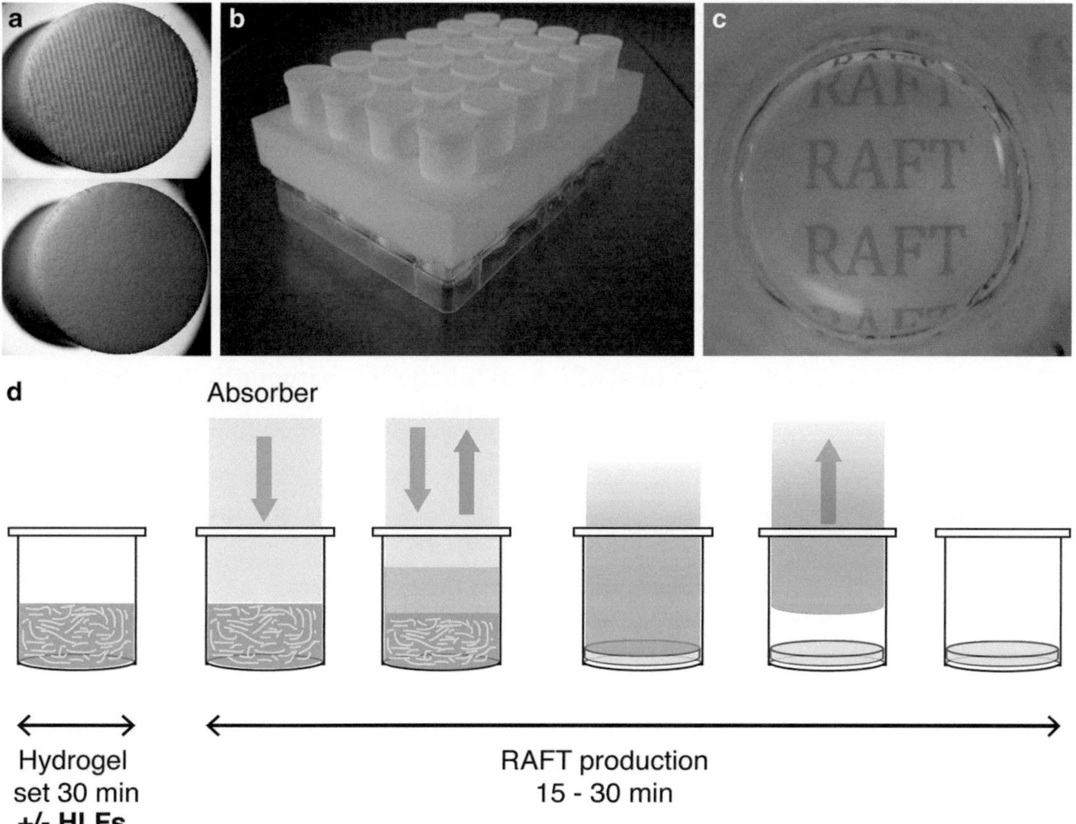

Fig. 5 RAFT production process. (**a**) Base of absorbers showing ridged and plain topographies. (**b**) Guide plate populated with absorbers on top of 24-well plate. (**c**) Acellular RAFT construct in the base of a 24-well plate. (**d**) Schematic summarizing the RAFT production process

3.5.4 Production of the RAFT 3D Constructs

1. Populate the guide plate with the correct number and configuration of sterile absorbers (ridged or plain) to match the hydrogels (*see* Fig. 5a) (*see* **Note 21**).
2. Open the lid of the plate heater and remove the lid of the cell culture plate.
3. Place the guide plate on top of the cell culture plate aligning the absorbers with the hydrogels and leave for 30 min (*see* Fig. 5b) (*see* **Note 22**).
4. Remove the guide plate and accompanying absorbers by lifting it straight up.
5. Discard the absorbers.
6. Add 500 μL of hLF medium to the surface of the RAFT constructs and replace the lid of the well plate.
7. Place the culture plate in a humidified incubator at 37 °C with CO_2 in air until addition of hLEC to the surface. An

example of an acellular RAFT construct is shown in Fig. 5c. (A schematic to describe the hydrogel formation and RAFT production process is shown in Fig. 5d).

3.5.5 Seeding of hLE onto the Surface of RAFT Constructs

1. Selectively trypsinize and discard 3T3-J2 feeder cells as in Subheading 3.2.
2. Trypsinize hLEC as in Subheading 3.2.
3. Count hLEC using a hemocytometer and dilute cell suspension to the correct density to add 560,000 cells to each RAFT construct in a 24-well plate (*see* **Note 15**).
4. Add a total of 1.5 mL of CECM (without EGF) (*see* **Note 23**) to each well and return the plate to the incubator at 37 °C and 5 % CO_2 in air.

3.5.6 Submerged Culture to Expand hLE on the Surface of RAFT

1. After 24 h, aspirate the CECM and any unattached cells and add 1.5 mL of fresh CECM.
2. Change medium three times per week.
3. Maintain RAFT constructs for 2 weeks in submerged culture.

3.5.7 Airlifting the Culture to Achieve a Stratified Epithelium on the Surface of RAFT

1. Remove the RAFT constructs from wells and place onto an insert in a 6-well culture plate using forceps and a flat spatula.
2. Add 800 μL of CECM to the well below the insert adding a drop of medium to the surface of the constructs and fill any empty wells with DPBS (*see* **Note 24**).
3. Change the medium three times per week, each time adding a drop of CECM to the surface of the constructs.
4. Maintain the RAFT constructs for 1 week in airlifting culture.

3.6 Wholemount Immunohisto-chemistry

3.6.1 Fixing

1. Fix RAFT constructs for 30 min using 4 % PFA (*see* **Note 25**).
2. Following fixation, wash the construct three times with DPBS (*see* **Note 26**) and, if required, store at 4 °C in DPBS, up to a maximum of 2 weeks.

3.6.2 Blocking

1. Prevent nonspecific binding by blocking with blocking buffer (*see* **Note 27**) for 1 h.
2. Wash once with DPBS (*see* **Note 26**).

3.6.3 Capture Antibody

1. Apply capture antibody in capture antibody solution at the required dilution (*see* Table 2) (*see* **Note 28**).
2. For the isotype negative control (*see* **Note 29**), apply the isotype control in capture antibody solution at the same concentration used for the capture antibody (*see* Table 2).

Table 2
Antibody and isotype controls supplier information and dilutions for immunohistochemistry

Antigen	Capture Ab manufacturer	Dilution for wholemount	Dilution for wax sections	Isotype control manufacturer
p63α	Cell Signaling Technology	1:50	1:100	New England Biolabs
Pax-6	Covance	1:100	1:100	
Ki67	Millipore	1:100	1:100	
CK3	Millipore	1:200	1:500	Cambridge Bioscience

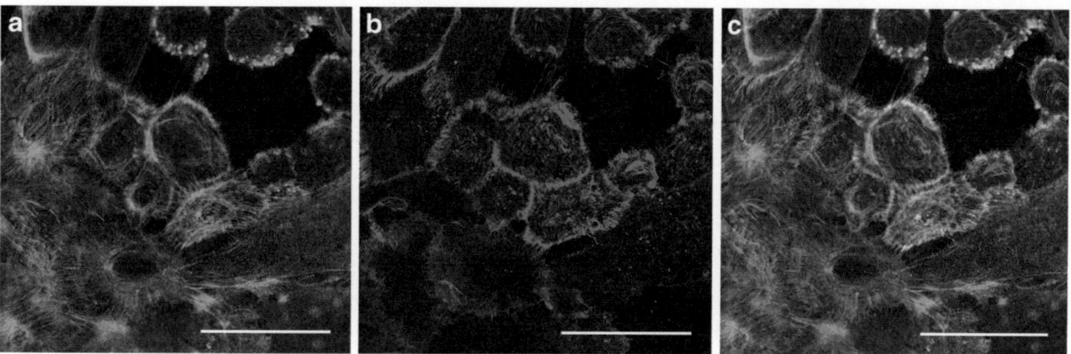

Fig. 6 Example of a wholemount-stained RAFT construct. (**a**) hLEC on RAFT stained for β1-integrin, putative stem cell marker. (**b**) hLEC on RAFT stained for phalloidin-FITC. (**c**) Merge. Scale bars: 50 μm

3. Incubate samples overnight at 4 °C.
4. Following the overnight incubation, wash samples with DPBS for 3 × 5 min (*see* **Note 26**).

3.6.4 Detection Antibody

1. Apply Alexa Fluor secondary antibody (1:500) (*see* **Note 30**) and phalloidin (1:1,000) (*see* **Note 31**) in DPBS.
2. Incubate samples for 1 h at RT in the dark.
3. Wash samples with DPBS for 3 × 5 min (*see* **Note 26**).

3.6.5 Mounting

1. Transfer the samples to glass slides and remove any excess liquid carefully by blotting with a tissue.
2. Apply one or two drops of Vectashield mounting medium to each sample and fix in place with a coverslip (*see* **Note 32**).
3. Seal using nail polish, carefully brushing the polish around the perimeter of the coverslip.
4. Leave to dry for 30 min before visualizing using a confocal fluorescence microscope. An example of RAFT wholemount staining is given in Fig. 6.

3.7 Immunohistochemistry on Paraffin Wax Embedded Sections

3.7.1 Fixing

1. Fix RAFT constructs for 30 min using 4 % PFA (*see* **Note 25**).
2. Following fixation, wash the construct three times with DPBS and, if required, store at 4 °C in DPBS, up to a maximum of 2 weeks.

3.7.2 Processing

1. Carefully wrap RAFT constructs in specimen wrapping paper using forceps and place into a tissue cassette.
2. Leave in formalin until transfer to the tissue processor, which processes the tissue using gentle agitation, as follows:
   - 70 % IMS for 5 min.
   - 90 % IMS for 30 min.
   - 100 % IMS for 1 h (×2).
   - 100 % IMS for 1.5 h (×2).
   - 100 % IMS for 2 h, xylene for 2 h (×2).
   - paraffin for 2 h (×2).
3. The tissue cassette should be left in paraffin until required.

3.7.3 Embedding

1. Open the tissue cassette and remove the RAFT construct.
2. Embed in paraffin wax and leave to cool for at least 1 h before sectioning (*see* **Note 33**).

3.7.4 Sectioning

1. Trim excess paraffin from the sides of the tissue cassette to ensure that the block fits in the microtome (*see* **Note 34**).
2. Cut 5 μm sections and float onto Superfrost ++ slides.
3. Allow excess water to drain away and transfer the slide to the hot plate for 15 min to ensure paraffin wax has melted and section is firmly stuck to the glass slide.
4. Store at RT until staining.

3.7.5 Dewaxing

1. Place labeled slides into a Coplin jar.
2. Dewax in xylene for 3 × 5 min.

3.7.6 Rehydrating

1. In a Coplin jar, perform the following steps:
   - 100 % IMS for 5 min.
   - 90 % IMS for 5 min.
   - 70 % IMS for 5 min.
   - Water for 5 min.
2. Wash slides in DPBS for 5 min at 95 °C in a water bath.

3.7.7 Antigen Retrieval

1. Perform antigen retrieval with sodium citrate buffer at 95 °C in a water bath for 20 min.
2. Wash for 3 × 5 min with DPBS.

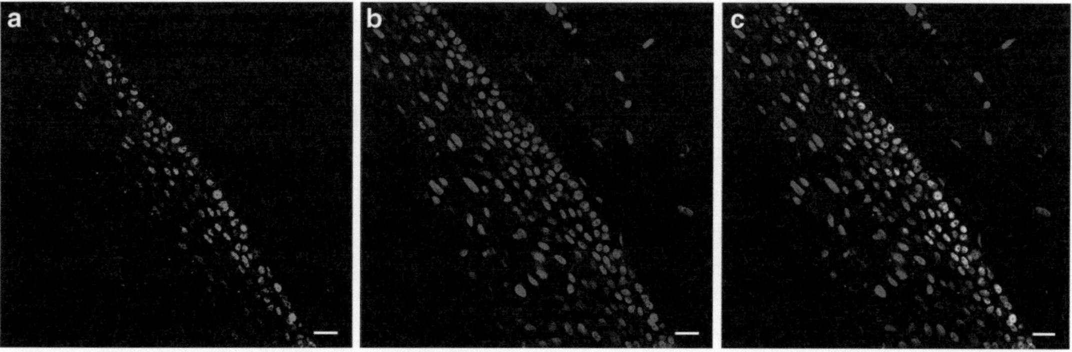

Fig. 7 Example of a stained RAFT section. (**a**) hLEC on RAFT stained for p63α, putative stem cell marker. (**b**) PI counter-staining of hLEC (*left*) nuclei on RAFT and entrapped hLF (*right*) nuclei. (**c**) Merge showing that only basal hLEC express p63α. Scale bars: 20 μm

3.7.8 Blocking

1. Prevent nonspecific binding by blocking in blocking buffer for 1 h.

3.7.9 Capture Antibody

1. Apply the capture antibody in capture antibody solution at the required dilution (*see* Table 2) (*see* **Note 30**).
2. To prepare an isotype negative control, add the isotype control in capture antibody solution at the same concentration used for the capture antibody (*see* Table 2) (*see* **Note 29**).
3. Incubate samples overnight at 4 °C.
4. Following the overnight incubation, wash samples with DPBS for 3 × 5 min.

3.7.10 Detection Antibody

1. Apply Alexa Fluor secondary antibody (1:500) (*see* **Note 30**) and phalloidin (1:1,000) (*see* **Note 31**) in DPBS.
2. Incubate samples for 1 h at RT in the dark.
3. Wash samples with DPBS for 3 × 5 min.

3.7.11 Mounting

1. Apply one or two drops of Vectashield to each sample and fix in place with a coverslip (*see* **Note 32**).
2. Seal using nail polish, carefully brushing the polish around the perimeter of the coverslip.
3. Leave to dry for 30 min before visualizing using a fluorescence microscope.

   An example of RAFT section staining is given in Fig. 7.

4 Notes

1. Cadaveric donor corneal rims with appropriate research consent are used to isolate hLE, hLF, and hCSSC. Tissue can be fresh postmortem or Optisol-stored tissue. Dispose of human tissue as defined in local policies.

2. Particular care must be taken with osmium tetroxide. The vapor pressure and the acute toxicity of osmium are high. Use under a fume cupboard and wear suitable protective clothing and gloves. Used osmium and washes should be stored for safe disposal.
3. Uranyl acetate is very toxic and radioactive. Use only in a fume cupboard and wear suitable protective clothing and gloves. Used uranyl acetate including washes should be stored for safe disposal.
4. Do not allow the tissue dry out.
5. Use this incubation time to prepare a fresh solution of thiocarbohydrazide.
6. It is important to trim at this point, in order to ensure that the specimen surface still fits in the cutting window of the 3View microtome during serial sectioning.
7. Refer to the instruction of the manufacturer for a detailed description of the loading process.
8. It may be necessary to top the feeder cells if they begin to detach during culture.
9. Tenon's capsule contains fibroblasts, which may contaminate the hCSSC culture if not removed.
10. Remove excess sclera and central cornea using a scalpel.
11. The pellet is very small so care must be taken not to dislodge/ remove it.
12. hCSSC can be left in hCSSC media for up to 2 days if necessary.
13. It is difficult to distinguish the hLEC from hCSSC at this stage of culture.
14. By the second day of culture, hLEC usually begin to grow in small colonies with the characteristic cobblestone morphology. hCSSC grow more sparsely but in close proximity to the epithelial colonies. hCSSC typically look small and square, with four points. Some keratocyte-like cells that are more dendritic in morphology also appear in early passages (Fig. 4c). It is necessary to passage hCSSC as soon as they appear, rather than waiting for confluence since they will become fibroblastic.
15. Cell densities have been optimized for this specific application; the final cell density in collagen construct must be determined by the end user depending on the application.
16. The reagent volumes include an excess to allow for solution lost when pipetting and transferring between different vessels.
17. If hLF is not required in the RAFT construct, add hLF culture media alone.

18. Keep all reagents ice-cold and perform all processes on ice as far as possible after neutralization. Gelling will commence if the solution is left to warm.
19. When dispensing the viscous collagen solution with serological pipettes, reverse pipette to increase accuracy.
20. It is possible to carry out the fibrillogenesis process in a 37 °C cell culture incubator.
21. If bioengineered limbal crypts in the RAFT construct are required, use ridged absorbers. If a flat surface is required, use plain absorbers. The process is identical for both.
22. Take care not to apply too much pressure to the hydrogels. Allow the absorbers to gently fall onto the surface of the gels. It is possible to carry out this process directly in the cell culture hood without the use of the plate heater.
23. Omission of EGF when initially cell seeding hLE cells aids attachment to the surface of RAFT constructs.
24. Adding DPBS to the empty wells helps to maintain a humid atmosphere and addition of a drop of medium to the surface of the constructs prevents drying of the cells on the construct.
25. Fixation reagents and incubation times may vary dependent upon the antibody used. Therefore, consult the relevant data sheet if antibodies other than those given in Table 2 are used before performing this step.
26. For all wash steps, be careful to expel liquid around the construct, and not directly on top. This prevents epithelial detachment.
27. For cell surface antigens Triton-X-100 detergent should be omitted from the blocking solution as the detergent may destroy cell membrane structure.
28. For capture antibodies other than those given in Table 2, the user should optimize the dilution.
29. The isotype control should be sourced from the same species as the host antibody, and should match the same antibody subtype. The purpose of the isotype control is to enable the end user to assess the level of background staining, thereby helping to differentiate between nonspecific and true antigen-antibody binding.
30. Secondary antibody selection is important to allow for antibody-antibody binding. The secondary antibody should be raised in the same species used for the blocking serum, but directed against the immunoglobulins of the primary antibody species; i.e., the appropriate secondary antibody to use in a protocol using goat serum and a rabbit primary antibody would be termed "goat anti-rabbit."

31. Phalloidin binds to the cells' actin filaments, allowing for visualization of the cells' cytoskeleton. Therefore, this reagent is a useful tool to aid localization of the target antigen within the cell.
32. Mounting medium prevents rapid photo-bleaching of fluorophores and provides an optimal refractive index enabling sharper images to be obtained. Excess liquid may alter the refractive index resulting in poorer quality images. Removal of excess liquid is advised.
33. Ensure a margin of at least 2 mm of paraffin on all sides of the gel to give it a good cutting support. If cracks are observed in the wax, re-melt the paraffin and repeat the embedding process, as the block will crack further during sectioning making obtaining sections impossible.
34. Cool each block facedown on an ice block for at least 10 min prior to sectioning to reduce friction between the blade and block. This also reduces the chance of cracking the block.

References

1. Thoft R, Friend J (1983) The X, Y, Z hypothesis of corneal epithelial maintenance. IOVS 10:1442–1443
2. Pellegrini G, Golisano O, Paterna P et al (1999) Location and clonal analysis of stem cells and their differentiated progeny in the human ocular surface. J Cell Biol 145:769–782
3. Schlotzer-Schrehardt U, Kruse FE (2005) Identification and characterization of limbal stem cells. Exp Eye Res 81:247–264
4. Shortt AJ, Secker GA, Munro PM et al (2007) Characterization of the limbal epithelial stem cell niche: novel imaging techniques permit in vivo observation and targeted biopsy of limbal epithelial stem cells. Stem Cells 25: 1402–1409
5. Denk W, Horstmann H (2004) Serial block-face scanning electron microscopy to reconstruct three-dimensional tissue nanostructure. PLoS Biol 2:329
6. Du Y, Funderburgh ML, Mann MM et al (2005) Multipotent stem cells in human corneal stroma. Stem Cells 23:1266–1275
7. Brown RA, Wiseman M, Chuo CB et al (2005) Ultrarapid engineering of biomimetic materials and tissues: Fabrication of nano- and microstructures by plastic compression. Adv Funct Mater 15:1762–1770
8. Levis HJ, Brown RA, Daniels JT (2010) Plastic compressed collagen as a biomimetic substrate for human limbal epithelial cell culture. Biomaterials 31:7726–7737
9. Levis HJ, Menzel-Severing J, Drake RA et al (2013) Plastic compressed collagen constructs for ocular cell culture and transplantation: a new and improved technique of confined fluid loss. Curr Eye Res 38:41–52
10. Levis HJ, Massie I, Dziasko MA et al (2013) Rapid tissue engineering of biomimetic human corneal limbal crypts with 3D niche architecture. Biomaterials 34:8860–8868
11. Mort RL, Douvaras P, Morley SD et al (2012) Stem cells and corneal epithelial maintenance: insights from the mouse and other animal models. Results Probl Cell Differ 55: 357–394

Chapter 16

Isolation and Characterization of Stem Cells in the Adult Mammalian Ovary

Seema Parte, Hiren Patel, Kalpana Sriraman, and Deepa Bhartiya

Abstract

Female mammals are born with a fixed pool of germ cells, which does not replenish during adult life. However, this has been recently challenged and adult ovaries produce oocytes throughout adult life just like sperm in the testes. Evidence is accumulating on the presence of ovarian stem cells, but the need for robust protocols to isolate, identify, further characterize, and subject them to various functionality tests is essential. Knowledge about the function and potential of ovarian stem cells is well demonstrated by various groups, but their true identity remains elusive because of the variability in the approaches used to identify them by different groups. In order to address this we have made attempts to compile our protocols to isolate, identify, characterize, and culture the stem cells using different animal models including human. Two distinct populations of stem cells exist in adult mammalian ovary, including very small embryonic-like stem cells (VSELs) and the progenitors termed ovarian germ stem cells (OGSCs). VSELs are relatively quiescent and undergo asymmetric cell division to give rise to OGSCs, which divide rapidly, occasionally form germ cell nests and undergo meiosis and differentiation into oocytes, which are surrounded by granulosa cells to assemble as primordial follicles.

Key words Ovary, Stem cells, VSELs, Oct-4

1 Introduction

The widely accepted concept in female ovarian biology is that women and other mammals produce a nonrenewable pool of germ cells, i.e., primordial follicles during fetal development, which does not replenish during adult life. This was challenged in 2004 by Tilly's group [1] and since then several investigators have also reported the existence of ovarian stem cells (*recently reviewed in* 2–4). Although unequivocal evidence exists about the presence of stem cells in the ovary surface epithelium (OSE), there is a need to establish consensus on methods to isolate and identify them at a single cell level, characterize further, and subject them to various functionality tests. This chapter describes our simplified approach

Ivan N. Rich (ed.), *Stem Cell Protocols*, Methods in Molecular Biology, vol. 1235,
DOI 10.1007/978-1-4939-1785-3_16,

to isolate, identify, characterize, and culture ovarian stem cells from the adult mammalian ovary. Our protocols to isolate ovarian stem cells differ from the recently compiled protocols for isolation of female germline/oogonial stem cells [5, 6], since we have reported the presence of two distinct populations of ovarian stem cells including the pluripotent stem cells and progenitors as discussed in relevant sections below.

The ovary is covered by a single layer of epithelial cells referred to as the ovarian surface epithelium (OSE). These epithelial cells share a common source of origin from the peritoneal mesothelial cells by differentiation and conversion of mesenchymal cells into epithelial cell type by extracellular matrix remodeling [7]. Ovarian surface epithelium forms a single layer of flat/cuboidal epithelial cells surrounding the ovarian surface and is attached tenuously to the underlying stromal layer and can be easily detached by mechanical means. The simple squamous to-cuboidal single-layered epithelial cell structure of the normal human OSE belies its complex biology. OSE plays an active role in follicular rupture, oocyte release, subsequent ovarian remodeling, and repair of follicle walls. It is a relatively less differentiated, uncommitted layer of cells that expresses both epithelial and mesenchymal markers, unlike most normal epithelia found in other tissues. It covers only a certain area in a functional ovary where it gets disrupted by regular ovulatory episodes; however, in resting ovaries, e.g., during anovulatory cycles, PCOS, during menopause, or sclerotic ovaries, the entire surface of the ovary is covered with OSE [8]. Ovarian stem cells have been shown to exist in the OSE by various investigators (including our group) in human, sheep, marmoset, rabbit, and mouse ovaries [1, 3, 9–14]. However, it is important to note that by using varied approaches, stem cells of variable cell sizes, gene signatures, and terminologies have been reported by other groups and their true identity remains elusive.

Use of Tilly's group DEAD box polypeptide 4 (DDX-4- also termed as mouse vasa homologue MVH), a germ cell-specific marker-based flow cytometry approach by Tilly's group to isolate and demonstrate the presence of oogonial stem cells resulted in controversy. The use of DDX-4 as a marker to isolate the putative stem cells remains controversial as it is generally accepted as a cytoplasmic protein [15], but the use of DDX-4 to isolate stem cells in their protocols was recently shown [5]. The stem cells described by his group are equivalent to SSCs in the testis [2] and are (5–8 μm in size) and express germ line markers [Prdm1, Dppa3, Ifitm3, Tert, Ddx4, and Dazl] [16]. Bukovsky et al. [9] first showed that ovarian surface epithelial stem cells were a bipotent source of germ as well as somatic granulosa cells. In vitro, OSE scraped from postmenopausal human ovary could develop oocyte-like structures of about 180 μm in the presence of a medium with phenol red (estrogenic stimuli). Ovarian epithelial stem cells underwent asymmetric division and contributed to new germ cells formation [17]. Virant-Klun's group [11, 12, 18, 19] reported the presence of tiny spherical cell types with embryonic-like characteristics and measuring 2–4 μm in diameter in the adult human ovaries (menopausal and premature ovarian failure). These putative ovarian stem cells expressed pluripotent stem cell markers (Oct-4, Nanog, Sox-2) and differentiated in culture to form oocyte-like structures measuring 90 μm in diameter, expressing germ cell-specific markers (c-kit, VASA, and ZP2) at the mRNA transcript level and could develop into embryoid body-like, and parthenote embryo-like structures, and neuronal phenotype cells.

Our group has reported the presence of two distinct populations of stem cells in OSE scrapings, which include very small embryonic-like stem cells (VSELs, 1–3 μm), slightly bigger progenitors (OGSCs, 4–7 μm) [10] and occasional germ cell nests/cysts (formed by rapid nuclear divisions and incomplete cytokinesis of OGSCs) [3, 14, 20]. The pluripotent VSELs with nuclear OCT-4 and cell surface SSEA-4 (in humans) give rise to OGSCs with cytoplasmic OCT-4, which undergo rapid clonal expansion and result in formation of cysts and further meiosis and differentiation to give rise to oocytes (Fig. 1).

Very small embryonic-like stem cells are a novel stem cell population present in both adult testis and ovary [10, 20, 21], are more primitive and give rise to the progenitors (SSCs and OGSCs) by asymmetric cell division. The progenitors undergo rapid proliferation and clonal expansion as chains in testes and as cysts in the ovaries. They undergo meiosis and further differentiation to produce haploid gametes [21, 22]. Thus, we report a novel stem cell population (VSELs), in addition to that by Tilly's group [2] as shown in Fig. 2.

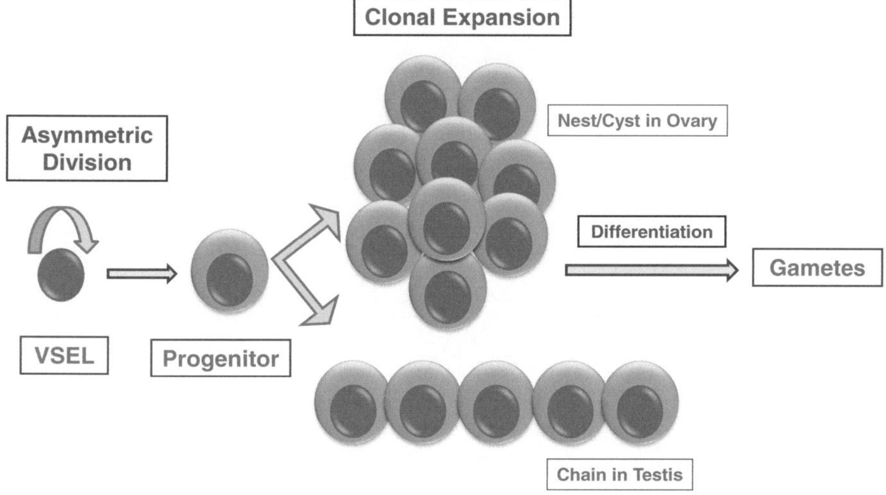

Fig. 1 Asymmetric cell division of stem cells in mammalian gonads: Very small embryonic-like stem cells (VSELs) are very small in size and undergo asymmetric cell division to self-renew themselves and to give rise to slightly bigger progenitors. These undergo rapid proliferation, clonal expansion meiosis, and further differentiation into gametes

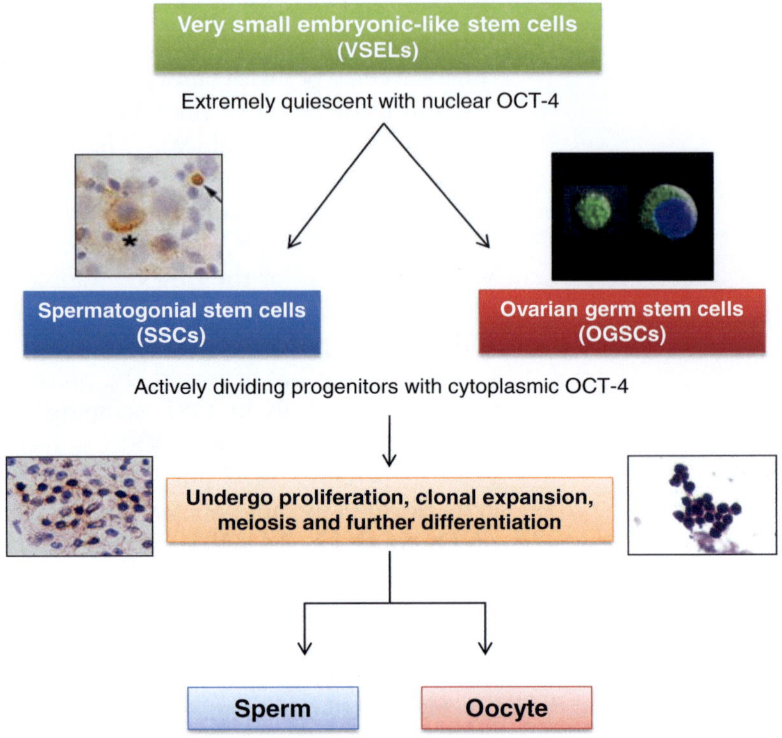

Fig. 2 Stem cell biology in mammalian gonads: VSELs exist in both testis and ovary; give rise to progenitors, which further differentiate into gametes. Note that the presence of relatively small VSELs with nuclear OCT-4 and slightly bigger progenitors in both testis and ovary with cytoplasmic OCT-4. Also rapid proliferation of testicular progenitors occurs as a chain in testis and as a nest in the ovary [3, 10, 14, 20–22]

We have studied the ovarian stem cells in scraped OSE cells. This method results in a heterogeneous population of stem cells as pointed out by Woods and Tilly [5], but provides a better understanding of stem cell biology that may account for postnatal oogenesis. Lei and Spradling [23] recently carried out elegant studies and concluded absence of stem cells in the adult mouse ovary since they did not observe any cysts. However, we have shown the presence of cysts in human, sheep, and mouse ovaries [3].

Very small embryonic-like stem cells are pluripotent stem cells which exist in various adult body tissues at the top of the hierarchy of tissue committed stem cells, including gonads, and serve as a backup pool to maintain homeostasis throughout life. Ratajczak's group reported VSELs for the first time [24] but doubts on their very existence were raised in recent literature [25, 26]. However, Ratajczak's group [27], has reviewed protocols for isolation of VSELs and also pointed out technical reasons that might have led to the recent controversy. We have established flow cytometry protocols to identify mouse ovarian VSELs [28] and these stem cells are relatively quiescent, survive oncotherapy, and can be stimulated to undergo oocyte-specific differentiation in vitro in response to follicle stimulating hormone. The approach of immuno-isolation of DDX4 positive stem cells by FACS described by Tilly's group does not pick up the VSELs as they report the presence of uniform sized oogonial stem cells in the range of 5–8 μm size which are possibly the progenitors. Their protocol also fails to detect the ovarian germ cell nests/cysts. Thus, every protocol has its own advantages and it is unfair to discard any one in favor of another [5]. Our work suggests that ovarian stem cell function is modulated by FSH through a novel FSH receptor isoform R3 [29]. We have reported a novel function of FSH on the ovarian stem cells lodged in the OSE [13]. This is in addition to the existing paradigm that FSH acts on the granulosa cells of growing follicles and that primordial follicle growth is independent of gonadotropin action [30]. We also have demonstrated postnatal oogenesis from ovarian stem cells leading to primordial follicle assembly in PMSG-treated mouse ovaries [31].

In order to identify and characterize stem cells from the OSE layer it is essential to first isolate the OSE cells followed by methods to enrich/isolate various populations of stem cells. OSE can be isolated mechanically through gentle scraping of ovaries for human, sheep, monkey, and rabbit and by controlled enzymatic digestion of mouse ovary. In this chapter we have described various methods for their isolation, identification, and characterization followed by a brief overview of exploring their differentiation potential in culture conditions by establishing primary cultures of OSE.

2 Materials

2.1 Isolation of Stem Cells from Adult Mammalian Ovaries

Human ovarian tissue can be obtained from peri-menopausal women with an age range of 40–60 years undergoing total abdominal hysterectomy or ovariectomy for various reasons other than ovarian pathology, infection or malignancy and young ovarian tissue from autopsy cases. Sheep ovaries are procured and transported from an abattoir in 0.9 % normal saline containing antibiotics (penicillin plus streptomycin) at ambient temperature to the laboratory for further processing. Monkey, rabbit, and mouse ovaries can be obtained from animal house facilities. All the procedures need approval from the Institutional Ethics Committee for human and animal research.

2.1.1 Mechanical Isolation of OSE from Adult Mammalian Ovary (Human, Sheep, Monkey, and Rabbit)

1. 0.9 % normal saline (NS)-containing antibiotics (penicillin 100 U/mL, streptomycin 100 μg/mL; Invitrogen).
2. 1X calcium and magnesium-free Dulbecco's phosphate-buffered saline (PBS) (Invitrogen).
3. Fetal bovine serum (FBS) (Invitrogen).

4. DMEM/F12 (high glucose, Sigma Aldrich).
5. 60 mm tissue culture dishes (BD Falcon).
6. 90 mm tissue culture dishes (BD Falcon).
7. 15 mL centrifuge tubes (Tarsons).
8. Centrifuge machine.
9. Surgical instruments: small forceps, scissors, surgical blade number 25 or a blunt scalpel blade.
10. Inverted microscope with Hoffman optics for viewing the scraped cell suspension.

2.1.2 Enrichment of Stem Cells from OSE by Immunomagnetic Method

1. OSE cells suspended in PBS plus 2 % FBS.
2. SSEA-4 antibody (Stem Cell Technologies Inc).
3. EasySep® SSEA-4 Selection Kit (Stem Cell Technologies Inc).
4. Purple magnet concentrator (Stem Cell Technologies Inc).
5. 5 mL round-bottom tubes (BD Biosciences).
6. Sorting medium: PBS containing 0.2 % FBS and 100 mM EDTA.

2.1.3 Flow Cytometry (Immunophenotyping) Analysis of Sheep Ovarian Stem Cells Using OCT-4

1. Ovary surface epithelium cell suspension.
2. 4 % paraformaldehyde (PFA, pH 7.4; Sigma Aldrich).
3. PBS with 1 % BSA.
4. DMEM/F12 with antibiotics.
5. Primary antibodies (Table 1).
6. 0.05 % NP40 (Sigma).
7. RBC Lysis buffer (BD Biosciences).
8. 5 mL round-bottom tubes (BD Biosciences).
9. 15 mL tubes (Tarsons,).
10. Hemocytometer.
11. Microscope.
12. Flow Cytometry Size Calibration Kit microspheres (Invitrogen).
13. BD FACS Aria Flow cytometer.
14. Centrifuge at 4 °C.

2.1.4 Enzymatic Digestion of Mouse Ovary to Obtain Stem Cells

1. DMEM-High Glucose media (Invitrogen).
2. Collagenase IV (Invitrogen).
3. Fetal Bovine Serum (Invitrogen).
4. Phosphate Buffered Saline (PBS-Sigma Aldrich).
5. 4 % Paraformaldehyde solution (PFA-Sigma Aldrich).
6. 1.5 mL Microcentrifuge tubes (Tarsons).

Table 1
Antibodies used to characterize ovarian stem cells

Antibody	Vendor	Dilution	Concentration for flow cytometry
OCT-4	Abcam, UK (ab19857) and Millipore, USA (AB3209)	1:100 1:300	1 μg/1 × 10^6 cells
SSEA-4	Millipore, USA	1:100	–
SSEA-1	Millipore, USA	1:100	–
CD 133	Miltenyi Biotec	1:50	–
FRAGILIS	Abcam, UK	1:50	–
STELLA	Millipore, USA	1:50	–
DAZL	Abcam, UK	1:100	–
GDF-9	Abcam, UK	1:100	–
VASA	R&D Systems, USA	1:100	–
SCP-3	Abcam, UK	1:50	–

2.1.5 Flow Cytometry of Mouse Ovarian Cells to Show Presence of VSELs

1. DMEM-High Glucose media (Invitrogen).
2. Collagenase IV (Invitrogen).
3. DNAse I (Sigma).
4. Fetal Bovine Serum (Invitrogen).
5. Phosphate Buffered Saline (PBS-Invitrogen) Mouse-specific FcR Block (Stem Cell Technologies Inc).
6. Flow Cytometry Size Calibration Kit microspheres (Invitrogen).
7. 2 and 1.5 mL centrifuge tubes.
8. 40 μm strainer/mesh filters (BD Biosciences).
9. Antibodies from BD Biosciences: FITC rat Anti-mouse SCA-1, PE rat Anti-mouse CD45, and APC mouse Lineage antibody cocktail.
10. Centrifuge.
11. BD FACS Aria Flow cytometer.

2.2 Detection and Characterization of Ovarian Stem Cells

2.2.1 Preparation of OSE Smears and H&E Staining

1. Ovary surface epithelium cell suspension.
2. 4 % PFA fixative.
3. Poly-L-lysine (Sigma Aldrich).
4. Glass slides (HiMedia).
5. Hematoxylin and Eosin stains.
6. Alcohol grades (30, 50, 70, 90, and 100 %).
7. 1 % Acid alcohol (add concentrated HCl in 70 % alcohol).

8. Scott's buffer.
9. Xylene.
10. Glass slides and coverslips.
11. DPX mountant.

2.2.2 Immuno-localization Studies

1. OSE cell smears on poly-L-lysine coated glass slides.
2. 4 % paraformaldehyde (PFA, pH 7.4; Sigma Aldrich).
3. 1× D-PBS (Invitrogen) or PBS (Sigma).
4. Triton X-100 (Sigma).
5. Blocking buffer (0.1 mM EDTA (Qualigens) + DPBS + 3 % BSA (Sigma)) or.
6. 20 % Serum for blocking.
7. Washing buffer (0.1 mM EDTA + DPBS + 0.5 % bovine serum albumin).
8. Primary antibody (Table 1).
9. Secondary antibodies: Alexa Fluor 488/568 labeled anti-mouse IgG, or anti- rabbit IgG (1:1,000) (Molecular Probes, Invitrogen).
10. 4′,6-Diamidino-2-phenylindole (DAPI) (Sigma; 300 nM).
11. Propidium iodide (PI, Sigma Aldrich; 1:10,000 dilution of 5 mg/mL, i.e., 0.5 μg/mL).
12. Vecta Mount medium (Vector Laboratories Inc).
13. Coverslips and nail-paint for sealing.

Confocal Laser Scanning Fluorescent microscope for viewing immuno-localization equipped with argon laser at $\lambda = 488$ nm, blue diode laser at $\lambda = 405$ nm and DPSS laser at $\lambda = 561$ nm for observing FITC, DAPI, and PI staining channels. Immuno-cytochemical localization by DAB method can be documented by viewing under an Upright light microscope (90i, Nikon) and representative fields at magnification X400 can be recorded.

2.2.3 RNA Isolation and c-DNA Synthesis

For transcription analysis, total RNA is isolated by the guanidinium-isothiocyanate-phenol chloroform method using TRIzol reagent. After checking purity of RNA, cDNA is prepared using iScript mix provided by BioRad as per standard manufacturer's instructions.

1. Ovary surface epithelium cells prior culture.
2. Cultured ovary surface epithelium cells.
3. TRIzol (Invitrogen).
4. Chloroform (Qualigens).
5. Isopropanol (Qualigens).
6. 70 and 100 % Ethanol (SD Fine Chemical, RNA grade).
7. iScript mix (Bio Rad).

8. Diethyl pyrocarbonate (DEPC) (0.1 % Sigma Aldrich)-treated water.
9. DNAse I.
10. Autoclaved Eppendorf tubes1.5 mL.
11. UV Spectrophotometer.
12. Refrigerated Microcentrifuge.
13. Micropipettors with aerosol-barrier tips (Axygen).
14. Pair of powder-free gloves.

2.2.4 RT-PCR

1. Forward and Reverse Primers (10 μM) (Table 2 stem cell-specific markers; Table 3 germ cell-specific markers [10, 13, 14, 28, 29, 31]).
2. DEPC (0.1 % Sigma Aldrich)-treated water.
3. cDNA sample.
4. PCR tubes (0.2 mL).
5. Thermal cycler machine (G-Storm, Labtech).
6. Pipettes for dispensing volumes from 100 to 0.1 μL.

2.3 Culture of Ovarian Stem Cells

1. Ovarian tissue sample.
2. DMEM/F12 (Sigma Aldrich).
3. Fetal bovine serum (FBS) (Invitrogen).
4. 35 mm culture dishes (BD Falcon).
5. Chambered slides (BD Falcon).
6. Inverted microscope with Hoffman optics for viewing cell suspension.
7. 5 % CO_2 Incubator at 37 °C (38.5 °C for sheep).
8. Sterile surgical instruments (forceps and scalpel).
9. Laminar flow with warm stage at 37 °C (K Systems; Kivex Biotech Ltd).

3 Methods

3.1 Preparation and Isolation of Stem cells

3.1.1 Mechanical Isolation of OSE from Adult Mammalian Ovary (Human, Sheep, Monkey, and Rabbit)

1. Rinse ovarian samples/ovaries gently but thoroughly in PBS to avoid any clumping of cells (4–6 times for 5 min each).
2. Hold ovaries in PBS or 20 % FBS + DMEM/F12 with forceps.
3. Dissect out extraneous tissue and hilum portion contributing to blood cells with sharp scissors.
4. Gently scrape surface of each ovary with the blunt edge of surgical blade superficially without applying pressure into a 35 mm Petri dish containing 20 % FBS + DMEM/F12.
5. Collect cells released in a 15 mL centrifuge tubes and sediment at 1,000 × g for 10 min at RT.

Table 2
Details of primers for pluripotent stem cell markers

Gene	Primer sequence	Amplicon size (base pair)
Human pluripotent stem cell markers		
Oct-4	*F*: GAAGGTATTCAGCCAAACGAC *R*: GTTACAGAACCACACTCGGA	315
Oct-4A	*F*: AGCCCTCATTTCACCAGGCC *R*: TGGGACTCCTCCGGGTTTTG	448
Nanog	*F*: TGCAAATGTCTTCTGCTGAGAT *R*: GTTCAGGATGTTGGAGAGTTC	285
Sox-2	*F*: ATGCACCGCTACGACGTGA *R*: CTTTTGCACCCCTCCCATTT	437
TERT	*F*: AGCTATGCCCGGACCTCCAT *R*: GCCTGCAGCAGGAGGATCTT	185
Sheep pluripotent stem cell markers		
Oct-4A	*F*: CAATTTGCCAAGCTCCTAAA *R*: TTGCCTCTCACTTGGTTCTC	290
Nanog	*F*: TTCCCTCCTCCATGGATCTG *R*: AGGAGTGGTTGCTCCAAGAC	501
Sox-2	*F*: TGATACGGTAGGAGCTTTGC *R*: CTTTTGCCCCTTTAGAGACC	362
Stat-3	*F*: TGGACAACATCATTGACCTG *R*: CTGCTGCTTGGTGTAAGGTT	239
Mouse pluripotent stem cell markers		
Oct-4A	*F*: CCATGTCCGCCCGCATACGA *R*: GGGCTTTCATGTCCTGGGACTCCT	235
Oct-4	*F*: CCTGGGCGTTCTCTTTGGAAAGGTG *R*: GCCTGCACCAGGGTCTCCGA	177
Nanog	*F*: CAGGAGTTTGAGGGTAGCTC *R*: CGGTTCATCATGGTACAGTC	223
Sca-1	*F*: AGAGGAAGTTTTATCTGTGCAGCCC *R*: TCCACAATAACTGCTGCCTCCTGA	**223**
Sheep differentiation markers		
Oct-4 (all isoforms)	*F*: GAGCCGAACCCTGAGGAGTCCC *R*: CAGCAGGGGCCGCAGCTTAC	225
FSH receptor transcripts		
FSH-R1	*F*: CATTCACTGCCCACAACTTTCATC *R*:TGAGTGTGTAATTGGAACCATTGGT	84
FSH-R3	*F*: TCTCCACTGCTGCACTGTTGGGCT *R*: ATTCAAATACAGGAAATAGAGAAA	382
Germ cell marker (human)		
c-Kit	*F*: AAGGACTTGAGGTTTATTCCT *R*: CTGACGTTCATAATTGAAGTC	345
Housekeeping genes		
Human Gapdh	*F*: GTCAGTGCTGGACCTGACCT′ *R*: CACCACCATGTTGCTGTAGC	255
Sheep Gapdh	*F*: GCC CAG AAC ATC ATC CCT G *R*: GGT CCT CAG TGT AGC CTA G	232
Mouse Gapdh	*F*: GTCCCGTAGACAAAATGGTGA *R*: TGCATTGCTGACAATCTTGAG	458

Table 3

Details of primers for human germ cell markers

Gene	Primer sequence	Amplicon size (base pair)
Early germ cell markers		
Oct-4	*F*: CCCCTGGTGCTGTGAAGCTGG *R*: CCCCAGGGTGAGCCCCACAT	124
C-kit	*F*: CCTGGGATTTTCTCTGCGTT *R*: ATTGGTCACTTCTGGGTCTG	376
Vasa	*F*: GAC TGC GGC TTT TCT CCT ACC *R*: TTT GGC GCT GTT CCT TTG AT	418
Primordial follicle transition markers		
AMH	*F*: CACCTGGAGGAAGTGACCTG *R*: CCACCGCTAACACCAGGTAG	202
Primary follicle oocyte markers		
Gdf-9	*F*: CTCCTGGAGACCAGGTAACAGGAAT *R*: TGCACACACATTTGACAGCAGAGG	291
Lhx8	*F*: CAAGCACAATTTGCTCAGGA *R*: GGCACGTAGGCAGAATAAGC	230
Housekeeping gene		
18S	*F*: GGAGAGGGAGCCTGAGAAAC *R*: CCTCCAATGGATCCTCGTTA	171

6. Resuspend cells in minimal volume of media and use cells for:
   - Observing under an inverted microscope to identify different cell types.
   - Enrichment of stem cells by immuno-magnetic method.
   - Stem cells analysis by flow cytometry.
   - Preparation of cell smears for immuno-localization.
   - For RNA extraction followed by RT-PCR and/or q-RTPCR.

3.1.2 Enrichment of Stem Cells from OSE by Immuno-magnetic Method

1. After overnight culture, collect OSE cells in a tube, gently pipette to obtain a single cell suspension and adjust cell count to 1×10^8 cells/mL or for rare cells, start with a cell concentration of 2×10^8 cells/mL (*see* **Note 1**). (*Subsequent steps are as per the manufacturer's instructions.*)
2. Incubate cells with species-specific FcR blocking antibody (100 μL/mL for human) followed by EasySep® SSEA-4 Selection Cocktail on ice for 15–20 min and subsequently with EasySep® Magnet Nanoparticles for 10–15 min.

3. Cell suspension is brought to a total volume of 2.5 mL by adding sorting medium. Cells are mixed in the tube by gently pipetting up and down 2–3 times. Place the tube (without cap) into the magnet. Set aside for 5–10 min.
4. With the tube in the magnetic field, the tube is inverted in one continuous motion and the supernatant fraction is poured off. Magnetically labeled cells remain inside the tube, held by the magnetic field (*see* **Note 2**).
5. SSEA-4 positive cells bound to EasySep® magnet are collected after two washes by repeating **steps 3** and **4**. Remove tube from magnet and resuspend cells in desired medium. The positively selected cells are now ready for use for immuno-localization and RT-PCR (Fig. 3).

3.1.3 Flow Cytometry (Immunophenotyping) Analysis of Sheep Ovarian Stem Cells Using OCT-4

1. Sheep ovaries are washed thoroughly and OSE cells are isolated (as described in Subheading 3.1) and a single cell suspension is prepared and resuspended in PBS with 1 % BSA (*see* **Note 3**).
2. Red blood cells (RBC) are lysed by incubating the cell suspension with lysis buffer in the dark at room temperature for 3 min. The cell suspension containing RBC lysis buffer is diluted to 10 mL and briefly vortexed twice for 2 s each followed by centifugation at $1,000 \times g$ for 10 min at 4 °C and washed twice with PBS. OSE cells are fixed with 2 % PFA for 10–15 min, rinsed with PBS and filtered through a 40 μm sieve and resuspended in PBS/1 % BSA (*see* **Note 4**).
3. Adjust cell concentration to 1×10^6 cells/mL in ice cold plain PBS and distribute cells equally in each vial.

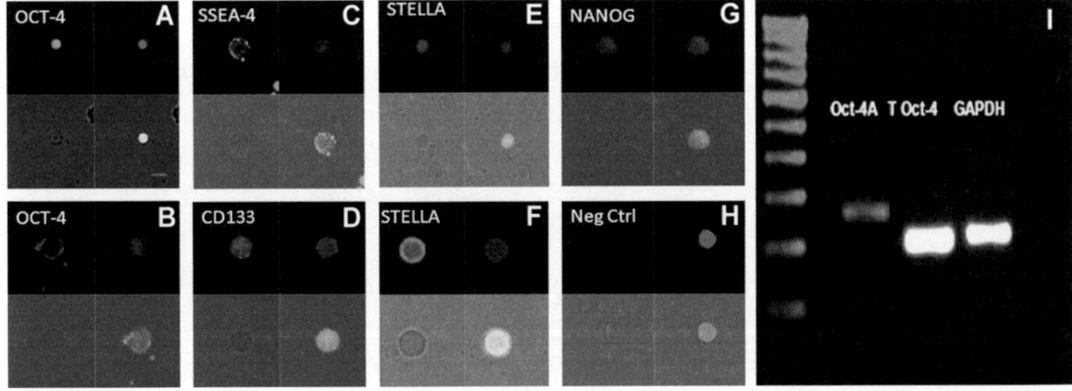

Fig. 3 Characterization of Sheep OSE cells post SSEA-4 based immunomagnetic cell separation: Immunomagnetically separated cells expressed both (**a**, **b**) nuclear and cytoplasmic OCT-4, (**c**) cell surface SSEA-4 and (**d**) CD133, (**e**, **f**) STELLA in both stem cells, (**g**) NANOG, by confocal microscopy. (**h**) Negative control was employed by omission of primary antibody. **a**–**h** is a composite of four panels representing fluorescent channels, bright field, and merged images of both by confocal microscopy. (**i**) RT-PCR for Oct-4A and Oct-4 and Gapdh in immunomagnetically SSEA-4 sorted cell fractions showed bands of expected size (290, 225, and 232 base pairs with no template control showing absence of band) [14]

4. Treat cells with 0.05 % NP40 followed by incubation with appropriate concentration of primary antibody for 1 h at 4 °C as optimized by the user, followed by washing and centrifugation at 1,000 × *g* for 10 min at 4 °C (*see* **Notes 5** and **6**).
5. Resuspend cells in the dilution of fluorochrome labeled secondary antibody in an appropriate volume of cold PBS and incubate in dark at 4 °C for 30–45 min followed by PBS wash two to three times.
6. Reconstitute the cells in 200 μL of PBS in 5 mL round-bottom tubes and leave in the dark at 4 °C until used. The cells are ready to be acquired on a flow cytometer and should be maintained on ice.
7. The unstained and OCT-4 stained cells are run on FACS Aria and the results are analyzed using FACS Diva software (BD Biosciences). Acquire unstained OSE cell suspension on an FSC-SSC plot and gate using standard flow cytometry to obtain cells in range of 2–9 μm (P1 and P2 in Fig. 4). From original P1 (2–4 μm) and P2 (4–9 μm) cell populations, select OCT-4+ cells (P4 and P8 respectively), of which further obtain DAPI+ nucleated cells (P6 and P10 respectively) on a DAPI-SSC plot. Calculate percentage of (OCT-4+/DAPI+) each individual population of stem cells or an average of both stem cell populations. Employ biological and technical replicates with appropriate negative controls. Experiments are repeated at least three times to get average stem cell populations with standard error (Fig. 4).

3.1.4 Enzymatic Digestion of Mouse Ovary to Obtain Stem Cells

Being very small in size, mouse OSE cannot be isolated by mechanical scraping. An enzymatic method is used to isolate the stem cells based on a method published by Symonds et al. [32]. It involves incubation of ovaries in a low concentration of collagenase for 30 min, which helps to partially digest the collagen fibers contained in the ovarian cortex/stroma and loosening of the OSE. These loosened OSE cells can be released into the medium through physical shear. The OSE cells are collected and studied for the presence of stem cells.

1. Excise the mouse ovaries from the surrounding tissue under a stereo microscope and rinse in DMEM-HG media.
2. Place the ovaries individually in a 1.5 mL tube with 0.1 mL (0.5 mL if the purpose is for RNA isolation) of DMEM-HG media containing 0.5 mg/mL of Collagenase Type IV.
3. Incubate at 37 °C for 30 min in a water bath (*see* **Notes 7** and **8**).
4. Vortex the tubes for exactly 2 min at high speed to release the OSE from the ovary.
5. Add half the volume of DMEM-HG media containing 15 % fetal bovine serum to stop the collagenase action (*see* **Notes 9** and **10**).

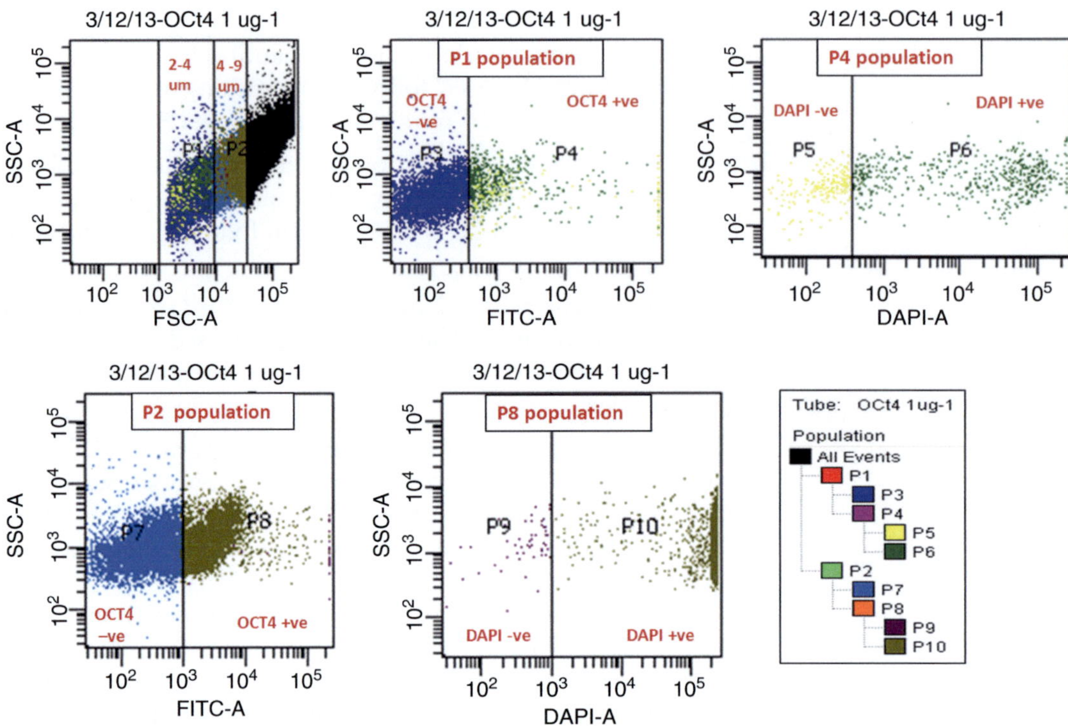

Fig. 4 Immunophenotyping of sheep ovarian stem cells: Immunophenotyping studies show the presence of two distinct populations of stem cells including 1.26 ± 0.19 % of (2–4 μm) cells and 6.86 ± 0.5 % of (4–9 μm) cells expressing OCT-4 within the OSE cell layer of sheep ovary. On average, 4.06 % cells of 2–9 μm size express OCT-4 obtained from scraped sheep OSE

6. Remove the ovaries and fix in 4 % PFA for further analysis.
7. Vortex the supernatant with isolated OSE cells again for 2 min to disaggregate epithelial sheaths.

3.1.5 Flow Cytometry of Mouse Ovarian Cells to Show Presence of VSELs

1. Excise the ovaries from the surrounding tissue and rinse in DMEM-HG media.
   Note: Two ovaries, usually from one animal are used per single tube.
2. Mince the ovaries in the enzyme mixture (DMEM-HG media containing 750 IU/mL of Collagenase IV and 1 μg/mL of DNAse I).
3. Incubate the minced ovaries in 1 mL of enzyme solution in a water bath at 37 °C for 20 min with intermittent mixing.
4. Resuspend well with a pipette to make a single cell suspension and stop the reaction with an equal volume of DMEM-HG media containing 20 % FBS. Filter through 40 μm cell strainer.
5. Centrifuge the filtrate at 1,000 × *g* for 10 min (*see* **Notes 11** and **12**).

6. Wash the pellet once with PBS.
7. Resuspend the pellet in an appropriate volume of PBS and count the cells using a hemocytometer.
8. Adjust the cell suspension to 1×10^6 cells/mL with PBS. (From this step onwards the cells are maintained on ice until flow cytomeric analysis.)
9. The cells are blocked with mouse-specific FcR blocking antibody (10 μL/mL) for 20 min and stained with FITC rat Anti-mouse SCA-1 (1 μg/10^6 cells), PE rat Anti-mouse CD45 (2 μg/10^6 cells) and APC mouse Lineage antibody cocktail (25 μL/mL of cells) on ice for 1 h. The cells without any antibody addition serve as unstained control used for setting gates.
10. Wash all samples with twice the volume of PBS and centrifuge for 10 min at $1{,}000 \times g$, 4 °C.
11. Resuspend cells in PBS and transfer them to 5 mL round-bottom tubes after passing through a 40 μm strainer/mesh filter to remove cell clumps. Keep on ice until analysis.
12. Beads of sizes 1–15 μm are prepared according to manufacturer's instructions.
13. Run the beads and samples in the flow cytometer.
14. Initially set the forward scatter (FSC) and side scatter (SSC) parameters to logarithmic scale and the threshold on the FSC parameter as mentioned in next step.
15. Run the mixture of predefined-sized microspheres (size calibration beads with standard diameters of 1, 2, 4, 6, 10, and 15 μm) and adjust the threshold for the machine to include 2 μm beads (a value of 2,000 in BD FACS Aria). This allows detection of VSELs, otherwise they will be discarded as debris since a default threshold picks up only cells greater than 5 μm.
16. Run the unstained and stained samples.

Analyze for SCA-1^+Lin$^-$CD45$^-$ cells using a sequential gating strategy. Briefly, select cells ranging from 2 to 10 μm size using beads as reference on an FSC-SSC plot (P1 in Fig. 5). From these cells, Lin$^-$ cells are selected from the SSC-APC plot (P2 in Fig. 5) followed by selection of CD45$^-$ in Lin$^-$ cells from a SSC-PE plot (P3 in Fig. 4). Finally, FITC positive cells in SSC-FITC plot represent VSELs (P5 in Fig. 5). All the gates are set using unstained samples.

3.2 Detection and Characterization of Ovarian Stem Cells

3.2.1 Preparation of OSE Smears and H&E Staining

1. Prepare the OSE smears on poly-L-lysine-coated glass slides.
2. Air-dry the smears and fix using freshly prepared 4 % PFA for 15 min at RT.
3. Wash the cell smears thrice with 1× PBS for 5 min. each.
4. Air-dry the slides and store at 4 °C until use.

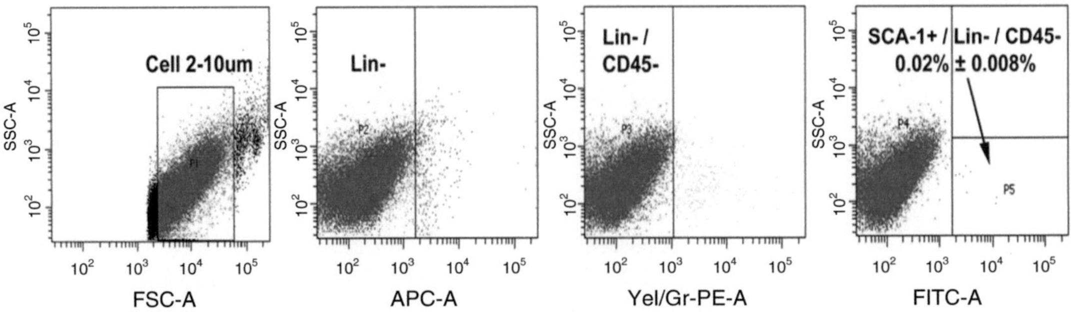

Fig. 5 Flow cytometry analysis of SCA-1+/Lin-/CD45- VSELs in mouse ovaries: Cells between 2 and 10 μm were gated using size calibration beads (i) followed by sequential selection for LIN negative population (ii), CD45 negative population (iii), and then SCA-1 positive population (iv) as mentioned in the protocol. The average percentage of SCA-1⁺/Lin⁻/CD45⁻ VSELs with standard deviation from minimum of four animals is reported

5. The cells smears are used for H&E staining and morphological analysis and immuno-fluorescence staining using specific markers for detailed characterization of the pluripotent stem cells.
6. During the preparation of the mouse OSE smears, the cells are fixed immediately after collagenase digestion. A certain degree of morphological change to the cells cannot be avoided and hence the cell shape may not appear normal.
7. OSE cell smears are rinsed with PBS two times for 5 min each (*see* **Note 13**) and then washed with tap water.
8. Stain the smears by dipping slides in Hematoxylin stain for 5 min followed by destaining of excess stain in 1 % acid alcohol and then place slides in water for 5 min.
9. Slides are immersed in Scott's buffer for 3–5 min to intensify the nuclear stain followed by water for 5 min.
10. Stain the slides with Eosin briefly by dipping for 15 s followed by increasing alcohol grades of 70, 90, and 100 %.
11. Excess alcohol is blotted and slides are air-dried and cleared in xylene for 2–5 min.
12. Mount slides permanently in DPX mountant with coverslips. The slides are ready to be viewed under the microscope (Fig. 6).

3.2.2 Immuno-localization Studies

1. Wash and hydrate the smears with 1X PBS/PBS twice for 5 min each.
2. Treat the cells with 0.3 % Triton X-100 for 10 min at room temperature for nuclear and cytoplasmic markers (step is avoided for cell surface markers) followed by wash with washing buffer twice for 5 min each.
3. Block nonspecific sites with blocking buffer/serum at room temperature for 1–1.5 h (*see* **Notes 14** and **15**).

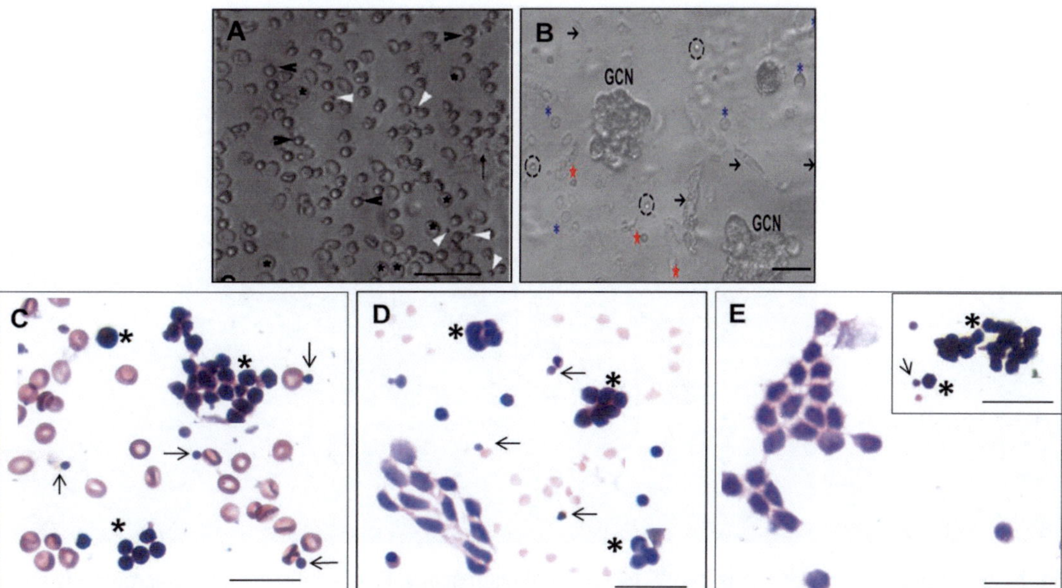

Fig. 6 Ovarian stem cells as smears prepared by scraping ovary surface epithelium: Presence of spherical putative stem cells of two distinct sizes and few germ cell nest-like structures (GCN) were observed under Hoffman optics in scraped OSE from (**a**) human (*white* and *black arrowhead*), (**b**) sheep (*circled*) ovary besides spindle-shaped epithelial cell sheets (*arrow*) and red blood cells (*black* and *red asterisks*). Hematoxylin and Eosin staining of (**c**) human (**d**) sheep and (**e**) mice OSE cell smears enabled identification of two spherical stem cell populations—VSELs and OGSCs (*arrow*) stained prominently with Hematoxylin stain having high nucleo-cytoplasmic ratio and rare germ cell nest/cyst-like structures (*asterisk*) with pale stained epithelial cells/cell sheets [3]. Scale bar = 20 μm

4. Incubate the cells with primary antibody diluted in blocking buffer/serum at 4 °C overnight (dilution of antibody is titrated as per sample to identify optimum dilution; (*see* **Notes 16** and **17**)). For directly tagged antibody, e.g., SCA-1-FITC, staining can be done in 1 day. The primary antibody is incubated at 4 °C for 2 h (**steps 2** and **6** of the protocol are omitted for SCA-1-FITC staining).
5. On the following day, remove the primary antibody and wash the cell smears with buffer three times for 5 min each.
6. Respective Alexa Fluor labeled secondary antibody diluted 1,000 times with washing buffer is added and incubated for 2 h at room temperature. Gently wash 5–7 times with washing buffer (5 min per wash).
7. Counterstain the cells with DAPI (1.47 μM) for 25 s (20 min to stain VSELs specifically) or with PI (0.5 μg/mL) for 20–30 s.
8. Mount smears in Vecta Mount medium and seal with nail varnish. Stained slides can be stored at −20 °C until observation. Slides are observed under a confocal microscope (Fig. 7).

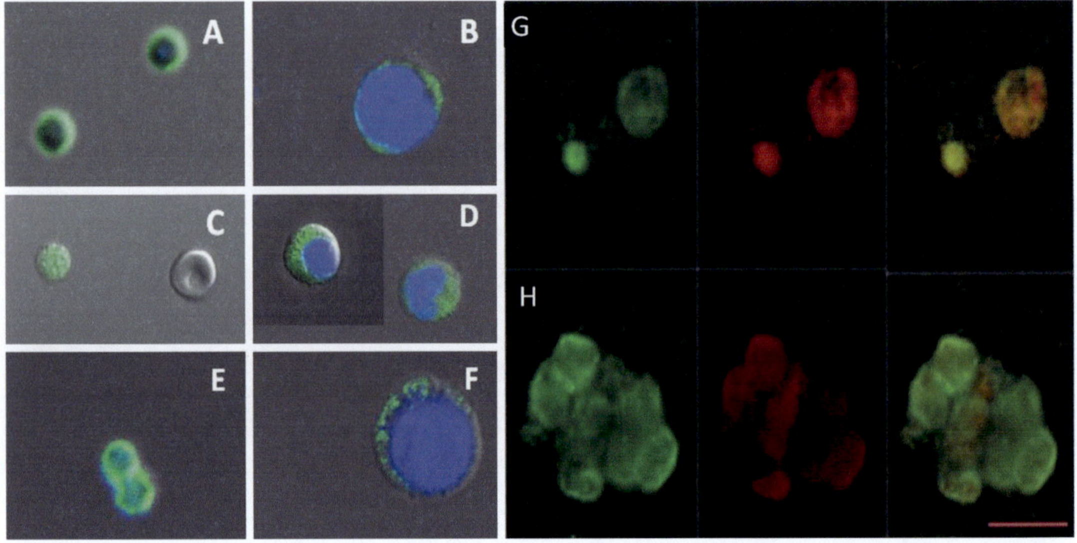

Fig. 7 Characterization of stem cells in OSE by confocal microscopy: Fluorescent confocal micrographs showing localization of pluripotent cell surface markers SSEA-4 in human (**a**) VSELs and (**b**) OGSCs, OCT-4 in (**c**) nucleus of VSELs and in (**d**) cytoplasm of OGSCs. Similarly SSEA-4 was also localized in (**e**, **f**) sheep VSELs and OGSCs. OCT-4 revealed similar localization pattern in sheep (**g**, **h**) ovarian stem cells and was also localized in cytoplasm of germ cell nest/cyst-like structures. Figures **a**, **b**, **e**, and **f** is at 882× magnification with 5× optical zoom and 520× magnification with 5× optical zoom in figures **c** and **d** [10, 29]. Scale bar =20 μm. Merged image of DAPI, FITC, and DIC in **a–f** and FITC, PI and both merged in **g**, **h**

3.2.3 RNA Isolation and c-DNA Synthesis

1. For RNA isolation, centrifuge the OSE cells at 1,000×*g* for 10 min.
2. Wash the cell pellet once with PBS and collect in TRIzol or extraction buffer of Arcturus Picopure RNA Isolation Kit. The latter is meant for isolating RNA from very low number of cells (*see* **Note 18**).
3. Frozen cells are brought to room temperature and pipetted multiple times to disrupt the cells and held in Trizol reagent for 5–10 min at RT (*see* **Note 19**).
4. Add chloroform to the sample tube and shake the contents vigorously and invert the tubes. Stand tubes at RT for 5 min. and centrifuge at 12,000 rcf for 15 min at 4 °C (*see* **Note 20**).
5. Collect upper phase carefully in a fresh autoclaved (nuclease free) tube and add chilled isopropanol in the tube (*see* **Note 21**).
6. Shake and stand tube at RT for 10 min. Place the tubes on ice followed by −20 °C. Tubes can be stored at −80 °C overnight for enhancing precipitation.
7. Continue to thaw the tubes and centrifuge at 12,000 rcf for 10 min. at 4 °C to pellet the RNA and supernatant is discarded.

8. Wash the pellet with 75 % ethanol and spin down the RNA pellet at 7,500 rcf for 10 min at 4 °C. Repeat **step 8** once more while discarding the supernatant.
9. Add 100 % ethanol to the sample tubes and spin down again at 7,500 rcf for 10 min at 4 °C. Collect and discard supernatant and air dry the RNA pellets (*see* **Note 22**).
10. Reconstitute RNA pellet in DEPC-treated water and solubilize RNA by incubation at 65 °C for 5–7 min. Snap chilled on ice and stored at −80 °C (*see* **Note 23**).
11. Add DNAse I-RNAse free to the RNA sample and incubate at 37 °C for 30 min. DNAse is heat inactivated at 70 °C for 5 min and snap chilled.
12. RNA is quantified with a UV spectrophotometer acquiring readings at 260 nm and using following formula:

$$\text{RNA concentration} = \text{O.D. at 260 nm} \times \text{dilution factor} \times \text{conversion factor}^{*}.$$

(*conversion factor for 1O.D. = 40 μg/μl).

3.2.4 RT-PCR

1. Thaw all components upon ice. Mix vigorously and centrifuge before use.
2. Reaction set up: Label the 0.2 mL PCR tubes on the sides as per convenience. All reactions are to be performed in duplicate and incorporate suitable positive and negative controls.
   (a) Add 1 μL of cDNA to each PCR tube and place tube on ice. Add 1 μL sterile water instead of cDNA in the negative tube (No template control-NTC) and RT negative or RNA.
   (b) Prepare the master mix as given below (For a 50 μL reaction).
   (c) Add 49 μL master mix into respective tube and mix-vortex each tube.
   (d) From each tube transfer 24 μL of the master mix-cDNA mixture into two low profile tubes (*see* **Notes 24–27**).

Components	Volume (μL)
SYBRGreen 2× Super mix	25
Forward primer (10 μM)	1
Reverse primer (10 μM)	1
Sterile water	22
Total	49

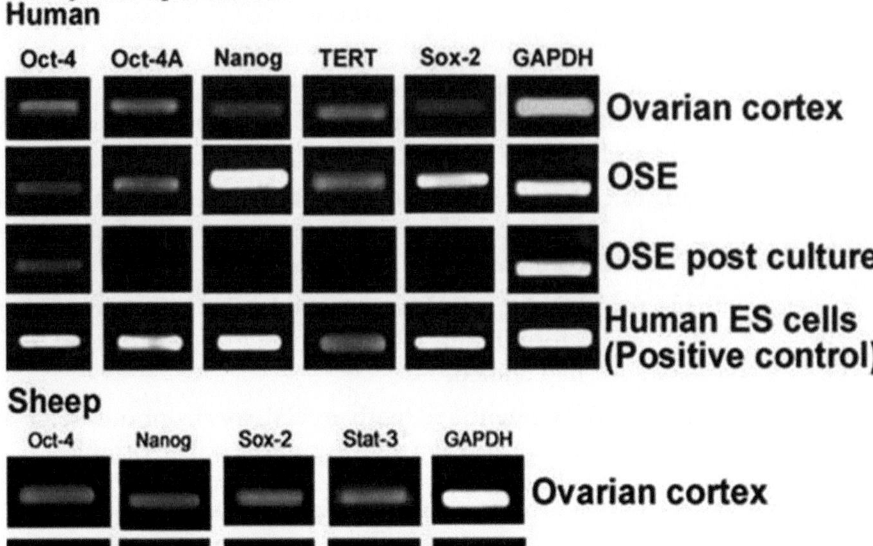

Fig. 8 Reverse transcriptase–polymerase chain reaction analysis of human and sheep ovarian cortical tissue with intact OSE, scraped OSE cells and OSE cell post culture: Expression of pluripotent transcripts for Oct-4 (315 bp), Oct-4A (448 bp), Nanog (285 bp), TERT (185 bp), and Sox-2 (437 bp) in human and Oct-4 (290 bp), Nanog (501 bp), Sox-2 (362 bp), and Stat-3 (239 bp) in sheep showed PCR products of expected band size. In-house derived human ES cells served as positive control. Transcripts for germ cell markers c-Kit (345 bp) and Oct-4 (315 bp) were observed post-culture and human testicular tissue was used as a positive control. Glyceraldehyde-3-phosphate dehydrogenase (GAPDH), a house-keeping control gene, was detected in all samples

3. Start the instrument and create a program based on annealing temperatures of the primers required. The program should be as follows:
   (a) Initial Denaturation: 95 °C for 5 min.
   (b) Denaturation: 95 °C for 10–30 s.
   (c) Annealing: optimum temperature for 20–30 s.
   (d) Elongation: 72 °C for 30 s.
   (e) Repeat **step b**, 39 times.
   (f) 95 °C for 30 s.
   (g) RT-PCR products are run on a 2 % agarose gel and imaged (Fig. 8).

3.3 Culture of Ovarian Stem Cells

1. Scrape the ovaries gently with sterile surgical blade in DMEM/F12 medium under laminar flow hood with warm stage at 37 °C (*see* **Note 28**).
2. After sedimentation at 1,000 × *g* for 10 min at room temperature the scraped OSE cell suspension is washed twice with DPBS and finally suspended in 250 μL of DMEM/F12

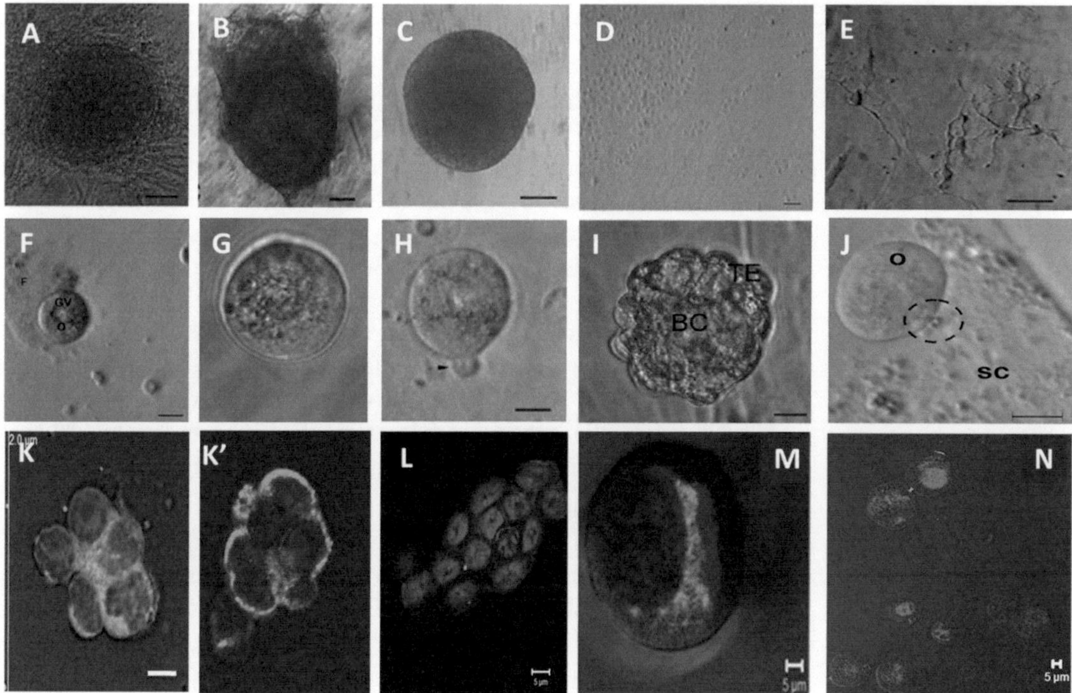

Fig. 9 Different cell types developed in OSE cultures by differentiation of stem cells: Stem cells in OSE differentiated into (**a**) ES cell-like colonies with distinct boundaries growing on a bed of fibroblasts, (**b**) alkaline phosphatase positive ES cell-like colonies, (**c**) embryoid body-like structure, (**d**) epithelial cell transition into mesenchymal type, (**e**) neural phenotype cells, (**f**) Oocyte-like structures [O] showing germinal vesicle [GV], (**g**) large oocyte-like structure, (**h**) Polar body formation (*arrowhead*), (**i**) Parthenote embryo-like structure showing typical [BC] blastocoel cavity and [TE] Trophoectoderm, (**j**) Cytoplasmic streaming with cross-talk (marked in *dashed circle*) between oocyte [O] and adjoining somatic fibroblasts [SC], Germ cell nest/cyst-like structures expressing pluripotent markers (**k**) cytoplasmic OCT-4, (**k′**) cell surface SSEA-4, germ cell-specific (**l**) VASA, (**m**) Live cell imaging of oocyte-like structures stained with MitoTracker FM dye to localize mitochondrial cloud/Balbiani body-like structure (Cultured cells were incubated with 350 nM Mitotracker green FM dye at 37°for ~1 h before live cell imaging) (N) negative control by omission of primary antibody. Scale bar = 50 μm in (**a**, **c**, **e**), 20 μm in (**b**, **d**, **f–j**), 5 μm in (**k–n**) [10, 14]. (**k′**, **l**) are sheep stem cells rest all are human ovarian stem cells

medium supplemented with 20 % FBS and antibiotic solution depending upon cell density.

3. The cell suspension is gently pipetted to loosen the epithelial cell sheets if present and to equally distribute the cells within each well of the culture plates.
4. 10 μL (representative volume) of the cell-suspension is observed under inverted microscope to confirm the cells obtained after scraping ovary surface.
5. Culture plates are returned to the incubator and cultures are incubated for 3 weeks with regular media changes every alternate day and monitored under inverted microscope with Hoffman optics on a warm microscope stage maintained at 37 °C (38.5 °C in case of sheep cultures) to observe the putative cells (Fig. 9) (*see* **Notes 29–31**).

4 Expected Results

1. The protocols described above result in the detection of ovarian stem cells, which comprise of two distinct populations including VSELs and OGSCs. The OGSCs are similar to the spermatogonial stem cells (SSCs) in the testis whereas the VSELs are a novel stem cell population present in both the ovary and testis (Fig. 2) [20]. Besides these two stem cell populations we also observe the presence of germ cell clusters (which arise due to rapid proliferation and incomplete cytokinesis), an essential feature of stem cell biology as argued recently by Lei and Spradling [22]. These are observed in various animal species, especially mouse ovaries after PMSG treatment [31], as well as in sheep OSE cultures treated with FSH [29].
2. Immuno-localization studies and RT-PCR analysis suggest that the VSELs express pluripotent (Oct-4A, Nanog, Sox-2, SSEA-4/1, Tert, Stat-3), primordial germ cells (Fragilis, Stella, Oct-4, Vasa, Dazl) and VSELs specific markers (CD133 in human and SCA-1 in mouse). Interestingly, the VSELs are pluripotent with nuclear OCT-4 whereas the OGSCs express cytoplasmic OCT-4. We have proposed that when a pluripotent stem cell is committed and undergoes differentiation, nuclear OCT-4 is no longer required and shifts to the cytoplasm and eventually gets degraded as cell becomes further differentiated. Similar staining patterns have been reported previously in the human testis [21]. The stem cells can be enriched by magnet-based separation employing SSEA-4 antibody or immunophenotyping by using anti-OCT-4 antibody in higher mammals. In mice, it can be identified by flow cytometry using the SCA-1+/Lin-/CD45- phenotype. The OSE cells in culture grow rapidly and spontaneously differentiate into oocyte-like structures as reported earlier [10]. Epithelial cells undergo EMT to form mesenchymal cells, which make up the granulosa-like somatic cells which surround the growing oocyte resulting in primordial follicle assembly [22]. The stem cells also give rise to other cell types such as embryonic stem cell-like colonies, embryoid body-like structures, parthenote embryo-like structures, and neuronal phenotype cells [10]. During spontaneous differentiation of ovarian stem cells in vitro, we have recently reported hallmark features like germ cell nests/cysts formation, Balbiani body formation (mitochondrial clouds) and cytoplasmic streaming reminiscent of fetal stage developing ovary [14].
3. VSELs are the primordial germ cells that persist into adulthood and also in the menopausal ovary. Menopause occurs as a result of a compromised somatic micro-environment where the stem cells are lodged and not due to a lack of stem cells. Moreover, the uncontrolled proliferation of VSELs (due to yet

not understood changes in the somatic niche) results in cancers. The presence of VSELs in OSE explains why 90 % of ovarian cancers are epithelial in nature and also exhibit the presence of OCT-4A [33]. The VSELs may be transformed into cancer stem cells. Both VSELs and CSCs share a common property of quiescence. The VSELs have been described as the lost pearls [34] and could bring about a paradigm shift in the basic understanding of ovarian biology and also pathology.

5 Notes

1. Ensure that cells are in a uniform suspension by vigorous pipetting more than five times. 15 mL conical tubes should be used instead of round bottomed tubes.
2. Do not shake or blot off any drops that may remain hanging from the mouth of the tube while pouring in the inverted position during immuno-magnetic cell separation.
3. Cells must be maintained at 4 °C throughout until acquisition in a flow cytometer.
4. The cell fixation protocol is unnecessary if viable cells are required. Alternatively, cells may be fixed after the immunostaining protocol is complete, if final acquisition on a flow cytometer is to be delayed. Fixation will help preserve the cell morphology and stabilize light scatter and inactivate most biohazardous agents. The concentration, type of fixative, and duration of fixation requires optimization depending on the markers being assessed and their localization. Fixatives such as acetone or methanol can also be used. Use of polystyrene/plastic tubes should be avoided with acetone.
5. The optimum concentration of primary antibody (i.e., least concentration with best signals above background levels) should be assessed prior to use. Cells may be fixed after incubation with primary antibody.
6. Cells should be incubated in the dark if a directly labeled antibody is used. Cells can be immediately analyzed on a flow cytometer or stored for few hours at 4 °C. The washing step can be repeated id sufficient cells are available. Multiple wash steps should be avoided if cells of interest are rare.
7. Enzymatic digestion step should be carried out with intermittent shaking at regular intervals to dissociate the tissue efficiently.
8. Multiple freeze-thaw cycles of enzymes should be avoided. Temperature fluctuations affect enzyme activity. Therefore, fresh aliquots of enzymes should be used for accuracy and reproducibility of results. Temperature variations should be avoided during incubation steps.

9. Care should be taken not to overexpose ovaries to enzyme action as OSE cells will be affected (morphology, viability, and epitopes) and the chances of the underlying cell layer dissociating, increases. Precise temperature should be maintained for optimal enzyme activity.
10. Since the cells are fixed immediately after collagenase digestion, a certain degree of morphological change to cells must be expected.
11. During multiple centrifugation steps followed by washes, supernatant should be gently decanted and not aspirated to avoid cell loss.
12. Each step of cell processing should be followed by a cell viability determination.
13. OSE cell smears can be rehydrated in alcohol grades (100, 90, 70, 50, and 30 %) for 3 min each similar to procedure followed for paraffin sections.
14. Normal goat, rabbit, or donkey blocking serum must be used depending upon the host species in which secondary antibody is raised. This step is very crucial as it counters nonspecific binding of secondary antibody and hence background staining.
15. Nonspecific signals can be reduced by increasing concentration of blocking solution and duration of blocking step.
16. The concentration of primary and secondary antibodies should be titrated to achieve optimum working concentration of the antibodies. Only one parameter should be changed at a time.
17. Batch or lot variations of antibody may yield variable results.
18. Cell/tissue samples can be snap frozen in Trizol/ extraction buffer at −80 °C until use. Sample volumes should not exceed 10 % of the volume of Trizol reagent used. Use 1 mL of Trizol reagent for 50 mg of tissue or 10^6 cells.
19. After harvesting cells from tissue/culture plates, cell samples should be maintained at 4 °C. The RNA yield will be lower if incubation in Trizol reagent occurs at RT prior to cell disruption.
20. Insoluble particulate material before chloroform addition should be removed by filtration (especially when extracting RNA from tissue) and ensure complete homogenization of cells/tissue. If this is not performed, the DNA may get trapped affect the results.
21. If poor phase separation is observed, add an additional half-volume of Trizol Reagent. Centrifugation steps should be followed strictly at 4 °C otherwise phase separation may not be complete and the RNA yield will be low.
22. Do not over dry RNA pellet and do not lyophilize or vacuum dry samples.

23. Do not store isolated RNA at –20 °C. Storage should be at –80 °C only to ensure optimal quality of RNA.
24. Housekeeping genes should always be incorporated in each PCR run.
25. Annealing temperature for each primer needs to be standardized.
26. Efficiency of all primers must be determined by serial dilution method (efficiencies should be between 90 and 110 %).
27. The Ct values generated can be analyzed using absolute quantification or a comparative delta Ct method depending on the nature of the study.
28. Tissue processing and cell harvest should be performed as soon as possible after tissue is acquired in order to avoid cell death. Scraping of ovary surface must be gentle otherwise cell death may occur and adversely affect cultures.
29. Volume of spent media removed should be replaced with an equal volume of fresh media pre-equilibrated at 37 °C. If the cell density is too high in a particular well, excess cells can be aspirated and reseeded into a fresh well until day 2 of culture. Plates should remain undisturbed for the first 2 days of culture.
30. Cell density should be monitored. If too few cells are present, proliferation will be slow. Overcrowding may also affect cell viability due to nutrient depletion and competition for adherence to plate surface.
31. If too many red blood cells are present simply shake culture plates gently during media change and the RBCs can be eliminated with subsequent media change.

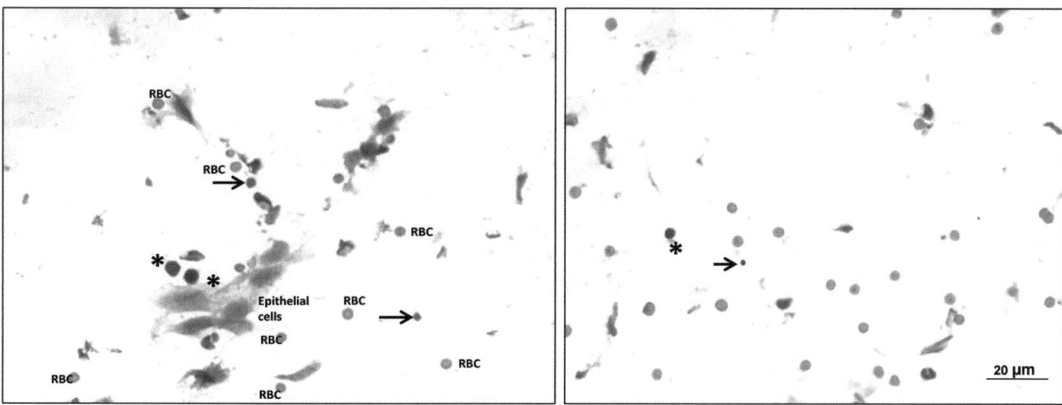

While this chapter was being published we have found that sheep ovaries, fixed overnight in neutral buffered formalin, can also be used to scrape surface epithelium and the H & E stained smears clearly show the presence of stem cells of two distinct sizes (VSELs are marked with arrow; OGSCs are marked with asterix)

Acknowledgements

Study was financially supported by Institute core support provided by Indian Council for Medical Research, Government of India, New Delhi. Research fellowship of SP from CSIR and KS from DST, Government of India, New Delhi, India are acknowledged.

References

1. Johnson J, Canning J, Kaneko T et al (2004) Germline stem cells and follicular renewal in the postnatal mammalian ovary. Nature 428:145–150
2. Woods DC, Tilly JL (2013) An evolutionary perspective on adult female germline stem cell function from flies to humans. Semin Reprod Med 31:24–32
3. Bhartiya D, Sriraman K, Parte S et al (2013) Ovarian stem cells: absence of evidence is not evidence of absence. J Ovarian Res 6:65 [Epub ahead of print]
4. Virant-Klun I, Stimpfel M, Skutella T (2012) Stem cells in adult human ovaries: from female fertility to ovarian cancer. Curr Pharm Des 18:283–292
5. Woods DC, Tilly JL (2013) Isolation, characterization and propagation of mitotically active germ cells from adult mouse and human ovaries. Nat Protoc 20:966–988. doi:10.1038/nprot.2013.047
6. Zou K, Yuan Z, Yang Z et al (2009) Production of offspring from a germline stem cell line derived from neonatal ovaries. Nat Cell Biol 11:631–636
7. Salamanca CM, Maines-Bandiera SL, Leung PC et al (2004) Effects of epidermal growth factor/hydrocortisone on the growth and differentiation of human ovarian surface epithelium. J Soc Gynecol Investig 11:241–251
8. Auersperg N, Wong AS, Choi KC et al (2001) Ovarian surface epithelium: biology, endocrinology, and pathology. Endocr Rev 22: 255–288
9. Bukovsky A, Svetlikova M, Caudle MR (2005) Oogenesis in cultures derived from adult human ovaries. Reprod Biol Endocrinol 3:17–30
10. Parte S, Bhartiya D, Telang J et al (2011) Detection, characterization, and spontaneous differentiation in vitro of very small embryonic-like putative stem cells in adult mammalian ovary. Stem Cells Dev 2:1451–1464
11. Virant-Klun I, Zech N, Rozman P et al (2008) Putative stem cells with an embryonic character isolated from the ovarian surface epithelium of women with no naturally present follicles and oocytes. Differentiation 76:843–856
12. Virant-Klun I, Rozman P, Cvjeticanin B et al (2009) Parthenogenetic embryo-like structures in the human ovarian surface epithelium cell culture in postmenopausal women with no naturally present follicles and oocytes. Stem Cells Dev 18:137–149
13. Parte S, Bhartiya D, Manjramkar DD et al (2013) Stimulation of ovarian stem cells by follicle stimulating hormone and basic fibroblast growth factor during cortical tissue culture. J Ovarian Res 6:20. doi:10.1186/1757-2215-6-20
14. Parte S, Bhartiya D, Patel H et al (2014) Dynamics associated with spontaneous differentiation of ovarian stem cells in vitro. J Ovarian Res 7:25
15. Castrillon DH, Quade BJ, Wang TY et al (2000) The human VASA gene is specifically expressed in the germ cell lineage. Proc Natl Acad Sci U S A 97:9585–9590
16. White YA, Woods DC, Takai Y et al (2012) Oocyte formation in mitotically active germ cells purified from ovaries of reproductive-age women. Nat Med 18:413–421
17. Bukovsky A, Caudle MR (2012) Immunoregulation of follicular renewal, selection, POF, and menopause in vivo, vs. neo-oogenesis in vitro, POF and ovarian infertility treatment, and a clinical trial. Reprod Biol Endocrinol 10:97
18. Virant-Klun I, Stimpfel M, Skutella T (2011) Ovarian pluripotent/multipotent stem cells and in vitro oogenesis in mammals. Histol Histopathol 8:1071–1082
19. Stimpfel M, Skutella T, Cvjeticanin B et al (2013) Isolation, characterization and differentiation of cells expressing pluripotent/multipotent markers from adult human ovaries. Cell Tissue Res 354(2):593–607
20. Bhartiya D, Unni S, Parte S et al (2013) Very small embryonic-like stem cells: implications in reproductive biology. Biomed Res Int. doi:10.1155/2013/682326

21. Bhartiya D, Kasiviswanathan S, Unni SK et al (2010) Newer insights into pre-meiotic development of germ cells in adult human testis using Oct-4 as a stem cell marker. J Histochem Cytochem 58:1093–1106
22. Bhartiya D, Sriraman K, Parte S (2012) Stem cell interaction with somatic niche may hold the key to fertility restoration in cancer patients. Obstet Gynecol Int 2012:921082. doi:10.1155/2012/921082
23. Lei L, Spradling AC (2013) Mouse primordial germ cells produce cysts that partially fragment prior to meiosis. Development 140: 2075–2081
24. Kucia M, Reca R, Campbell FR et al (2006) A population of very small embryonic-like (VSEL) CXCR4(+)SSEA-1(+)Oct-4+ stem cells identified in adult bone marrow. Leukemia 2:857–869
25. Abbott A (2013) Doubt cast over tiny stem cells. Nature 499:390
26. Miyanishi M, Mori Y, Seita J et al (2013) Do pluripotent stem cells exist in adult mice as very small embryonic stem cells? Stem Cell Rep 1:198–208
27. Ratajczak MZ, Zuba-Surma E, Wojakowski W et al (2014) Very small embryonic-like stem cells (VSELs) represent a real challenge in stem cell biology: recent pros and cons in the midst of a lively debate. Leukemia 28(3):473–484. doi:10.1038/leu.2013.255
28. Sriraman K, Bhartiya D, Bhutda S (2013) Mouse ovarian very small embryonic-like stem cells resist chemotherapy and initiate oocyte-specific differentiation *in vitro* in response to follicle stimulating hormone. Mol Rep Dev. (under revision)
29. Patel H, Bhartiya D, Parte S et al (2013) Follicle stimulating hormone modulates ovarian stem cells through alternately spliced receptor variant FSH-R3. J Ovarian Res 6:52
30. Siegel ET, Kim HG, Nishimoto HK et al (2013) The molecular basis of impaired follicle-stimulating hormone action: evidence from human mutations and mouse models. Reprod Sci 20:211–233
31. Bhartiya D, Sriraman K, Gunjal P et al (2012) Gonadotropin treatment augments postnatal oogenesis and primordial follicle assembly in adult mouse ovaries. J Ovarian Res 5:32
32. Symonds DA, Tomic D, Miller KP et al (2005) Methoxychlor induces proliferation of the mouse ovarian surface epithelium. Toxicol Sci 83:355–362
33. Ahmed N, Thompson EW, Quinn MA (2007) Epithelial mesenchymal inter conversions in normal ovarian surface epithelium and ovarian carcinomas: An exception to the norm. J Cell Physiol 213:581–588
34. Zuba-Surma EK, Kucia M, Ratajczak J et al (2009) "Small stem cells" in adult tissues: very small embryonic-like stem cells stand up! Cytometry A 75:4–13

Chapter 17

Clonal Culture of Adult Mouse Lung Epithelial Stem/Progenitor Cells

Jonathan L. McQualter and Ivan Bertoncello

Abstract

Clonal culture of stem cells is crucial for their identification, and the characterization of the cellular and molecular mechanisms that regulate their proliferation and differentiation. In the adult mouse lung, epithelial stem/progenitor cells are defined by the phenotype CD45^{neg} CD31^{neg} EpCAM^{pos} CD104^{pos} CD24^{low}. Here we describe a tissue dissociation and flow cytometry strategy for the detection and isolation of adult mouse lung epithelial stem/progenitor cells, and a three-dimensional colony-forming assay for their clonal culture in vitro.

Key words Epithelial colony-forming unit assay, Lung stem cells, Flow cytometry, Tissue dissociation

1 Introduction

The adult lung is a regenerative organ that contains numerous populations of epithelial stem and progenitor cells (EpiSPC) that contribute to the maintenance of the epithelium lining the conducting airways and the gas-exchanging alveolar bed throughout adult life. In the upper airways, in vitro colony-forming assays and in vivo lineage tracing studies have shown that basal cells (NGFR^{pos}, CD104^{pos}, Krt14^{pos}, and/or Krt5^{pos}) act as lineage-restricted airway progenitor cells, giving rise to club, ciliated and goblet cells in the trachea and proximal airways [1–4]. Basal cells can be isolated from the mouse trachea and human airways based on the expression of NGFR and CD49f (integrin α6), and can be grown and propagated in vitro in Matrigel as spheroids [3, 5].

In the bronchiolar airways of the adult mouse lung, a subset of EpiSPC (CCSP^{pos} CyP450^{neg}) cells that colocalize with neuroepithelial bodies in the distal airways have been shown to give rise to mature club cells and ciliated cells in vivo [2, 6–8]. Similarly, at the bronchoalveolar duct junction, a subset of EpiSPC termed bronchoalveolar stem cells (CCSP^{pos} SP-C^{pos}) have been shown to

Ivan N. Rich (ed.), *Stem Cell Protocols*, Methods in Molecular Biology, vol. 1235,
DOI 10.1007/978-1-4939-1785-3_17,

proliferate in response to both bronchiolar and alveolar injury [9]. In the alveoli, it is accepted that type I alveolar cells are descended from type II alveolar cells. A subset of Type II alveolar cells has also been shown to have the capacity for renewal [10, 11].

Importantly, we and others have demonstrated that EpiSPC can be isolated from the adult mouse distal lung by flow cytometry on the basis of their $EpCAM^{hi}$, $Sca\text{-}1^{low}$, $CD104^{pos}$, $CD49f^{pos}$, and $CD24^{low}$ expression [9, 12–15]. By culturing these cells in a three-dimensional colony-forming assay, we have shown that the adult mouse lung contains a renewing multipotent epithelial stem cell population with the capacity to differentiate into epithelial cells of the airway (club, ciliated and goblet cells), and alveolar (type I and type II) cell lineages.

In comparison, the development of robust assays for the identification and characterization of analogous stem/progenitor cell populations in the adult human distal lung lags far behind. Human studies are typically constrained by the limited availability of normal human lung tissue in sufficient quantities to permit isolation of rare cell subpopulations (*see* **Note 1**). The lack of concordance in cell separative strategies and cell culture assays in studies of regenerative cells in the human lung makes it impossible to establish the identity and relationship of human lung epithelial stem/progenitor cells described by various laboratories (*see* **Note 2**). Consequently, there is no consensus as yet about the immunophenotypic signature profile of human bronchiolar and bronchioalveolar stem/progenitor cells, or the relative merits of in vitro clonal assays used to measure their proliferative and differentiative potential.

The protocol described in this chapter has been developed for the isolation and characterization of adult mouse lung bronchiolar and bronchoalveolar stem/progenitor cells. Adaptations of this protocol may be useful for characterizing these cells in the future.

2 Materials

Prepare all solutions using ultrapure water (prepared by purifying deionized water to attain a sensitivity of 18 MΩ cm at 25 °C). Dispose of all waste materials as per environment, health, and safety regulations.

2.1 General Components

1. Hank's balanced salt solution (HBSS): 140 mg/L $CaCl_2$, 100 mg/L $MgCl_2{\cdot}6H_2O$, 100 mg/L $MgSO_4{\cdot}7H_2O$, 400 mg/L KCl, 60 mg/L KH_2PO_4, 350 mg/L $NaHCO_3$, 8,000 mg/L NaCl, 48 mg/L Na_2HPO_4, 1,000 mg/L d-Glucose, pH 7.4. Store at 4 °C (*see* **Note 3**).
2. Centrifuge: 400 × *g*, 4 °C

3. 15 and 50 mL sterile polypropylene tubes.
4. FACS buffer: HBSS, 0.2 % bovine serum albumin (BSA).

2.2 Tissue Dissociation Components

1. Dissecting equipment: forceps, scissors, and single-sided razor blade.
2. 30 mm petri dish.
3. 18 gauge and 21 gauge needles.
4. 20 mL syringes.
5. Liberase solution: For stock solution, reconstitute 50 mg Liberase TM Research Grade (Roche) in 10.4 mL sterile HBSS to make a stock solution at 25 Wunsch U/ml. Aliquot and store at −20 °C (*see* **Note 4**).
6. Red cell lysis buffer: 1,000 mg/L $KHCO_3$, 8,024 mg/L NH_4Cl, 37 mg/L EDTA, pH 7.4. Store at 4 °C.
7. Liberase wash buffer: HBSS, 5 % fetal bovine serum (FBS). Store at 4 °C.
8. 40 μm cell strainer (*see* **Note 5**).
9. Thermomixer (Eppendorf): 50 mL tube block (*see* **Note 6**).

2.3 Cell Depletion Components

1. Dynabeads Biotin Binder (Life Technologies): 4×10^8 beads/mL.
2. DynaMag-15 Magnet (Life Technologies): Compatible with 15 mL tubes.
3. Tube rotator.
4. Centrifuge: 400 rcf, 4 °C.
5. Depletion antibody cocktail: Biotinylated anti-mouse antibodies directed against CD31 (clone 390 or MEC13.3), CD45 (clone 30-F11), and TER119 (clone TER-119) antigens.

2.4 Flow Cytometry Components

1. Positive selection antibody cocktail: FITC-conjugated anti-mouse CD104 (clone 346-11A), Alexa Fluor 647-conjugated anti-mouse EpCAM (clone G8.8), Brilliant Violet 421-conjugated anti-mouse CD24 (clone M1/69), PECy7 Streptavidin (*see* **Note 7**).
2. Viability dye: Propidium Iodide (PI).
3. 5 mL (12 × 75 mm) FACS tubes.
4. 5 mL (12 × 75 mm) FACS tubes with 35 μm cell strainer cap.
5. Cell Sorter: 100 μm nozzle, 20 psi.
6. Collection Buffer: DMEM/F12 media, 10 % FBS. Store at 4 °C.

2.5 Epithelial Colony-Forming Unit Assay Components

1. Mlg 2908 mouse lung fibroblast cell line (ATCC CCL-206): Harvested during log-phase growth (*see* **Note 8**). Alternatively, freshly isolated primary EpCAM^{neg}, Sca-1^{pos} primary lung stromal cells can be used to support the growth of EpiSPC colonies (*see* **Note 9**).
2. Growth factor reduced Matrigel (Corning).
3. Millicell cell culture inserts (Merck Millipore): 0.4 μm pore size, hydrophilic PTFE, 30 mm diameter.
4. 6 well flat bottom tissue culture plate.
5. CFU-Epi medium: DMEM/F12, penicillin, streptomycin, glutamax (Gibco), insulin, transferrin, selenium, 10 % FBS, 2 μg/mL Heparin sodium salt (STEMCELL Technologies).
6. Tri-gas incubator: 5 % v/v O_2, 10 % v/v CO_2, 85 % v/v N_2 (*see* **Note 10**).

3 Methods

Carry out all procedures on ice (or at 4 °C) and in a sterile biological safety cabinet or a laminar-flow hood unless otherwise specified.

3.1 Dissociation of Mouse Lung Tissue

1. Dilute 40 μL of the Liberase stock solution in 4 mL of sterile HBSS for each mouse lung (*see* **Note 11**) in a 50 mL tube and preheat to 37 °C.
2. Exsanguinate deceased mouse (*see* **Note 12**) by severing the major arteries behind the intestines. Open the thoracic cavity and excise the lungs. Remove the extralobular airways and place the lung lobes in a 50 mL tube containing 30 mL of HBSS. Shake to wash out excess blood and transfer the lungs into a fresh 50 mL tube containing 30 mL of HBSS.
3. Transfer the lungs into a sterile petri dish, and finely mince the lungs using a single-sided razor blade. Transfer the minced tissue into a 50 mL tube and add 4 mL of the preheated Liberase solution per lung. Place the tube in the Thermomixer and agitate (750 rpm) at 37 °C for 30 min.
4. Triturate the sample with an 18-gauge needle attached to a 20 mL syringe until the tissue passes freely through the needle. Place in the Thermomixer and agitate (750 rpm) at 37 °C for a further 15 min (*see* **Note 13**).
5. Triturate the sample with a 21-gauge needle attached to a 20 mL syringe until the tissue passes freely through the needle. Strain the tissue digest through a 40 μm cell strainer into a 50 mL tube to remove tissue debris and cell clumps. Top up the tube to 50 mL with wash buffer and centrifuge at 400 × *g*, 4 °C for 5 min. Remove supernatant, resuspend the cell pellet

in 50 mL Liberase wash buffer and centrifuge at 400 × *g*, 4 °C for 5 min. Remove supernatant and resuspend the cell pellet in FACS buffer for cell counting.

6. Count the cells and calculate cell concentration (*see* **Note 14**).
7. Aliquot approximately 100,000 cells for each fluorescence compensation tube and unstained control tube for FACS setup (*see* **Note 15**).

3.2 Depletion of Hematopoietic and Endothelial Cells (*See* Note 16)

1. Resuspend the cell pellet in 50 mL FACS buffer and centrifuge at 400 × *g*, 4 °C for 5 min and discard the supernatant. Resuspend the cell pellet at 2.5×10^7 cells/mL in FACS buffer containing the depletion antibody cocktail: biotinylated anti-CD31 (1/250), biotinylated anti-CD45 (1/250) and biotinylated anti-TER119 (1/250). Incubate on ice for 20 min.
2. Prepare Dynabeads by mixing 5 μL of Dynabeads per 1×10^6 cells (*see* **Note 17**) with 10 mL FACS buffer in a 15 mL tube. Mix on the tube rotator at 4 °C for 5 min then load the tube on the magnet for 1 min. Discard the supernatant and resuspend the Dynabeads in 10 mL of FACS buffer, mix on the tube rotator at 4 °C for 5 min then load onto the magnet for 1 min. Discard the supernatant and store washed Dynabeads on ice.
3. Wash labeled cells to remove unbound antibody by resuspending in FACS buffer and centrifuge at 400 × *g*, 4 °C for 5 min. Discard the supernatant.
4. Resuspend the cell pellet in FACS buffer at 1×10^7 cells/mL and mix with washed Dynabeads. Incubate 4 °C for 20 min with gentle rotation on the tube rotator.
5. Load the tube on the magnet for 3 min. Transfer the supernatant containing the unbound cells to a 50 mL tube. Wash the bound Dynabeads with 10 mL FACS buffer and leave the tube on the magnet for 3 min. Collect the supernatant and combine with the other unbound cells. Top up tube to 50 mL with FACS buffer and centrifuge at 400 × *g*, 4 °C for 5 min. Resuspend cell pellet in FACS buffer for cell counting.
6. Count the cells and calculate cell concentration.
7. Aliquot 200,000 cells into 5 ml tubes for fluorescence minus one (FMO) controls. Store on ice.

3.3 Flow Cytometry

1. Resuspend cells in a 50 mL tube at 2.5×10^7 cells/mL in FACS buffer containing the positive selection antibody cocktail: FITC anti-CD104 (1/100), Alexa Fluor 647 anti-EpCAM (1/250), Brilliant Violet 421 anti-CD24 (1/100) and PECy7 Streptavidin (1/500). Incubate on ice in the dark for 20 min.
2. Top up the tube to 50 mL with FACS buffer and centrifuge at 400 × *g*, 4 °C for 5 min. Repeat this wash step two more times.

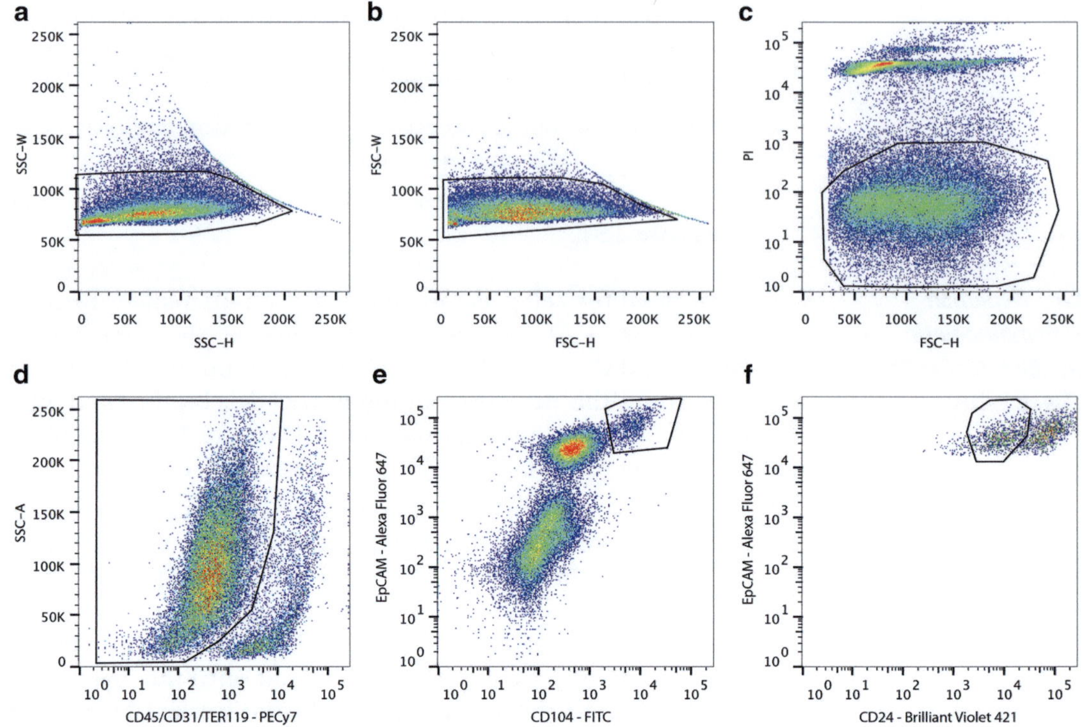

Fig. 1 Flow cytometry gating strategy for adult mouse lung EpiSPC. Setup sequential gates for (**a**) SSC-W doublet exclusion, (**b**) FSC-W doublet exclusion, (**c**) viability, (**d**) CD45^{neg} CD31^{neg} TER119^{neg}, (**e**) EpCAM^{pos} CD104^{pos}, and (**f**) CD24^{low} cells

3. Resuspend cells at 1×10^7 cells/mL in FACS buffer containing 1 μg/mL PI. A monodispersed cell suspension is prepared by pipetting the cells through the fitted cell strainer cap of a 5 mL FACS tube to ensure the elimination of cell aggregates.
4. Setup individual FMO control tubes for each antibody by staining cells with all antibodies except replace one of the antibodies in each tube with the relevant fluorochrome-matched isotype control antibody (*see* **Note 18**).
5. Set up flow cytometer with 100 μm nozzle at 20–30 psi. Set voltages and fluorescence compensation using unstained and single color controls.
6. Isolate epithelial stem/progenitor cells by setting up sequential gates for selection of single (SSC-W vs. SSC-H, FSC-W vs. FSC-H), viable (PI vs. FSC-H), nonhematopoietic, nonendothelial (PECy7 vs. SSC-H), EpCAM^{pos} CD104^{pos} (Alexa Fluor 647 vs. FITC), CD24^{low} (Alexa Fluor 647 vs. Brilliant Violet 421) cells as shown in Fig. 1.
7. Collect cells in 5 mL collection tubes containing 1 mL collection buffer.

3.4 Cell Culture

1. Centrifuge the sorted cells at 400 × *g*, 4 °C for 5 min. Discard the supernatant and resuspend the cell pellet in 1 mL chilled CFU-Epi medium. Store on ice.
2. Take a small aliquot for cell counting and calculate the cell concentration.
3. Mix sorted epithelial stem/progenitor cells with Mlg 2908 cells so that the final cell concentrations are at 2×10^4 cells/mL and 2×10^6 cells/mL, respectively. Centrifuge at 400 × *g*, 4 °C for 5 min.
4. Resuspend cell pellet in chilled Matrigel diluted at 1:1 ratio with CFU-Epi medium so that the final concentration of epithelial stem/progenitor cells is 2×10^4 cells/mL, and Mlg 2908 cells is 2×10^6 cells/mL. Mix the Matrigel cell suspension (*see* **Note 19**).
5. Place 30 mm Millicell inserts in a 6 well culture plate.
6. Add 3 × 25 μL drops of Matrigel cell suspension atop of the filter membrane of a Millicell insert. Allow maximum distance between drops to avoid amalgamation of drops. Incubate cultures at 37 °C for 5 min to allow Matrigel to set.
7. Add 1200 μl of CFU-Epi medium around the insert in each well.
8. Incubate cultures at 37 °C, 5 % O_2, 10 % CO_2, 85 % N_2 and change media three times weekly.
9. Score colonies using a stereo-microscope under phase contrast as shown in Fig. 2 (*see* **Note 20**).

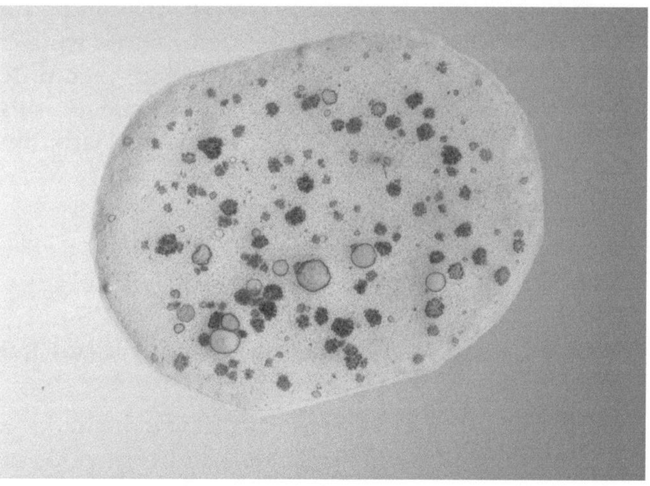

Fig. 2 Representative phase contrast image of CFU-Epi. Day 14 image (×5 magnification) of EpiSPC grown in 3-D matrigel coculture with supporting Mlg 2098 cells

4 Notes

1. Resected normal human lung tissue is usually obtained from patients undergoing diagnostic or surgical procedures for a variety of disease indications as well as lungs deemed unsuitable for transplantation. Factors which confound the development of robust assays for the identification, characterization, and quantitation of human airway and alveolar stem/progenitor cells include the age of the donor, the stage of lung disease progression, the patients' medical condition and previous treatment, and their smoking status. Reproducible sampling of lung tissue at precise regions of the airways along the proximal-distal lung axis is often also beyond the investigator's control further confounding assay reproducibility and the interpretation of assay readouts.
2. To date human studies have focused on the characterization of basal cells of the trachea and proximal airways, with only a few studies looking at EpiSPC in the human distal lung. Kajstura et al. propagated unfractionated human lung cells in liquid culture prior to sorting c-kit^{pos} cells from this expanded cell population which appeared to regenerate bronchiolar and alveolar epithelium as well as vasculature when injected into an injured mouse lung [16]. However, the validity of this study has been questioned by experts in the field and must therefore be considered with great caution. On the other hand, Oeztuerk-Winder et al., have described an E-cadherin ^{pos}Lrg6^{pos} lung stem cell cohort able to regenerate damaged bronchioalveolar epithelium following bleomycin-induced lung injury in mice, as well as regenerate bronchioalveolar tissue when transplanted under the kidney capsule whereas c-kit^{pos} cells isolated in this study were not able to do so [17]. These studies have all used different cell culture media, different medium supplements, different cytokine cocktails, and different cell culture conditions to characterize the proliferative and differentiative potential of candidate stem/progenitor cell populations.
3. Dulbecco's Phosphate-Buffered Saline (DPBS) (Life Technologies) can be used as a substitute for HBSS: 100 mg/L $CaCl_2$, 100 mg/L $MgCl_2{\cdot}6H_2O$, 200 mg/L KCl, 200 mg/L KH_2PO_4, 8,000 mg/L NaCl, 2,160 mg/L $Na_2HPO_4{\cdot}7H_2O$, pH 7.4.
4. Liberase TM Research Grade (Roche) contains highly purified Collagenase I and II blended in a precise ratio with a medium concentration of Thermolysin. We use this enzyme mix because the high purity provides higher lot-to-lot consistency. However, other sources of Collagenase are equally effective for tissue dissociation using this protocol.

5. As a cheaper alternative to cell strainers, sterilized 40–80 μm Nylon mesh can be used to strain cells.
6. Any apparatus that maintains a constant 37 °C with agitation can be used. Agitation is important for the penetration of Liberase into the tissue fragments.
7. Different fluorochrome-conjugates can be used. The optimal combination of fluorochromes will depend on the laser and filter configuration of the flow cytometer.
8. The Mlg 2908 cell line is maintained as a monolayer culture in DMEM/F12, penicillin, streptomycin, glutamax (Gibco), 10 % FBS. To obtain cells in log-phase growth we typically seed 5×10^4 cells in a T75 tissue culture flask and harvest with 0.25 % Trypsin 3 days later. However, the growth kinetics of Mlg 2908 cells will vary with cell batch, passage number, and incubation conditions. Investigators should therefore carefully monitor the growth properties of the Mlg 2908 cell line to ensure that cells are routinely harvested in log-phase growth to provide optimal support for EpiSC colony growth.
9. Primary lung stromal cells can be isolated by flow cytometry from the mouse lung based on their $CD45^{neg}$ $CD31^{neg}$ $EpCAM^{neg}$ $Sca\text{-}1^{pos}$ immunophenotypic signature profile, and are seeded directly in the CFU-Epi assay as an alternative to the Mlg 2908 cell line as previously described [13, 18]. These cells must be used fresh and seeded at 2×10^6 cells/mL in this assay. They can also be expanded in monolayer culture in DMEM/F12, penicillin, streptomycin, glutamax (Gibco), 10 % FBS but require the addition of the TGF-β inhibitor SB431542 (10 μM) to maintain their epithelial supportive activity [19].
10. We use a tri-gas incubator set at low oxygen tension (5 % v/v O_2, 10 % v/v CO_2, 85 % v/v N_2) which has been shown to be optimal for the growth of stem/progenitor cells at clonal density in vitro [20]. However, CFU-Epi can be grown under standard oxygen tension (10 % v/v CO_2 in air) but the cloning efficiency may be lower.
11. Up to five lungs can be pooled per 50 mL tube for tissue dissociation. We have found that tissue dissociation efficiency is compromised if more than five lungs per tube are processed.
12. Mice should be killed in accordance with ethical codes of practice governing animal experimentation at individual institutes, and only as approved by institutional animal ethics committees.
13. When tissue digestion is complete no chunks of pink lung tissue should be seen. However, clumps of extracellular matrix will be visible as white strands in the suspension.
14. A Sysmex KX-21N, or similar automated cell counter can be used to count cells. Alternatively a hemocytometer can be used with Trypan Blue to exclude nonviable cells.

15. To avoid unnecessary use of valuable samples, this step can be excluded and CompBeads (BD) can be used for fluorescence compensation.
16. We routinely use Dynabeads to deplete contaminating hematopoietic and endothelial cells by negative immunomagnetic selection prior to cell sorting. Alternatively, cells can be labeled with fluorochromes-conjugated anti-CD45 and anti-CD31 antibodies for excluding hematopoietic and endothelial cells flow cytometrically in the sort gating strategy. However, because $CD45^{pos}$ and $CD31^{pos}$ cells are major contaminants, this will significantly extend the cell sorter time required to isolate EpiSPC.
17. The Dynabead to cell ratio used in this protocol is two beads per cell. To improve depletion, this step may be repeated with half the volume of Dynabeads (one bead per cell).
18. FMO controls should be used to set gates for positive events. The level of background fluorescence (nonspecific staining) is determined by FACS analysis of cells labeled with relevant isotype control antibodies.
19. Matrigel must be kept on ice as it will begin to gel at slightly elevated temperatures. It is best to use pre-cooled pipettes, tips, and tubes when preparing Matrigel for use. It is also important to avoid creating bubbles when mixing Matrigel. This can be achieved by holding the tube in the centre of a vortex to create a swirling motion.
20. Cultures can be scored in real-time or at the end of the culture period. Colonies typically emerge after 5 days and can be maintained for up to 3 weeks.

Acknowledgements

This work was supported by grants from the Australian National Health and Medical Research Council (NHMRC) and the Australian Stem Cell Centre.

References

1. Rawlins EL, Okubo T, Xue Y et al (2009) The role of Scgb1a1+ Clara cells in the long-term maintenance and repair of lung airway, but not alveolar, epithelium. Cell Stem Cell 4:525–534
2. Hong KU, Reynolds SD, Watkins S et al (2004) In vivo differentiation potential of tracheal basal cells: evidence for multipotent and unipotent subpopulations. Am J Physiol Lung Cell Mol Physiol 286:L643–L649
3. Rock JR, Onaitis MW, Rawlins EL et al (2009) Basal cells as stem cells of the mouse trachea and human airway epithelium. Proc Natl Acad Sci U S A 106:12771–12775
4. Hegab AE, Ha VL, Gilbert JL et al (2011) Novel stem/progenitor cell population from murine tracheal submucosal gland ducts with multipotent regenerative potential. Stem Cells 29:1283–1293

5. Tesei A, Zoli W, Arienti C et al (2009) Isolation of stem/progenitor cells from normal lung tissue of adult humans. Cell Prolif 42:298–308
6. Hong KU, Reynolds SD, Watkins S et al (2004) Basal cells are a multipotent progenitor capable of renewing the bronchial epithelium. Am J Pathol 164:577–588
7. Giangreco A, Reynolds SD, Stripp BR (2002) Terminal bronchioles harbor a unique airway stem cell population that localizes to the bronchoalveolar duct junction. Am J Pathol 161: 173–182
8. Reynolds SD, Giangreco A, Power JH et al (2000) Neuroepithelial bodies of pulmonary airways serve as a reservoir of progenitor cells capable of epithelial regeneration. Am J Pathol 156:269–278
9. Kim CF, Jackson EL, Woolfenden AE et al (2005) Identification of bronchioalveolar stem cells in normal lung and lung cancer. Cell 121: 823–835
10. Chapman HA, Li X, Alexander JP et al (2011) Integrin alpha6beta4 identifies an adult distal lung epithelial population with regenerative potential in mice. J Clin Invest 121:2855–2862
11. Barkauskas CE, Cronce MJ, Rackley CR et al (2013) Type 2 alveolar cells are stem cells in adult lung. J Clin Invest 123:3025–3036
12. Teisanu RM, Chen H, Matsumoto K et al (2011) Functional analysis of two distinct bronchiolar progenitors during lung injury and repair. Am J Respir Cell Mol Biol 44:794–803
13. McQualter JL, Yuen K, Williams B et al (2010) Evidence of an epithelial stem/progenitor cell hierarchy in the adult mouse lung. Proc Natl Acad Sci U S A 107:1414–1419
14. Chen H, Matsumoto K, Brockway BL et al (2012) Airway epithelial progenitors are region specific and show differential responses to bleomycin-induced lung injury. Stem Cells 30: 1948–1960
15. Lee JH, Kim J, Gludish D et al (2012) SPC H2B-GFP mice reveal heterogeneity of surfactant protein C-expressing lung cells. Am J Respir Cell Mol Biol 48:288–298
16. Kajstura J, Rota M, Hall SR et al (2011) Evidence for human lung stem cells. N Engl J Med 364:1795–1806
17. Oeztuerk-Winder F, Guinot A, Ochalek A et al (2012) Regulation of human lung alveolar multipotent cells by a novel p38alpha MAPK/ miR-17-92 axis. EMBO J 31:3431–3441
18. McQualter JL, Brouard N, Williams B et al (2009) Endogenous fibroblastic progenitor cells in the adult mouse lung are highly enriched in the sca-1 positive cell fraction. Stem Cells 27:623–633
19. McQualter JL, McCarty RC, Van der Velden J et al (2013) TGF-beta signaling in stromal cells acts upstream of FGF-10 to regulate epithelial stem cell growth in the adult lung. Stem Cell Res 11:1222–1233
20. Wion D, Christen T, Barbier EL et al (2009) PO(2) matters in stem cell culture. Cell Stem Cell 5:242–243

Chapter 18

Culture and Characterization of Mammary Cancer Stem Cells in Mammospheres

Eleonora Piscitelli, Cinzia Cocola, Frank Rüdiger Thaden, Paride Pelucchi, Brian Gray, Giovanni Bertalot, Alberto Albertini, Rolland Reinbold, and Ileana Zucchi

Abstract

Mammospheres (MMs) are a model for culturing and maintaining mammary gland stem cells (SCs) or cancer stem cells (CSCs) ex situ. As MMs recapitulate the micro-niche of the mammary gland or a tumor, MMs are a model for studying the properties of SCs or CSCs, and for mapping, isolating, and characterizing the SC/CSC generated lineages. Cancer stem cells share with normal SCs the properties of self-renewal and the capacity to generate all cell types and organ structures of the mammary gland. Analysis of human tumor samples suggests that CSCs are heterogeneous in terms of proliferation and differentiation potential. Mammospheres from CSCs likewise display heterogeneity. This heterogeneity makes analysis of CSC generated MMs challenging. To identify the unique and diverse properties of MM derived CSCs, comparative analysis with MMs obtained from normal SCs is required. Here we present protocols for identifying and enriching cells with SC features from a cancer cell line using the LA7CSCs as a model. A comprehensive and comparative approach for identifying, isolating, and characterizing MMs from SCs and CSCs from human breast is also introduced. In addition, we describe detailed procedures for identifying, isolating, and characterizing mammary gland specific cell types, generated during MM formation.

Key words Stem cells, Cancer stem cells, Mammospheres, CellVue Maroon dye, ALDEFLUOR, LA7CSCs, Tubules and cysts

1 Introduction

Considerable and convincing research supports that a rare population of cells from tumors has overlapping features with stem cells (SCs) [1]. These cancer stem cells (CSCs) have the capacity to initiate heterogeneous tumors and invasion and metastatic potential [1]. Cancer stem cells, like normal SCs generate a daughter SC and diverse cell types that are in different stages of differentiation and with different proliferation rates in a tumor mass or in a mammosphere (MM) [2]. Evidence that MMs are generated by self-renewing SCs or CSCs and can regenerate Evidence that MMs are

Ivan N. Rich (ed.), *Stem Cell Protocols*, Methods in Molecular Biology, vol. 1235,
DOI 10.1007/978-1-4939-1785-3_18, © Springer Science+Business Media New York 2015

generated by self-renewing SCs or CSCs and can regenerate serially suggests that SCs or CSCs could be maintained in vitro suggested that SCs or CSCs could be maintained in vitro [3]. Recent protocols for identifying and isolating SCs from normal breast tissue have provided new opportunities to identify features unique to CSCs by comparative analysis [4–6].

Evidence supports that cancer lines, such as the LA7CSCs, may also contain a rare population of cells with features of stem cells and can be maintained in adherent or in non-adherent culture conditions. Evidence that CSCs can also be maintained in non-adherent cultures was shown with cancer cell lines, such as LA7CSCs. Single LA7CSCs recapitulate all the cell types and 3D structural architecture of the mammary gland in vitro and form heterogeneous tumors when injected into NOD/scid mice [7, 8]. An advantage of using a CSC line is that factors that influence CSC self-renewal, differentiation, and the dynamics of tumor formation can be studied at the single stem cell level [8]. Our experience suggests that passaging established CSC lines, often results in the loss of the CSCs due, in part, by not maintaining and routinely enriching the population of cells with SC features.

The first part of the chapter is a protocol for enrichment and routine maintenance of CSC lines based on serial regeneration followed by MM isolation and validating that the lines exhibit SC properties. Protocols are then described for the comparative analysis of human MMs from SCs or CSCs based on differences in dye dilution and ALDH activity [5, 6].

Using fluorescent dye retention and the ALDEFLUOR assay allows for the identification of distinct cell types from MMs: cells that have SC features and MM regeneration capacity, luminal cells that have highest ALDH activity and cyst formation potential, and cells with myoepithelial properties. The cysts generated by luminal cells recapitulate the alveolar structures of the mammary gland and often are indistinguishable from MMs, especially when cysts are in their early phase of formation. The acinar structure forming cells have a more homogeneous rate of dye dilution and highest ALDH activity compared to other cells from SC or CSC generated MMs. Dye dilution and ALDH enzymatic activity in MMs from CSCs is variable between tumor samples, supporting the notion that CSCs are heterogeneous in proliferation and differentiation potential.

2 Materials

2.1 Cell Culture Reagents

1. Dulbecco's Modified Eagle's Medium (DMEM, Life Technologies).
2. Ham's F12 (Life Technologies).
3. DMEM-GlutaMAX (Life Technologies).

4. Serum Replacement (SR) (Life Technologies).
5. Human recombinant Insulin (Sigma).
6. Hydrocortisone (Sigma).
7. Nonessential amino acids, minimal essential medium (NEA-MEM) (Life Technologies).
8. B27 supplement without Vitamin A (Life Technologies).
9. Human recombinant Epidermal Growth Factor (EGF, Sigma).
10. Human recombinant Fibroblast Growth Factor-basic (bFGF) (ImmunoTools).
11. Heparin (Sigma).
12. 0.05 % Trypsin-EDTA (1×) (Life Technologies).
13. Fetal Bovine Serum (FBS) (Sigma).
14. HEPES (Sigma).
15. Collagenase from *Clostridium histolyticum* (Sigma).
16. Hyaluronidase (Sigma).
17. 2-Mercaptoethanol (Life Technologies).
18. Mitomycin-C (Sigma).
19. Dimethyl sulfoxide (DMSO) (Sigma).

2.2 Differentiation Matrix

1. Rat-tail collagen.
2. Growth Factor Reduced Matrigel (Becton Dickinson).
3. Triiodothyronine (Sigma),
4. Apo-transferrin (Sigma),
5. Cyclic adenosine monophosphate (c-AMP) (Sigma).

2.3 Cell Membrane Labeling

1. CellVue® Maroon Cell Labeling Kit (MTTI).

2.4 ALDH Activity

1. ALDEFLUOR kit (Stem Cell Technologies).

2.5 Antibodies

1. E-cadherin (Santa Cruz).
2. CD24 (Santa Cruz).
3. Cytokeratin 18 (Sigma).
4. Secondary antibodies (Invitrogen Molecular Probes): Alexa Fluor 594-conjugated rabbit anti-goat IgG (H+L); Alexa Fluor 488-conjugated goat anti-mouse IgG (H+L); Alexa Fluor 555-conjugated goat anti-mouse IgG (H+L).

2.6 Equipment and Tissue Culture Plates

1. Low-attachment plates (Thermo Fisher).
2. Adherent multi-well plates (VWR-PBI).

3. 40 and 100 μm cell strainers (Thermo Fisher).
4. MoFlO XDP cell sorter (Beckman Coulter).
5. Tissue culture incubator.

2.7 Preparation of Tissue Culture Reagents

Medium 1 for culturing of MMs from LA7CSCs: DMEM-GlutaMAX supplemented with 0.5×B27 without vitamin-A, 50 ng/mL recombinant insulin, 10 ng/mL EGF, 10 ng/mL bFGF, and 4 μg/mL heparin.

Medium 2 for MM or adherent cell culture dissociation: 0.05 % Trypsin-EDTA (1×). The reaction is stopped with sera containing medium.

Medium 3 for adherent culturing of LA7CSCs: DMEM containing 5 % FBS supplemented with 150 U/mL collagenase and 50 U/mL hyaluronidase.

Medium 4 for human tissue enzymatic digestion: DMEM containing 5 % FBS supplemented with 150 U/mL collagenase and 50 U/mL hyaluronidase.

Medium 5 for freezing dissociated cells: 90 % FBS and 10 % DMSO.

Medium 6 for culturing MMs from human SCs or CSCs: Ham's F12/DMEM-GlutaMAX (1:1) containing 10 % SR, 1× NEA-MEM, 1 μg/mL insulin, 0.5 μg/mL hydrocortisone, 10 ng/mL EGF, and 4 ng/mL bFGF.

Medium 7 for Matrigel cultures: Ham's F12/DMEM-GlutaMAX (1:1), 10 % SR, 1 % NEA-MEM, 1 μg/mL insulin, 0.5 μg/mL hydrocortisone, 100 ng/mL EGF, 10 nM triiodothyronine , 25 ng/mL apotransferin, 10 nM cAMP, 0.1 nM estradiol, 4 ng/mL bFGF, with half of the medium replaced with fresh media every 2–3 days.

Unless otherwise specified, all the media are pre-warmed to 37 °C.

3 Methods

3.1 Guidelines for Maintaining and Enrichment of Cells with SC Features from Mammary Cancer Cell Lines

Certain cancer cell lines contain cells with CSC properties, especially when recently derived from a tumor or when a deliberate effort is made to maintain in these lines cells with stem cell features. Maintenance and enrichment of a population of cells with stem features are based on selecting SCs by their functional properties rather then expression of candidate SC markers. For instance, LA7 or cancer cell lines from a commercial distributor may require that cells with CSC features are enriched and then subcloned to generate a CSC line. While CSC lines can be passaged indefinitely, the number of CSCs decreases with passaging and may even disappear. Once the cells with SC features are lost in the cell line, they cannot be reestablished by propagation. The protocol for enrichment and routine maintenance of a CSC line is based on

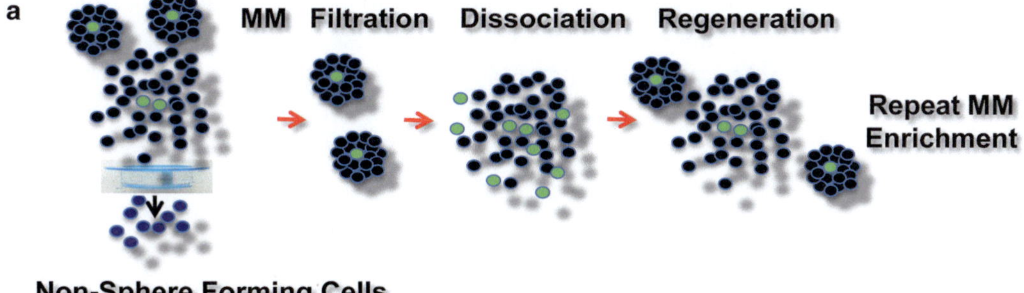

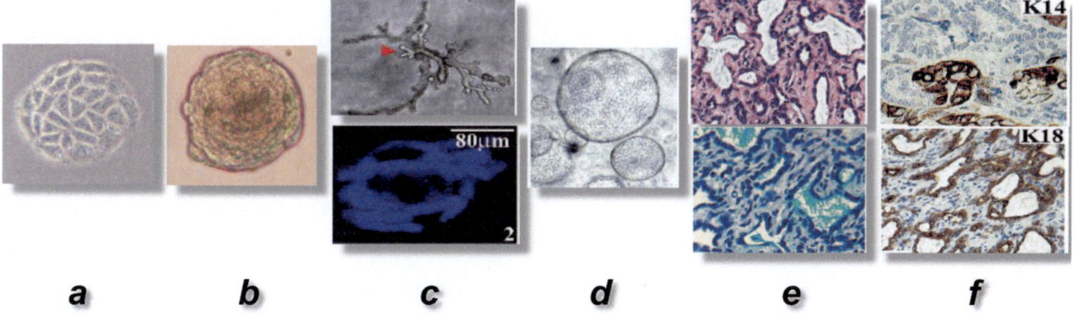

Fig. 1 Generation of MMs from the LA7CSC line. (**a**) Scheme for enrichment of CSCs and routine maintenance of a CSC line, (**b**) A single LA7CSC generates holoclones in adherent conditions (a "in italic"), MMs in suspension (**b**), tubules (**c**, upper panel phase contrast and lower panel confocal of Hoechst nuclear dye 33342 stained cells showing lumen of a tubule), and acini (**d**), and a heterogeneous tumor with tubules (f, upper panel H&E) with secretion capacity (f, bottom panel alcian blue staining), and differentiated cells expressing cytokeratin 14 (**e**, upper panel) and cytokeratin 18 (**e**, lower panel) (Magnifications 10× (a), 4× (b), 20× (c), 20× (d), 40× upper and 20× lower (e), 20× upper and 40 lower (Images a, c, d, e and f are Copyright (2007 and 2008) National Academy of Sciences, U.S.A. [4, 6])

serially regenerating and collecting MMs by filtration and is schematically represented in Fig. 1a.

3.1.1 Generation of MMs from LA7CSC or Other CSC Lines

While rat LA7CSCs are used as an example for generating MMs from a cancer cell line, the protocols presented are also suitable for other cancer cell lines containing CSCs.

1. LA7CSCs are plated in low-attachment plates in *Medium 1* at a density of 500 cells per 2.5 mL of media or as a single cell/well in a 96-well plate.
2. MMs are regenerated every 7 days by collecting them with a cell strainer with a pore size of 40 μm to remove single cells or small cell aggregates.
3. MMs are enzymatically dissociated into single cells with *Medium 2* for approximately 10 min. Efficiency of MM dissociation is monitored with a microscope.

It is essential that MMs are generated from single cells as cell aggregates result in the eventual loss of the CSC pool. LA7CSC MMs can be regenerated for at least 60 passages, with each passage being 7 days [7–9]. The stem cell properties of LA7CSCs can also be maintained indefinitely under adherent culture conditions [7, 9].

3.1.2 Propagation of LA7CSC Line as Adherent Cultures

1. Self-renewing LA7CSCs are propagated as adherent cultures in *Medium 3* at a density of 1,000 cells per cm^2.
2. Cells are enzymatically dissociated in *Medium 2* for approximately 5 min to obtain single cells and re-plated at the same density every 2–3 days. Re-passaging time should be strictly maintained as cell confluence or formation of large colonies results in the gradual loss of the CSC line.

3.1.3 Evaluation of Cancer Cell Lines for CSC Properties

Single cells from established CSC lines are evaluated for their capacity to form holoclones in adherent cultures (Fig. 1b panel a), MMs in suspension (panel b), and for their ability to produce a 3D branching morphogenesis in vitro (tubules, panel c; cysts, panel d). In addition, they exhibit the capacity to differentiate into all mammary gland cell types in vitro and in tumors in NOD/scid mice (Fig. 1b, panels e and f).

1. Tubules and cysts, representative of ductal and alveolar structures, respectively, are generated in 10–14 days when LA7CSCs or dissociated MM cells are seeded at a concentration of 300 cells in 500 μL of collagen (isolated from rat-tail collagen) in a single well of a 24-well plate.
2. After collagen solidification, the cells are fed with *Medium 3* and maintained by replacing half of the media every 2–3 days.

3.2 Isolation of SCs and CSCs from Normal or Tumor Human Breast for MM Cultures

As it is desirable to compare the functional properties and gene expression profiles of normal SCs and CSCs, and since normal tissue is often available from cancer patients, methods for the isolation of both mammary gland normal SCs and CSCs are presented. The following protocols describe the isolation of both mammary gland normal SCs and CSCs. Large surgical material from normal breast can also be obtained by mammoplastic reduction procedures (Fig. 2a).

3.2.1 Preparation of Normal or Tumor Human Breast Tissue

1. Surgical material is placed into a container with 100 % FBS. Containers with FBS can be prepared ahead of time and left at room temperature (RT) at the surgical theater as active proteins such as growth factors in the sera are not essential at this step. The tissue filled containers should have as little "empty air space" as possible. Sterility is not essential at this point since the tissue needs to be sterilized before being cut.
2. Large tissue is cut into smaller pieces to allow for sterilization in small batches.

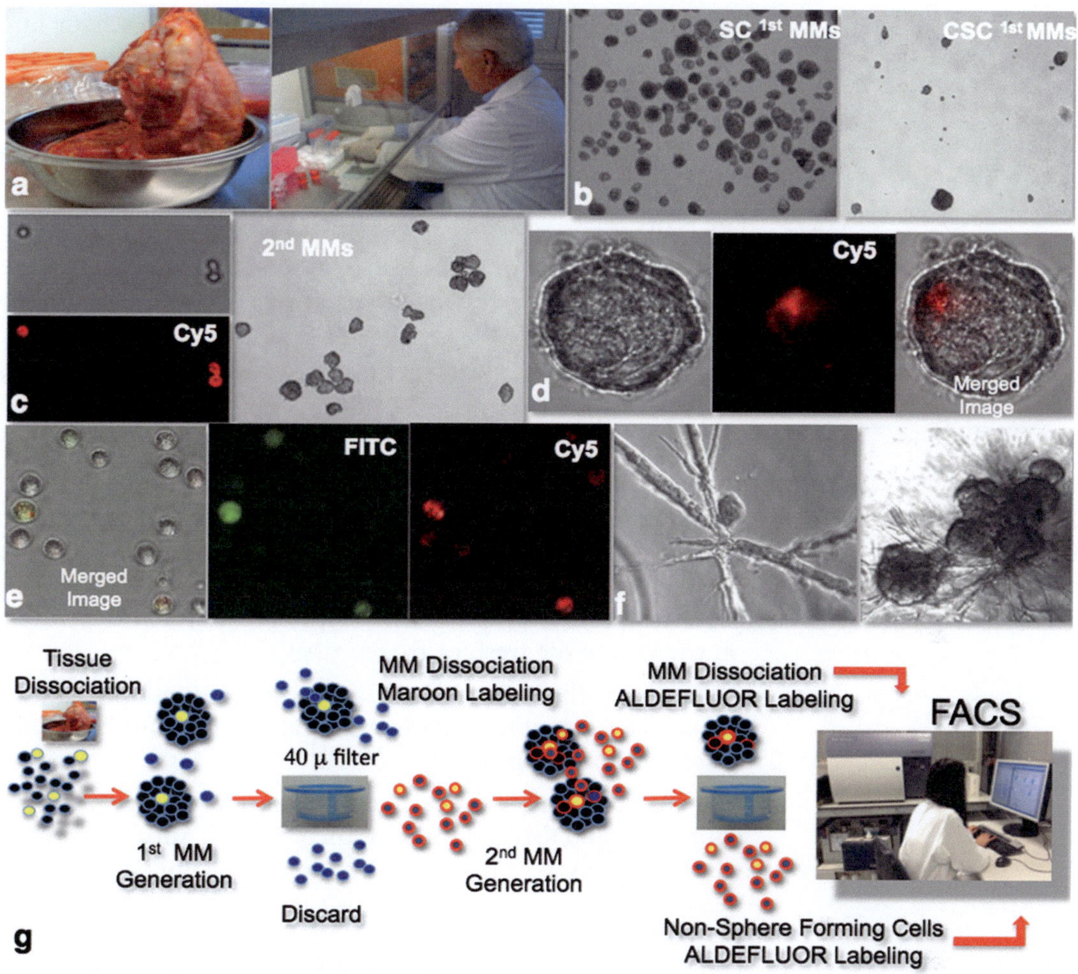

Fig. 2 Generation of MMs from human breast SCs and CSCs. (**a**) Tissue preparation and enzymatic dissociation. (**b**) 1st generation MMs from SCs (*left*) and CSCs (*right*). (**c**) CellVue Maroon (Cy5 filter) stained single cells from 1st generation dissociated MMs (*left*) and 2nd generation MMs formed from stained single cells (*right*). (**d**) Maroon dye dilution in 2nd MMs. (**e**) ALDEFLUOR staining (FITC filter) of 2nd generation dissociated MMs generated from Maroon labeled SCs. (**f**) Tubular (*left*) and alveolar (*right*) 3D structures generated from MMs formed from single SCs. (**g**) Scheme for labeling and collecting SCs and non-sphere forming cells from MMs (Magnifications: 4× (b), 20× (*left*) and 4× (*right*) (c), 20× (d), 40× (e), 20× (f)

3. The tissue is dipped into 80 % ethanol (EtOH) for a few seconds, washed in sterile PBS and then placed into the cutting solution. It is important that the tissue is not cut too small or punctured before sterilization (*see* **Note 1**).
4. Cutting solution consists of 100 % FBS if the material is to be stored frozen. If the cut tissue is to be used immediately for preparing single cells for MM formation then cutting should be performed with *Medium 4* (*see* Subheading 3.3).

5. For freezing sterilized tissue, cut the tissue with a scissor into sizes of 1–2 mm^3. Cutting should be performed quickly in a small diameter (2.5 cm) but long tube (12 cm) with a small medium volume. Do not cut the tissue in an open plate with a scalpel, since this will cause excess aeration of the cells. Expect variability in tissue density and firmness from sample to sample (*see* **Note 2**).
6. Just before freezing, adjust FBS to at least 5 tissue volumes and add DMSO to a final concentration of 10 % and mix quickly and well.
7. Freezing times (at –80 °C) should be controlled by use of thick wall styrofoam or isopropanol freezing containers.
8. Long-term storage should be in liquid nitrogen.

3.2.2 Collagenase Digestion of Normal or Tumor Tissue

1. The protocol for single cell preparation is performed in two steps over 2 days, but can be performed in one working day (see below). Steps for day 1 are performed with all solutions pre-warmed to 37 °C and take approximately 1 h depending on the size of the fresh or frozen tissue.
2. If starting with frozen material, the sample should be thawed as quickly as possible in a water bath at 37 °C. If fresh material is used, skip to **step 5**.
3. The thawed tissue is transferred into a new sterile tube containing fresh DMEM not less then 5× the tube volume in which the cells were frozen in. The freezing solution and tissue mass should be diluted as fast as possible when the cells are thawed. It is easier and faster to estimate the freezing tube volume then the actual volume of freezing solution and tissue mass. DMEM is used instead of PBS to monitor pH conditions throughout all processing steps.
4. After transfer, seal and place the tube into a water bath and induce a gentle rocking motion for 15 min, allowing for the DMSO of the freezing solution to diffuse out of the cells. (DMSO is a potent differentiating agent for SCs/CSCs.)
5. The tissue is then poured onto a strainer with a sieve pore size small enough not to allow the tissue to flow through and washed three times on the sieve with 5 tissue volumes of DMEM.
6. The tissue is cut into smaller pieces to allow the enzymes for digestion to penetrate. Appropriate cut size will determine whether single SCs or CSCs will be obtained for generating MMs. The cut size should be small enough so that cells after enzymatic digestion can exit the fibrous tissue mesh with as little mechanical disruption of the tissue as possible.
7. The tissue is enzymatically dissociated in *Medium 4* overnight in non-adherent plates.

To obtain single cells in one working day, dissociation is performed in 4–6 h with increase in the amount/concentration of enzymes.

3.2.3 Extraction of Single Cells from Enzymatically Digested Normal or Tumor Human Breast Tissue

The step for preparing single cells requires about 2 h. Generating good MMs depends on forming spheres from single SCs or CSCs, obtained after complete tissue digestion and good cell filtration.

1. The digested tissue from Subheading 3.2.2 is pipetted into a long tube with a narrow diameter tube (to limit excess aeration of the cells) and the cells are gently extracted from the tissue with a slow up and down pipetting motion with a serological pipette. The pipetting is repeated at least 40 times.
2. First Filtration step to remove fibers: tissue debris and fibers are strained away from single cells using a large sieve with a large (100 μm) pore size to prevent clogging. The flow through which contains single cells, SCs or CSCs is centrifuged at 200 × *g* in 15 or 50 mL tubes at 16 °C for 20–40 min depending on the amount of digested tissue. Long centrifugation time is required for normal tissue due the presence of fibers and lipids. For centrifugation, use long tubes to obtain a greater separation of cells from the tissue fibers. The tubes should be filled completely with media with little air space inside as possible to minimize aeration of the cells. The supernatant is discarded and the pellet is gently resuspended with 10 mL of DMEM at RT.
3. Second Filtration step to collect single cells from cell aggregates: if the collagenase digestion was performed adequately, the supernatant is relatively easy to filter to obtain single cells. Filtration should be performed gently using large volumes of DMEM and large mesh surface areas. It is often necessary to perform serial steps of filtration with decreasing mesh pore size strainers, from 100 to 40 μm.
4. After filtration the cells are centrifuged at 200 × *g* for 10 min and resuspended with *Medium 5* for freezing or *Medium 6* for generating MMs (*see* Subheading 3.3 and **Notes 3–6**). A large collection of frozen single cells can be prepared at this step for future use.

3.3 Generation of MM from Normal or Cancer Tissue from Human Breast

A schematic representation of the complete procedure is illustrated in Fig. 2g.

1. Single SCs or CSCs freshly derived or from frozen stocks are resuspended in *Medium 6*. Cells from frozen stocks must first be washed to remove the DMSO by resuspending the frozen stocks in 5 volumes of fresh DMEM.
2. Cells are gently mixed and centrifuged at 200 × *g* for 8 min, and resuspended in *Medium 6*.

3. Cells are seeded in low-attachment plates at a clonogenic density of 500 cells/mL in 15 cm diameter plate (*see* **Note 7**).
4. MMs are formed after 5–7 days and take approximately the same time to form if generated from normal SCs or CSCs.
5. MM dissociation to obtain single cells is performed with *Medium 2*. The dissociation time is 20 min or less. While dissociation of MMs into single cells is the ideal wanted outcome, in reality this may result in cell damage and lower yield of the number of SCs or CSCs. The dissociation time should be therefore optimized. MMs are dissociated by intermittent pipetting and returning them back to the incubator.

3.4 Comparative Analysis of MMs Generated from Normal SCs and CSCs

First generation MMs obtained from a pool of single cells derived from human normal or tumor breast tissue are usually interspersed with a heterogeneous population of 3D structures of various sizes and morphologies (Fig. 2b). This heterogeneity may represent 3D structures derived from cell aggregates, or cells at various stages of differentiation and with different proliferation capacities. Regeneration of MMs from cells derived from 1st generation MMs display significantly greater structural and proliferative homogeneity (Fig. 2c, right panel and Fig. 6a, left panel). Therefore, SCs or CSCs are enriched not directly from single cells of dissociated tissue, but from 1st generation dissociated MMs. A comprehensive approach for comparative analysis of MMs from SCs and CSCs from human breast is described below. The analysis is based on Maroon labeling of SCs and CSCs and performing the ALDEFLUOR assay on dissociated MMs.

3.4.1 Maroon Dye Labeling of 1st Generation MM Dissociated Cells

The CellVue Maroon fluorescent dye has ideal properties for isolating, tracking, and lineage determination of SCs. Additionally, Maroon is used for characterization of MMs generated from SCs or CSCs. Maroon is a very stable and robust membrane labeling dye and does not interfere with the commonly used FITC (fluorescein isothiocyanate) fluorochrome or with green fluorescent protein (GFP) detection. As a membrane labeling dye, dilution or retention of Maroon is a measure of the rate of proliferation of a cell and its descendant cells. Maroon analysis and fluorescence-activated cell cytometry profiling during SC self-renewal and differentiation provides insight into cell division cycles of specific cell types generated by a SCs or CSCs. As SCs have a low proliferation rate [6], SCs display the highest Maroon cell dye retention during MM formation (Fig. 2d; Fig. 3a, region R5 from Fluorescence Activated Cell Sorting (FACS) image.

1. First generation MMs are dissociated into single cells in *Medium 2* (*see* Subheading 3.3, **step 5**).
2. Single cell suspensions are labeled with 2 μM of the CellVue Maroon for 5 min at RT (Fig.2c, left panel).
3. Cells are blocked with 100 % FBS for 1 min.

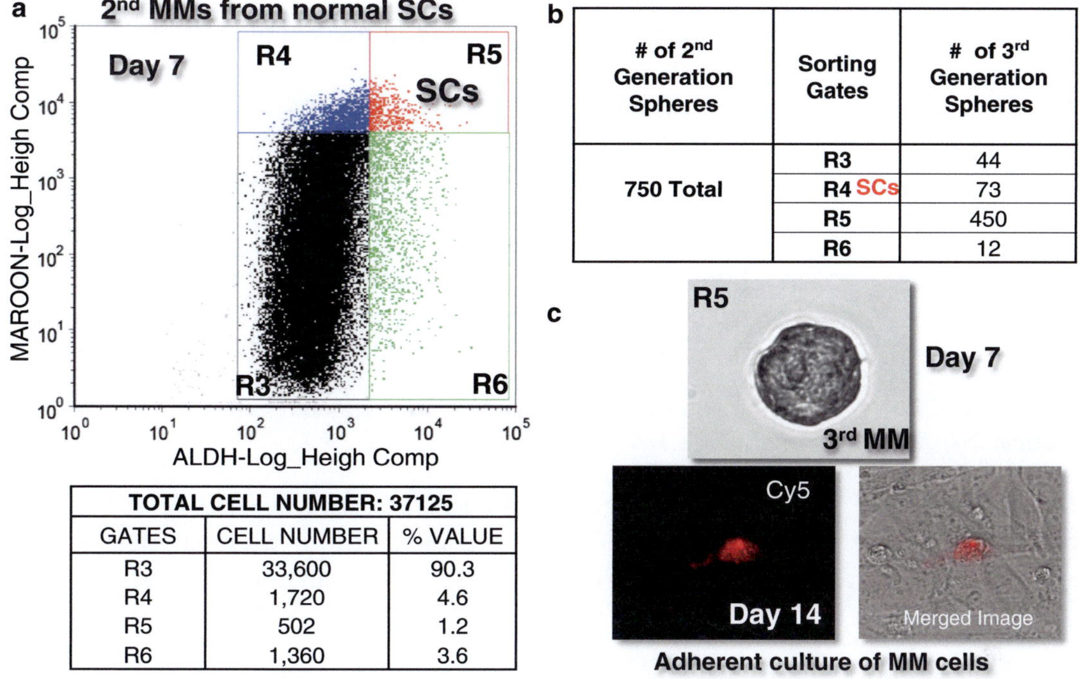

TOTAL CELL NUMBER: 37125		
GATES	CELL NUMBER	% VALUE
R3	33,600	90.3
R4	1,720	4.6
R5	502	1.2
R6	1,360	3.6

# of 2nd Generation Spheres	Sorting Gates	# of 3rd Generation Spheres
750 Total	R3	44
	R4 SCs	73
	R5	450
	R6	12

Fig. 3 Characterization of MMs generated from human normal breast SCs. (**a**) Representative FACS profile of dissociated 2nd generation MMs. Single cells derived from 2nd generation dissociated MMs obtained from Maroon stained SCs were labeled for the ALDEFLUOR assay. Region R5 contains the SC population displaying highest Maroon dye retention and highest ALDH activity. Table shows the percentage of cells in each region from the total number of dissociated cells used for the FACS input. (**b**) Number of MMs regenerated when sorted cells were placed back into MM forming conditions. (**c**) 3rd generation "MM from region R5 cells. MMs when placed onto human foreskin derived fibroblasts cells show that SCs retain the original Maroon dye staining and quiescent state" (Magnification (**c**) 10x upper and 40x lower

4. Cells are washed for 10 min with DMEM supplemented with 10% FCS twice and plated in suspension at clonogenic densities in Medium 6.
5. Labeled single cells are then used to form 2nd generation spheres (Fig. 2c, d). Dye dilution is due to cell proliferation and is monitored during sphere formation using fluorescence microscopy and a Cyanine5 (Cy5) specific detection filter.

Since LA7CSCs, or SCs or CSCs isolated from human mammary tissue express ALDH proteins [10], the ALDEFLUOR assay is combined with Maroon staining to isolate SCs (Figs. 2e and 3, and *see* Subheading 3.4.2).

3.4.2 ALDEFLUOR Labeling of Dissociated MM Cells

1. MMs generated after 7 days from Maroon stained single cells are filtered with a 40 μm pore size cell strainer and dissociated in Medium 2.
2. Single cells are then labeled with the ALDEFLUOR kit and placed at 37 °C in the tissue incubator for 40 min.

3. Using the buffer contained in the kit, the cells are washed to stop the reaction.
4. Cell sorting by FACS analysis is performed. SCs are sorted for highest dye retention (Maroon) and highest ALDH activity. Selecting the size of the gate that contains candidate SCs is based on the observation that the number of MMs formed at the 2nd and 3rd generations is approximately the same. Therefore, the number of cells sorted for the SC region (R5) is approximately equal to the number MMs used for FACS. If the number of spheres to be used for FACS is not known *see* **Note 8**.

3.5 SC and CSC Isolation from MMs

The protocols for generating and characterizing MMs from normal CSCs are identical for SCs.

3.5.1 SC Isolation from MMs

1. Single ALDEFLUOR labeled cells (detected by FITC) from dissociated MMs generated from Maroon stained cells (detected with a filter for allophycocyanin (APC)) are sorted with FACS with a nozzle diameter of 100 μm and back pressure of 20 psi.
2. Cells are sorted into four regions, with the region containing the SCs or CSCs (Figs. 3a 6b left panel) determining the boundaries of the other regions. SCs are defined with highest Maroon dye retention and highest ALDH activity (Fig. 3a–c and *see* **Note 8**).

3.5.2 CSC Isolation from MMs

CSCs may represent a heterogeneous population of cells. Alternative approaches therefore are utilized for their isolation (*See* Somatic Stem Cells: Methods and Protocols, Methods in Molecular Biology, vol. 879). CSCs, depending on the tumor sample, may not display highest dye retention or ALDH activity. Therefore, each region of sorted cells from dissociated MMs is plated under non-adherent conditions to evaluate their capacity for MM formation. The cell population from the region that generates MMs contains the CSCs.

The profiles of CSC MMs obtained after FACS, display significantly contrasting Maroon dye retention and ALDEFLUOR staining for different tumor grades (Ileana Zucchi unpublished data). For instance, MMs generated from CSCs obtained from aggressive tumors contain CSCs and cells derived from CSCs that display a predominantly homogeneous and fast rate of cell proliferation compared to SC MMs (Fig.7a right panel). In contrast, MMs generated from CSCs obtained from low aggressive tumors contain CSCs and cells derived from CSCs that display a heterogeneous (fast and low) rate of proliferation and few cells that are maintained in a quiescent state of proliferation as in MMs from normal SCs (regions R4 and R5 in Fig. 3a and Fig. 6b, left panel). While normal SCs are maintained in a quiescent state of cell proliferation during MM formation (Fig. 3 region R5), MMs generated from aggressive and low aggressive tumors may not contain quiescent CSCs. Extensive tumor sample analyses suggest that MMs that display homogeneous Maroon dye dilution and therefore similar cell cycle division rates are correlated with CSCs that have impaired differentiation potential (Ileana Zucchi unpublished observation).

3.6 Characterization of Cells from Dissociated MMs

1. Single ALDEFLUOR labeled cells from dissociated MMs generated from Maroon stained SCs or CSCs from the same patients or from different tumor grades are used for FACS (*see* **Note 9**).
2. Dye dilution of cells from dissociated SC MMs and CSC MMs is compared. SC MMs display a larger number of cells with dye retention compared to CSC MMs (Fig. 3a, and Fig. 6a, right panel and 6b, left panel). Protocols from Subheading 3.7 and Fig. 6b (right panel), suggest that cells from CSC MMs due to their aberrant high proliferation may delay or inhibit their differentiation (see Section 3.7).
3. Dye dilution of cells from dissociated CSC MMs from aggressive and non-aggressive tumors is compared. CSC MMs from aggressive tumors display a larger number of cells with uniformly faster dye dilution compared to CSC MMs from aggressive display a larger number of cells with uniformly faster dye dilution compared to CSCs MMs from low or non-aggressive tumors (Fig. 6a right panel, and Fig. 6b left panel).

3.7 Characterization of Non-sphere Forming Cells (NSFCs) Derived from MMs

The following protocols were developed to enrich and characterize cells from MMs that have no capacity to regenerate MMs. As MMs can generate all cell types and the 3D architecture of the mammary gland, non-sphere forming cells (NSFCs in Fig. 2g) represent the SC-descendent cells that are at various stages of differentiation and are generated by asymmetric self-renewal of SCs during MM formation. Two cell types can be identified; cells with luminal or myoepithelial characteristics. Cells with highest ALDH activity and high Maroon dye retention express cytokeratin 18 (K18), CD24, and CDH1 and have colony formation potential in adherent conditions on infant foreskin fibroblast cells or on FBS coated (Fig. 4 panel b). The K18 expressing cells also have the capacity to generate 3D acinar structures, suggesting that the cells have a luminal or alveorlar lineage-restricted phenotype. Our experience shows that, in contrast to SC MMs, CSC MMs derived from tumors (and in particular aggressive tumors) often do not generate (or generate few) acinar forming cells (Fig. 6b, right panel, *see* that region R6 contains few or no acinar forming cells) suggesting that depending on the tumor grade, CSCs may display inhibition or delay in their differentiation potential.

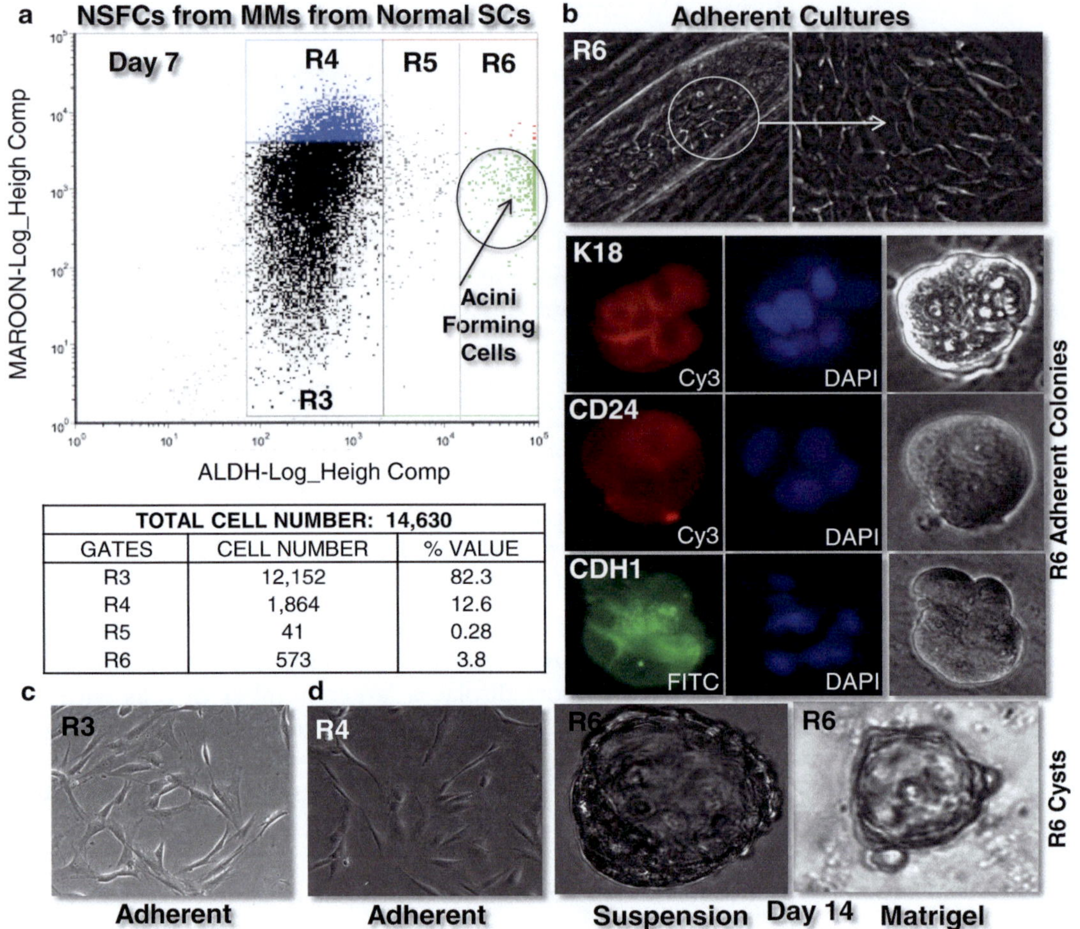

TOTAL CELL NUMBER: 14,630		
GATES	CELL NUMBER	% VALUE
R3	12,152	82.3
R4	1,864	12.6
R5	41	0.28
R6	573	3.8

Fig. 4 Characterization of non-sphere forming cells from MMs. (**a**) Representative FACS profile of *N*on-*S*phere *F*orming *C*ells (NSFCs in Fig. 2g) generated from SCs. Table shows percentage of cells in each region from the total number of NSF cells used for the FACS input. (**b**) While NSFCs do not form MMs, a subpopulation of cells (R6) have colony formation capacity (*top*) (on human foreskin derived fibroblast cells), express high levels of cytokeratin 18, CD24 and CDH1, and have the capacity to form 3D acinar structures in suspension conditions or in Matrigel (*bottom*). Cells from regions R3 (c) and R4 (d) have a fibroblast like morphology and no colony formation potential. (Magnifications: (b) top 10x left and 40x right, immuno-staining images 40x, bottom 10x, (c) and (d) 10x)

3.7.1 Identification of Identification of Acinar Structure Forming Cells Generated from SCs or CSCs

1. Single cells from 1st generation MMs are labeled with Maroon and allowed to reform into 2nd generation MMs.
2. NSFCs are separated by filtration from the 2nd generation MMs (Fig. 2g).
3. NSFCs are labeled for the ALDEFLUOR assay and FACS is performed (Figs. 4a and 6b right panel).
4. Most of the cells are collected from regions R3 and R4 as for SC dissociated MMs. R5 and R6 cells were chosen to have highest ALDH activity (Figs. 4a and 6b, right up panel). Only cells from region R6 generate 3D acinar structures (Fig. 4b).

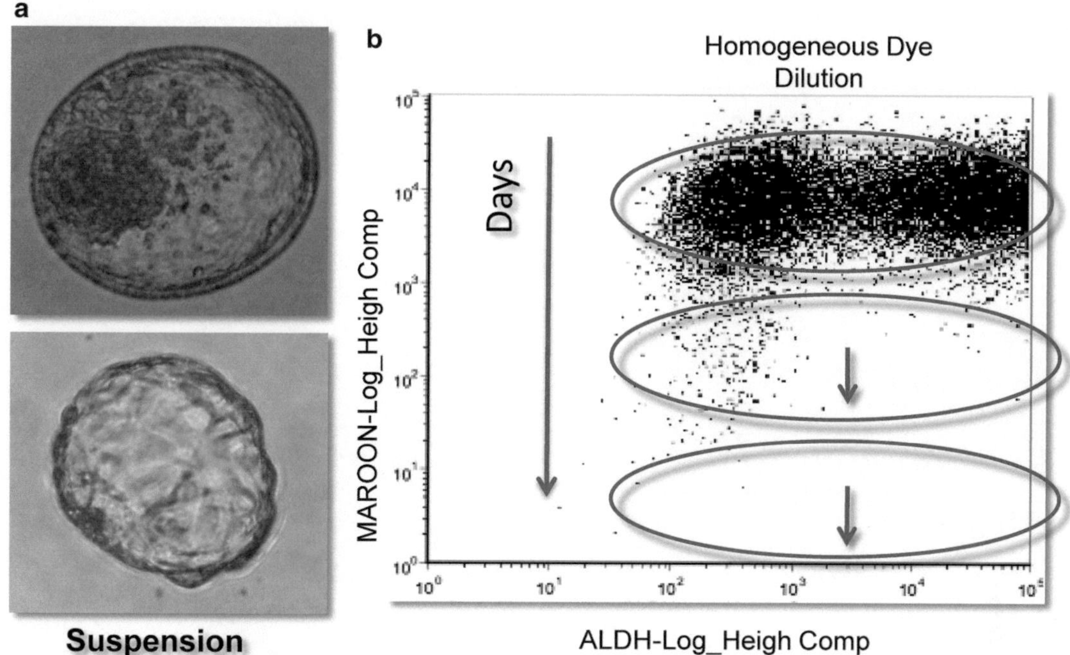

Fig. 5 Analysis of cells from dissociated acinar structures. (**a**) Acinar-like structures generated from region R6 of NSFCs from Fig. 4. (**b**) Representative FACS profile of single cells from acini dissociated at day 14. R6 cells were allowed to form acinar structures in Medium 6 in suspension conditions. At day 14 acini were dissociated and single cells were labeled for the ALDEFLUOR assay. Ovals indicate that Maroon dye is homogeneously diluted with cell division with acinar structure size Magnification: (a) 4x upper and 10x lower

Acinar forming cells are always generated from SC MMs. Depending on the properties of the tumors, CSC MMs can generate no, few or a significant number of acinar structure forming cells (Fig. 6b, right panel, *see* that region R6).

Since acinar structures can form in the same cultures with MMs, they are often mistaken as MMs, especially when acinar structures are in the early phase of formation. Protocols were developed to distinguish acinar structures from MMs based on differences in dye dilution and ALDH activity (Figs. 4a and 5b). In acinar structure formation, cells display a more homogeneous rate of Maroon dye dilution compared to MM cells. Dye dilution in MMs is more heterogeneous due to MMs containing cells with different states of differentiation and proliferation that are generated from multipotent SCs or CSCs. Since acinar structures are generated from a homogeneous population of cells with the same differentiation potential, dye dilution is more homogeneous. Highest activity of ALDH is found in acinar structure forming cells compared to MM cells. Therefore, when high ALDH activity similar to that found in acinar structure forming cells is detected in MM cultures by FACS this suggests that the suspension cultures are "contaminated" with 3D acinar structures resembling MMs.

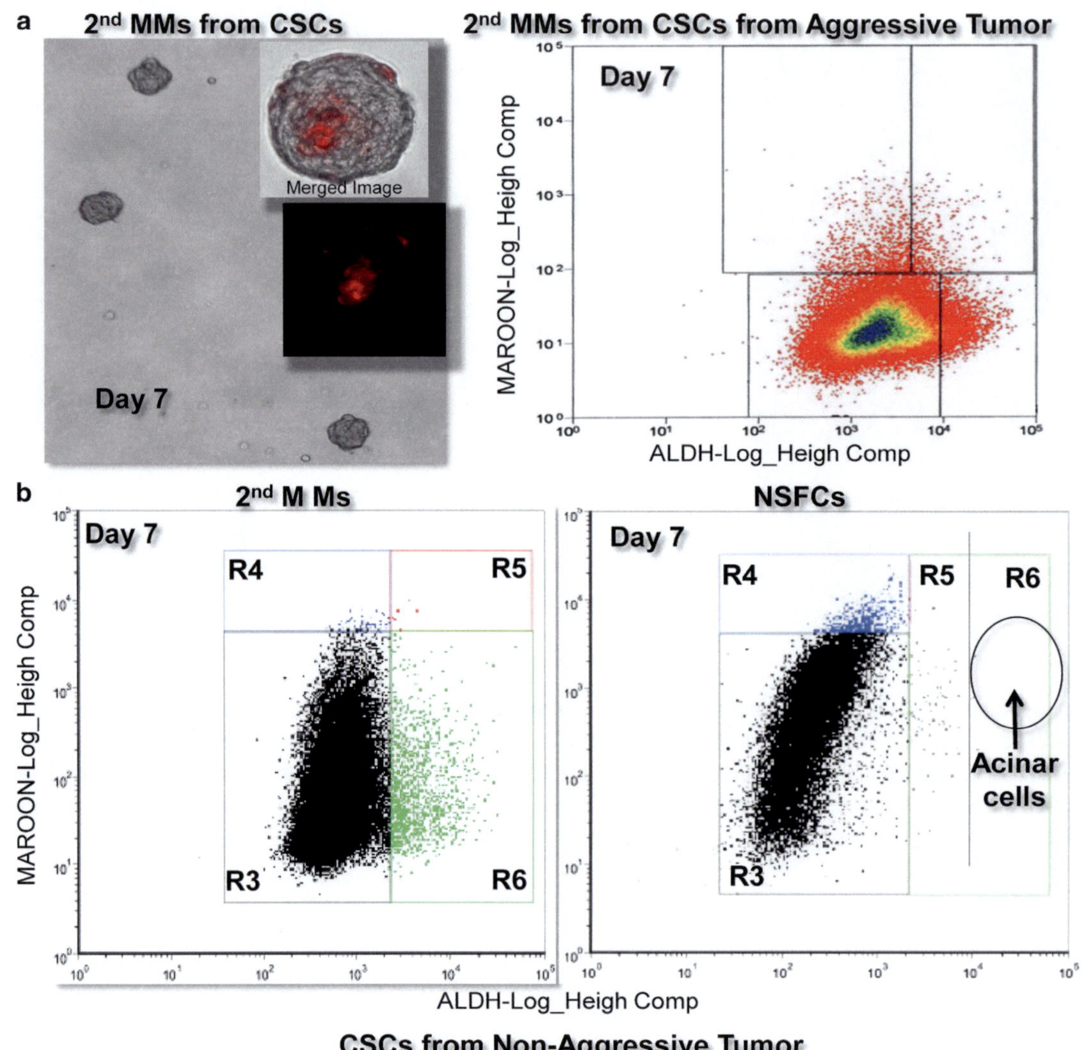

Fig. 6 Characterization of MMs and NFSCs generated from human breast CSCs. (**a**) 2nd generation MMs from CSCs derived from a non-aggressive tumor (*left*) and representative FACS profile of dissociated MMs generated from CSCs derived from an aggressive tumor (*right*). (**b**) Representative FACS profile of dissociated MMs (*left*) and NSFCs (*right*) both generated from CSCs derived from a non-aggressive tumor. Cells from CSC MMs display greater dye dilution compared to cells from normal SC generated MMs (Fig. 3a). Few or no cyst forming cells (R6) are often detected from the NSF cells of CSC MMs

3.8 Comparative analysis of Acinar Structures with MMs by FACS

1. NSFCs collected as flow through from MM filtration (Fig. 2g) contain Maroon stained and unstained cells. NSFCs are labeled with the Aldefluor kit and used for FACS. Cells that display highest ALDH activity and high Maroon retention are contained in region R6 (Fig. 4a) and the only NSFCs that

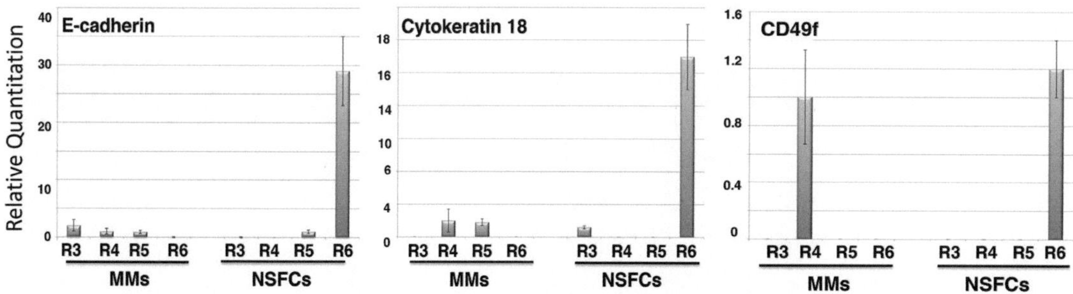

Fig. 7 Comparative mRNA expression analysis of cells from regions obtained from dissociated MMs and NSFCs sorted by FACS. Quantitative PCR expression values for cells from MMs and NSFCs are relative to the expression level of each gene, detected in R4 (the SCs containing region) from MMs cells. Endogenous control is HPRT1. Bars represent a confidence of 95%. When no expression is shown, no expression was detected

generate acinar structures. Acini are generated in suspension in *Medium 6*.

2. Acinar structures are apparent at 12–16 days of culture (Fig. 4b bottom panels and Fig. 5a).
3. Acinar structures are collected with a 40 μm filter, dissociated with Medium 2, and used for FACS (Fig. 5b).

3.9 Expression Analysis of FACS MM Cells and NSFCs

Gene expression analysis shows that the levels of CDH1 and K18 are significantly lower in MM cells than in NSFCs (Fig. 7). Highest ALDH activity is found not in SCs (Region R5) or any MM cells (Regions R3, R4 and R6) but in K18 expressing cells (Region R6) derived from NSFCs. The K18 expressing cells have acinar formation potential. NSFCs from region R4 (Fig. 4a), with no or low ALDH activity but highest dye retention, have no colony (Fig. 4d) and no acinar structure formation potential (Figs. 7 and 8).

1. Cells from all regions from FACS of MMs and NSFCs are used for RNA isolation.
2. Quantitative-PCR analysis is performed for Quantitative-PCR analysis is performed for e-cadherin (CDH1), cytokeratin 18 (K18), cytokeratin 14 (K14), and integrin alpha 6 (CD49f).

3.10 Evaluation of Media for Generating MMs

Medium 6 for generating MMs was established for generating "good" MMs that display highest regenerating potential, and minimize 3D structural artifacts and loss of SC/CSC self-renewal. The reader should be aware that various tissue culture medium compositions available commercially or from the literature generate a heterogeneous population of sphere-like structures from normal SCs and CSCs. It is important to note that an increase in MM-like

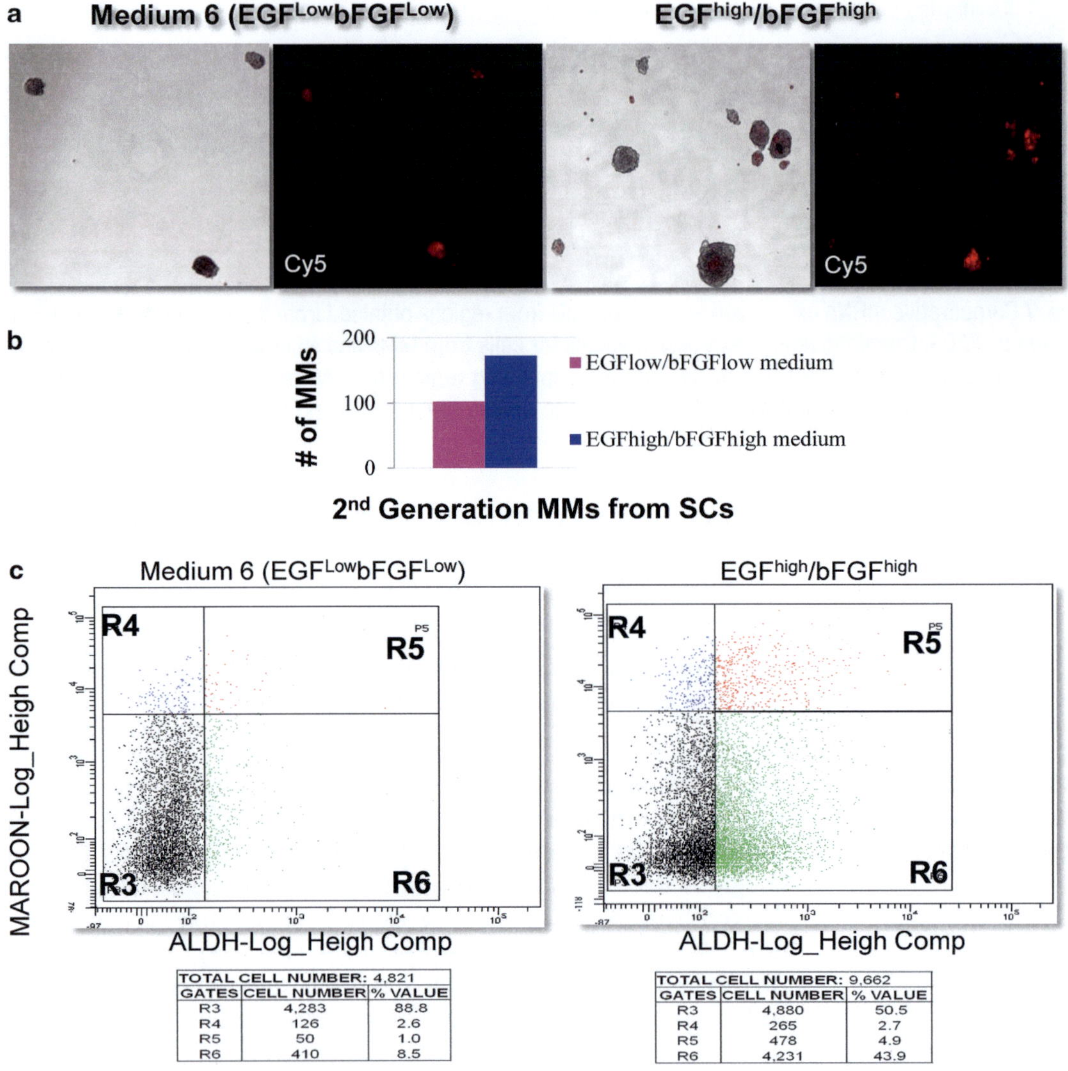

TOTAL CELL NUMBER: 4,821		
GATES	CELL NUMBER	% VALUE
R3	4,283	88.8
R4	126	2.6
R5	50	1.0
R6	410	8.5

TOTAL CELL NUMBER: 9,662		
GATES	CELL NUMBER	% VALUE
R3	4,880	50.5
R4	265	2.7
R5	478	4.9
R6	4,231	43.9

Fig. 8 Growth factor concentration effects on MM formation and number. Increase in tissue culture levels of EGF (20 ng/mL) and basic FGF (20 ng/mL) results in: (**a**) increase 3D structure heterogeneity. (**b**) Increase number of 3D structures. (**c**) Cells from dissociated 3D structures display variability in dye retention and ALDH1 enzymatic activity with increase in EGF and bFGF

structures may not represent a real increase in MM numbers. An example is observed with MM media containing higher levels of both EGF and basic FGF (Fig. 8). Increase in these factors results in addition to more sphere-like structures, structures with heterogeneous (Fig. 8a, b), variability in dye dilution and ALDH1 (Fig. 8c). Whether increases in sphere like structures by addition or increase in the levels of specific factors represent an actual increase in SC or CSC self-renewal and true MM numbers needs to be determined by the researcher.

4 Notes

1. Sterilization results in outer cells and infectious agents being killed while leaving the inner cells of the tissue mass intact. The amount of time for leaving the tissue in EtOH depends on the size, thickness, and contours of the tissue. Tissue pieces cut too small can result in the entire cell mass being killed with sterilization.
2. For cutting, use long and narrow tubes with a small cutting medium volume to prevent excess movement and shredding of the tissue. Use long-blade, high carbon or titanium-coated scissors for fast cutting.
3. Use of FBS is not recommended (unless tested) as it consists of components of unknown concentrations that may induce loss of self-renewal, accelerated differentiation, or epithelial to mesenchyme transition of SCs or CSCs. For this reason serum replacement (and defined factors at specific concentrations) is recommended when SCs, CSCs, or MMs are to be cultured.
4. Allow the pellet to resuspend itself slowly for 5 min at RT in a seal tube to prevent excessive aeration of the cells. Pelleted cells should not be immediately forced into a suspension by pipetting them.
5. The post centrifugation step is critical. If centrifugation results in a supernatant that is not relatively clear, this indicates that fibers were not sufficiently digested and have variable buoyancy. Heterogeneous fiber sizes interfere with density separation of the media and prevent single cells from pelleting to the bottom of the tube.
6. Failure to strain the cells properly to obtain single cells can result in the generation of 3D sphere-like artifacts or cell aggregates. It can also result in clogging of the strainer with SCs or CSCs being trapped in the sieve mesh and the tissue fibers. Generally larger surface area strainers result in less cell loss.
7. Cell plating number may not be the same for all diameter plates as small diameter plates promote the cells to collect to the center of plates, resulting in aggregate formation.
8. Selecting the size of the FACS region that contains candidate SCs is based on the calculation that a sphere of 7 days consists of approximately 50 cells and a population of 50,000 cells from dissociated MMs contains 1,000 SCs. The region containing SCs is therefore selected to have 2 % of all of the sorted cells. This 2 % is chosen from the region in which cells have both the highest dye retention (Maroon) and highest ALDH activity. Spheres generated from cells of regions R3, R4, and R6 suggest that more SCs could be included into region 5 if the gate (e.g., Region 4) was shifted slightly.

9. MMs regenerate from normal SCs or CSCs for about 5–6 passages with each passage being 7 days. MMs from CSCs in contrast to MMs from normal SCs, often contain more than one stem cell with MM regeneration potential suggesting that CSCs may under both asymmetric and symmetric self-renewal during MM formation. This number appears to be dependent on the properties of the tumor. We have not observed that either normal SCs or CSCs display immortality, indefinite self-renewal, and indefinite MM regeneration capacity.

Acknowledgements and Financial Support

We like to acknowledge the expert technical assistance of Sig. Emanuele Canonico of the FRACTAL-Servizio di Citometria Istituto Scientifico San Raffaele, Milano and the always supportive Loredana Ansalone for her administrative help. We thank the support Dr. Martin Götte from the Universitätsklinikum of Münster, Germany and the staff of the Department of Pathology of Leno (BS) for their excellent technical support. We acknowledge and are grateful to the IZ grant support: the N.O.B.E.L. Grant and Cariplo Progetti-Internazionali n. 2008–2015 (Fondazione CARIPLO), MIUR-FIRB Newton grant RBAP11BYNP and Fierce grant RBAP11BYNP. R.A.R is supported with a fellowship from the Fondazione CARIPLO Progetti-Internazionali. This work is dedicated to the memory of a dear friend, Sabina Cascella who died of cancer when she was still young and full of life.

References

1. Reya T, Morrison SJ, Clarke MF et al (2001) Stem cells, cancer, and cancer stem cells. Nature 414:105–111
2. Clevers H (2005) Stem cells, asymmetric division and cancer. Nat Genet 37:1027–1028
3. Dontu G, Wicha MS (2005) Survival of mammary stem cells in suspension culture: implications for stem cell biology and neoplasia. J Mammary Gland Biol Neoplasia 10:75–85
4. Al-Hajj M, Wicha MS, Benito-Hernandez A et al (2003) Prospective identification of tumorigenic breast cancer cells. Proc Natl Acad Sci U S A 100:3983–3988
5. Ginestier C, Hur MH, Charafe-Jauffret E et al (2007) ALDH1 is a marker of normal and malignant human mammary stem cells and a predictor of poor clinical outcome. Cell Stem Cell 1:555–567
6. Pece S, Tosoni D, Confalonieri S et al (2010) Biological and molecular heterogeneity of breast cancers correlates with their cancer stem cell content. Cell 140:62–73
7. Zucchi I, Sanzone S, Astigiano S et al (2007) The properties of a mammary gland cancer stem cell. Proc Natl Acad Sci U S A 104:10476–10481
8. Zucchi I, Astigiano S, Bertalot G et al (2008) Distinct populations of tumor-initiating cells derived from a tumor generated by rat mammary cancer stem cells. Proc Natl Acad Sci U S A 105:16940–16945
9. Cocola C, Sanzone S, Astigiano S et al (2008) A rat mammary gland cancer cell with stem cell properties of self-renewal and multi-lineage differentiation. Cytotechnology 58:25–32
10. Zucchi I, Dulbecco R (2002) Proteomic dissection of dome formation in a mammary cell line. J Mammary Gland Biol Neoplasia 7:373–384

Chapter 19

Isolation and Culture of Primary Glioblastoma Cells from Human Tumor Specimens

Sascha Seidel, Boyan K. Garvalov, and Till Acker

Abstract

Cultured tumor cells are a central tool in cancer research and have provided fundamental insights in tumor biology. Recent evidence, however, indicates that classically established cell lines from different tumors, including glioblastoma, do not fully reflect the genotypes and phenotypes of the respective primary tumors. By contrast, primary cells, isolated from human tumor samples and maintained in serum-free spheroid cultures at low passage under defined growth factor conditions, reproduce key aspects of tumor cell physiology much more faithfully. Among the tumor cell characteristics that are better represented in primary glioblastoma cell cultures is the self-renewal and differentiation potential of the tumor cells. Indeed, a large body of evidence from the past decade indicates that glioblastomas and other tumors are composed of a hierarchy of heterogeneous types of cells, which are generated and maintained by cells that share characteristics of stem cells. This cancer stem cell/tumor initiating cell population is optimally preserved and maintained in primary glioblastoma cultures. Here, we describe a method for the isolation and culture of primary tumor cells from human glioblastomas in serum-free conditions, which allows the routine generation and proper maintenance of tumor cells as spheroid cultures. Such primary tumor cultures can serve as a model of choice for the study of the mechanisms behind key aspects of glioblastoma biology, including tumorigenicity, stem cell hierarchy, invasion, and therapeutic resistance.

Key words Glioblastoma, Brain tumor, Primary glioblastoma cells, Brain tumor initiating cells, Brain tumor stem cells, Glioblastoma stem cells, Cancer stem cells, Tumor spheres, Primary culture

1 Introduction

Glioblastomas are the most common and most malignant brain tumors in adults, with a median patient survival time of less than 15 months [1]. Therefore, an improved understanding of the cellular and molecular mechanisms controlling glioblastoma growth and invasion has been the focus of intense research. Among the most important conceptual advances in the study of glioblastomas over the last decade has been the realization that these tumors are organized in a hierarchy of heterogeneous types of tumor cells that derive from neoplastic cells with stem cell characteristics [2, 3]. Glioblastoma is among the first solid tumor types for which the

Ivan N. Rich (ed.), *Stem Cell Protocols*, Methods in Molecular Biology, vol. 1235,
DOI 10.1007/978-1-4939-1785-3_19,

hierarchy (cancer stem cell) model was demonstrated and has remained a prototype system for the study of this model [4], which is now thought to be applicable to various other solid tumors [5, 6].

Traditionally, the cell biology of glioblastoma has been studied on the basis of cultured tumor cells that were extensively passaged and grown in adherent monolayers in the presence of serum, an approach commonly used also for other tumors. While such tumor cells can be tumorigenic and are convenient to handle and manipulate, they usually do not fully reproduce the characteristic invasive growth phenotype of glioblastomas in xenograft models. Moreover, they show drastically altered gene expression patterns in vitro compared to the original tumor [7, 8]. Importantly, they are not well suited for the study of the self-renewal (stem cell) capacity of glioblastoma cells, as serum induces profound differentiation, which may lead to a loss of the cancer stem cell fraction. Therefore, much recent research in glioma biology has focused on a cell culture system of primary tumor cells derived from patient tumors and cultured for a limited number of passages as spheroids in suspension, in the presence of defined growth factors but in the absence of serum [2, 9, 10], conditions that were originally established for the cultivation of neural stem cells [11]. Such cultures retain much greater similarity to the primary tumor from which they are derived [7] and their use has enabled the discovery and characterization of the key features of glioblastoma stem cells. For example, the primary glioblastoma cell culture system was applied to examine the role of microenvironmental stimuli on glioblastoma cells. It was shown that the cancer stem cell phenotype of glioblastoma cells is induced and maintained in specific niches, including a hypoxic niche [12–15] and a perivascular niche [10, 14, 16]. It was further shown that glioblastoma stem cells possess enhanced resistance to chemotherapy and radiotherapy [17, 18], a property that likely underlies their ability to drive tumor relapse after treatment. Interestingly, glioblastoma stem cells derived from primary cultures revealed an unexpectedly broad differentiation potential in vitro and in vivo, being able to contribute not only to the generation of the various types of tumor cells but also to the formation of the tumor vasculature, through transdifferentiation into endothelial cells and pericytes [19–21]. Furthermore, primary glioblastoma cells were employed to elucidate the role of inflammatory responses on tumor growth and the interaction between glioblastoma stem cells and immune cells [22–24].

In this chapter, we describe a method that has been routinely used in our laboratory for isolating primary tumor cells from human glioblastomas and their cultivation in vitro. The cells generated in this manner are maintained as spheroids in serum-free suspension cultures with defined growth factors. These conditions, which are analogous to the ones used for the culture of physiological stem cells, have also proven to be particularly suitable for the study of the cancer stem cell and brain tumor initiating phenotype of glioblastoma cells, as well as for many other features of glioblastoma biology.

2 Materials

2.1 Tumor Dissection

2.1.1 Preparation of Buffers and Solutions

1. Phosphate buffered saline (PBS) without calcium and magnesium, sterile (PAA).
2. 10× Hanks Balanced Salt Solution (HBSS) without calcium and magnesium, sterile (Gibco).
3. Eagle's Balanced Salt Solution (EBSS) without calcium and magnesium, sterile (Gibco).
4. D-Glucose solution (300 mg/mL), sterile filtered.
5. Sodium chloride solution (0.15 M), sterile filtered.
6. Bovine serum albumin (BSA), cell culture tested (Sigma).
7. Amphotericin B (250 μg/mL) (Sigma).
8. Gentamicin (50 mg/mL) (PAA).
9. HEPES buffer, 1 M, sterile (PAA).
10. Ultrapure water (18.2 MΩ).
11. Papain (Worthington).
12. DNase I (Sigma).
13. Trypsin (Sigma).
14. Collagenase I (Worthington).
15. Hyaluronidase (Sigma).
16. Red Blood Cell Lysis Buffer (Roche Diagnostics).
17. 0.22 μm sterile syringe filters.
18. Sterile syringes.
19. 0.22 μm Stericup-GV filters, 150 and 500 mL (Millipore).
20. Autoclaved glass bottles 500 mL.
21. Sterile conical tubes.

2.1.2 Tissue Dissociation

1. Sterile 10 cm petri dishes (Greiner Bio-One; *see* **Note 1**).
2. Sterile scalpel with a #21 blade.
3. Autoclaved single edged razor blades.
4. Autoclaved forceps.
5. Sterile cell strainers, 70 and 100 μm.
6. Sterile 15 and 50 mL conical tubes.
7. Sterile plastic pipettes.
8. Water bath at 37 °C.

2.2 Cell Culture

1. Dulbecco's Modified Eagle's Medium (DMEM)/F-12+ GlutaMAX (Gibco).
2. B27 Supplement minus vitamin A (Gibco; *see* **Note 2**).
3. Recombinant human EGF (PeproTech).
4. Recombinant human bFGF (PeproTech).
5. Accutase solution (PAA).
6. Cryo-SFM cryopreservation medium (Promo Cell).
7. Sterile 94 mm petri dishes (Greiner Bio-One).
8. Cell strainer 40 μm.
9. Freezing container (e.g., Mr. Frosty).

3 Methods

3.1 Tissue Preparation

In the following section, we describe a method for the isolation and the establishment of primary glioblastoma cell lines from human tumor specimens. All steps of tissue dissociation should be carried out under sterile conditions in a laminar flow cell culture hood to reduce the risk of contamination of the primary culture.

3.1.1 Preparation of Buffers and Solutions

Wash Buffer:

1. Add 450 mL PBS without calcium and magnesium to an autoclaved Erlenmeyer flask.
2. Dissolve 25 g of cell culture tested BSA in the PBS.
3. Add 5 mL amphotericin B (250 μg/mL).
4. Add 500 μL gentamicin (50 mg/mL) (*see* **Note 3**).
5. Fill up to 500 mL with PBS without calcium and magnesium.
6. Mix well and filter-sterilize with a 0.22 μm Stericup Filter into an autoclaved glass bottle, store the Wash Buffer at 4 °C.

Papain stock solution:

1. Dissolve the Papain Suspension (100 mg Papain, Worthington) in 5 mL EBSS without calcium and magnesium (final concentration 20 mg/mL = 500 U/mL).
2. Filter-sterilize with a 0.22 μm Syringe Filter.
3. Store the Suspension at 4 °C.

DNase I Stock Solution:

1. Dissolve 100 mg (500 U/mg) DNase I in 5 mL sterile 0.15 M NaCl.
2. Prepare 50 μl Aliquots and store them at −20 °C.

Dissociation solution:

1. Add 400 mL ultrapure water to an autoclaved Erlenmeyer flask.
2. Add 50 mL 10× HBSS without calcium and magnesium.
3. Add 9 mL D-glucose solution (300 mg/mL).
4. Add 7.5 mL 1 M HEPES buffer.
5. Mix carefully and adjust the pH to 7.4.
6. Add ultrapure water to a final volume of 500 mL.
7. Filter-sterilize with a 500 mL 0.22 μm Stericup filter into an autoclaved glass bottle.
8. Prepare 5 mL aliquots in sterile 15 mL conical tubes and 50 mL aliquots in sterile 50 mL tubes and store them at −20 °C.

Attention: 100 mL of dissociation solution is required for the Dissociation Medium 2. If the dissociation solution is prepared freshly do not freeze the whole amount.

Dissociation Medium 1:

Prepare freshly for every tumor dissection!

1. Thaw one 5 mL aliquot of Dissociation solution.
2. Transfer 4.75 mL of Dissociation solution to a new sterile conical 15 mL tube.
3. Add 200 μL of the papain stock solution.
4. Activate the above mixture at 37 °C for 30 min prior to use.
5. Thaw one 50 μL aliquot of the DNase I stock solution on ice.
6. Add the DNase I aliquot and mix carefully.

Dissociation Medium 2:

1. Pour 100 mL of pre-cooled Dissociation solution into an autoclaved Erlenmeyer flask on ice.
2. Add 70 mg of collagenase I to a final concentration of 0.7 mg/mL.
3. Add 70 mg of hyaluronidase to a final concentration of 0.7 mg/mL.
4. Add 100 mg of trypsin to a final concentration of 1 mg/mL.
5. Mix carefully until all the enzymes are completely dissolved.
6. Filter-sterilize with a 150 mL 0.22 μm Stericup filter into an autoclaved glass bottle.
7. Prepare 5 mL aliquots in sterile 15 mL conical tubes and store them at −20 °C.

Prior to Tumor dissociation:

1. Thaw one 5 mL aliquot of the above mixture.
2. Thaw one 50 μL aliquot of the DNase I stock solution on ice.

3. Add the DNase I aliquot and mix carefully to obtain Dissociation Medium 2.

Growth Factor Stock Solutions:

1. Reconstitute lyophilized recombinant human EGF and recombinant human bFGF following the supplier's recommendations in 5 mM Tris, 0.1 % BSA, pH 7.6 to a final concentration of 20 μg/mL.
2. Prepare 500 μL aliquots in sterile 1.5 mL tubes.
3. For long term storage keep the aliquots at −20 °C, the aliquot in use can be stored at 4 °C for up to 1 week.

Tumor Sphere Culture Medium

1. Take 480 mL DMEM/F-12 + GlutaMAX.
2. Add 5 mL amphotericin B (250 μg/mL).
3. Add 500 μL gentamicin (50 mg/mL) (*see* **Note 3**),
4. Add 2.5 mL HEPES Buffer (1 M).
5. Add 10 mL B27 Supplement minus vitamin A (*see* **Notes 2** and **4**).
6. Mix well, protect the bottle from light by wrapping it in aluminum foil.
7. Store the medium at 4 °C.

3.1.2 Tissue Dissociation

1. Transfer the tumor specimen into a sterile petri dish using autoclaved forceps (*see* **Notes 5** and **6**).
2. Wash the tumor specimen two times with wash buffer and discard the buffer by sucking it off.
3. Cut the tissue into small fragments using a sterile scalpel with a #21 blade (*see* **Note 7**).
4. Chop/cut the tumor fragments with a sterile single edged razor blade until it is well minced (*see* **Note 7**).
5. Transfer the tumor tissue to a sterile 15 mL conical tube by flushing the dish with the 5 mL activated Dissociation Medium 1 and pipetting it with a sterile pipette (*see* **Note 8**).
6. Incubate the tube in a 37 °C water bath for 30 min. During the incubation, take the tube out of the water bath every 5 min and mix gently with a 5 mL pipette (around 20 times). Avoid bubbling (*see* **Note 8**).
7. Centrifuge the tube for 5 min at 300 × *g*.
8. Suck off the supernatant carefully (the pellet is not solid!) (*see* **Note 9**).
9. Resuspend the pellet in 5 mL Dissociation Medium 2. (*see* **Note 8**).

10. Incubate the tube in a 37 °C water bath for 30 min. During the incubation take the tube out of the water bath every 5 min and mix gently with a 5 mL pipette (around 20 times). Avoid bubbling (*see* **Note 8**).
11. Centrifuge the tube for 5 min at 300 × *g*.
12. Suck off the supernatant carefully (the pellet is not solid!) (*see* **Note 9**).
13. Resuspend the pellet in 2 mL of Red Blood Cell Lysis Buffer and incubate the tube for 10 min at room temperature (*see* **Note 10**).
14. Add 8 mL of wash buffer.
15. Centrifuge the tube for 5 min at 300 × *g*, and prepare a conical 50 mL tube with a 100 μm cell strainer.
16. Carefully remove the supernatant and resuspend in 10 mL wash buffer.
17. Filter the suspension through the cell strainer (*see* **Note 11**).
18. Wash the filter with 10 mL wash buffer.
19. Centrifuge the tube for 5 min at 300 × *g*, in the meantime prepare a conical 50 mL tube with a 70 μm cell strainer.
20. Filter the suspension through the cell strainer (*see* **Note 11**).
21. Centrifuge the tube for 5 min at 300 × *g*.
22. Resuspend the cell pellet in Tumor Sphere Culture Medium.

3.2 Cultivation of Primary Glioblastoma Cells

This section describes how to establish and maintain the primary tumor cells, obtained and plated during the previous procedure.

1. Count the cells from **step 22** in the previous section (Subheading 3.1.2).
2. Plate them at a density of 2×10^6 cells in 10 mL of Tumor Sphere Culture Medium in a 10 cm petri dish.
3. Add 10 μL (1:1,000) of the EGF and bFGF stock solutions to a final concentration of 20 ng/mL.
4. After 24 h, collect the cells with a plastic pipette, transfer them to a 15 mL conical tube, and centrifuge at 190 × *g* for 3 min (*see* **Note 12** and Fig. 1a).
5. Resuspend the pellet in 10 mL Tumor Sphere Culture Medium and transfer them to the same petri dish from which they were collected.
6. Add 10 μL (1:1,000) of the EGF and bFGF stock solutions to a final concentration of 20 ng/mL.
7. The culture medium should be refreshed twice a week by transferring the cells with a plastic pipette to a 15 mL conical tube and centrifuging at 190 × *g* for 3 min.

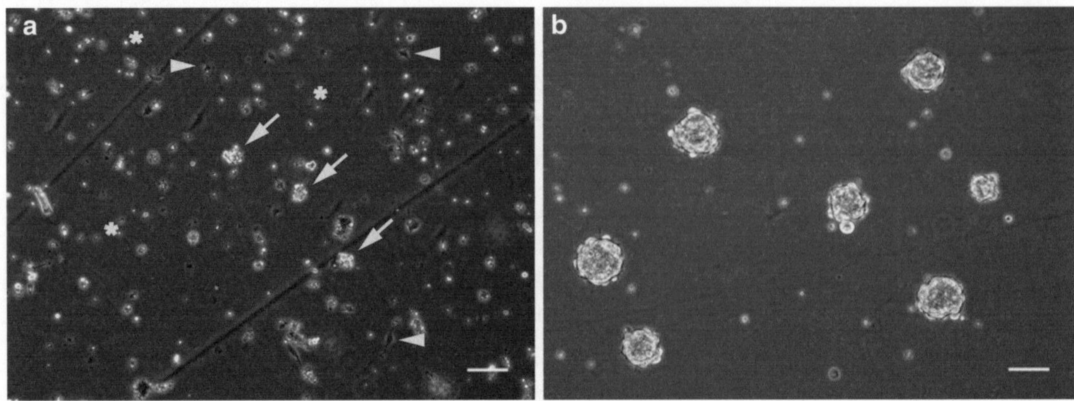

Fig. 1 Primary glioblastoma cells in culture. (**a**) Isolated glioblastoma cells 12 h after tumor dissection. The culture consists of tumor cells, which partly already start to form tumor spheres (*arrows*), some cell debris (*asterisks*), and non-tumor cells that adhere to the petri dish (*arrowheads*). (**b**) Tumor Spheres of primary glioblastoma cells 5 days after splitting. Scale bars 100 μm

When refreshing the medium for the first time aspirate half of the old medium and replace it with an equal volume of fresh medium (i.e., refresh half of the medium). Afterwards re-plate the cells in the original dish.

When refreshing the medium for the second time, aspirate all the medium and replace it with fresh medium. Thereafter, re-plate the cells in a new dish. Always add 10 μL (1:1,000) of the EGF and bFGF stock solutions to a final concentration of 20 ng/mL after changing the medium.

The obtained culture of primary glioblastoma cells should form three-dimensional spheroids suspended in the medium (tumor spheres) around 24–48 h after the dissociation (Fig. 1). The frequency of sphere formation depends on the characteristics of the primary cell line, e.g., the number of brain tumor initiating cells in the original tumor specimen and their self-renewal capacity (*see* **Notes 13–15**).

3.2.1 Splitting of Tumor Spheres

Depending on their size, tumor spheres of primary glioblastoma cells should be split every 7–10 days.

1. Centrifuge the tumor spheres for 3 min at 190 × *g*.
2. Aspirate the medium and resuspend the pellet in 1 mL of Accutase solution.
3. Incubate the spheres for 15 min at 37 °C.
4. In the meantime prepare a sterile 50 mL conical tube with a 40 μm cell strainer.
5. Mix the cell suspension with a 1,000 μL pipette tip until there are no visible spheres left.
6. Add 9 mL of Tumor Sphere Culture Medium and filter through the cell strainer.

7. Centrifuge the cells for 3 min at 190 × *g*.
8. Remove the supernatant and resuspend the pellet in 10 mL of Tumor Sphere Culture Medium.
9. Plate the cells at a density of 2×10^6 cells in 10 mL of Tumor Sphere Culture Medium in 10 cm petri dishes.
10. Add 10 μL (1:1,000) of the EGF and bFGF stock solutions to a final concentration of 20 ng/mL.

3.2.2 Cryopreservation of Primary Glioblastoma Cells

The primary glioblastoma cells should be used at low passage number, as there is a risk of genetic alterations and phenotypic shifts over the course of prolonged culture with multiple passaging. Therefore, a substantial number of aliquots should be cryopreserved at the earliest passage when they can be sufficiently expanded. Subsequently, fresh frozen aliquots should be periodically thawed and cultured for a limited number of passages and new fresh aliquots at low passages should be frozen.

1. Split and seed the cells as described in Subheading 3.2.1.
2. Spin down the spheres 24 h after the re-seeding for 3 min at 190 × *g*.
3. Resuspend the pellet of one petri dish in 2 mL of Cryo-SFM cryopreservation medium.
4. Transfer the cells suspension to cryo vials at 1 mL per vial.
5. Transfer the vials to a freezing container at room temperature.
6. Freeze the cells slowly in the freezing container for 24 h at −80 °C.
7. Transfer the cells to liquid nitrogen for long-term storage.

4 Notes

1. Petri dishes used must *not* have a specially treated surface for cell culture. As the primary glioblastoma cells are maintained as spheroids in suspension, the surface of the dish should prevent, rather than promote the attachment of the cells.
2. The B27 Supplement is light sensitive. Bottles containing medium with B27 should be wrapped in aluminum foil to prevent exposure to light.
3. The addition of antibiotics and antimycotics to the buffers and cell culture media is optional, but can reduce the risk of contamination, especially if it is not clear whether the tissue samples were collected under fully sterile conditions.
4. The stability of the B27 Supplement can be affected by long-term storage at 4 °C. Therefore, if the throughput of Tumor

Sphere Culture Medium is low, add B27 Supplement minus vitamin A from frozen aliquots stored at -20 °C directly to the medium in the culture dish. In this case, the B27 Supplement should be diluted 1:50.

5. The yield and the quality of the isolated primary cells strongly depend on the quality of the provided tumor material. We recommend to contact your department of neurosurgery and discuss with them an optimal way for the collection and transfer of the tumor material. At the same time we recommend to obtain the clinical information about the patient regarding potential viral infections (e.g., HIV, Hepatitis A/B/C), which would preclude using the tumor specimen for research purposes in a routine lab environment. The tumor material should be processed as fast as possible after the resection. We suggest that the material should be collected either in a sterile 50 mL conical tube filled with 20 mL cold PBS, or in a sterile screw lid sample container on sterile gauze soaked with PBS to prevent the tissue from drying. Until the cell isolation procedure is performed, the tumor material should be kept on ice. If you process more than one tumor sample at a time, we recommend using a separate set of instruments for tissue dissection to prevent cross-contamination.
6. It is important to obtain ethical approval from the responsible Institutional Review Board (IRB) before working with human tissue samples. We strongly recommend the wearing of suitable personal protective equipment (laboratory coat, protective goggles, gloves) during the tissue dissociation, as all human tissue is potentially infectious and sometimes insufficiently tested for infectious agents such as hepatitis virus or HIV. For the same reasons it is recommended that the laboratory personnel performing the primary cell isolation undergo immunization against hepatitis A/B.
7. Mincing the tissue well can take time. Take particular care during this step when handling scalpels and razor blades to prevent injuries, as the tumor material has to be seen as potentially infectious.
8. The dissociation procedure described here is suitable for most tumor samples with a size of up to 0.5 cm^3. Nevertheless, different tumors samples may have different textures, cell density, amounts of necrosis and may originate from different brain regions. Therefore, for large tumor samples, it might be necessary to increase the amount of Dissociation Media 1 and 2, or the incubation time for the cell dissociation.
9. During the dissociation procedure care should be taken while removing the supernatant during the centrifugation steps, especially when aspirating the supernatant with a pump. The pellet usually is not solid and homogenous. The cells are

located at the very bottom of the conical tube and are overlaid with debris. The latter appears as the upper layer of the pellet and, especially during the first dissociation step, can be quite viscous and sticky due to released DNA and extracellular matrix proteins. Hence, there is the danger of aspirating the cell pellet together with the supernatant and the debris. Therefore, the supernatant must be aspirated slowly and carefully.

10. Depending on the tumor sample it could be necessary to repeat the Red Blood Cell Lysis step. If the cell pellet appears to be still reddish after the washing step, repeat the incubation with Red Blood Cell Lysis Buffer for 5 min.

11. Depending on the amount and density of the tumor, the cell suspension may be quite thick. In such cases, pipette directly onto the cell strainer with increased pressure or press the suspension through the cell strainer with the pipette tip.

12. At this stage the culture could still contain a small amount of debris and non-tumor cells, which usually adhere to the surface of the petri dish, whereas the tumor cells float in suspension, some of them already forming small spheres (Fig. 1a). The non-tumor cells will not proliferate under the culture conditions used for the primary tumor cells and are going to die out. The debris will usually be washed out during the medium change the day after tumor dissection.

13. The primary glioblastoma cells isolated using the described protocol show self-renewal capacity and stem cell characteristics. This is reflected, for example, by their ability to form spheres in culture (Fig. 1), which express neural stem cell markers such as nestin (Fig. 2a). In addition, they express established brain tumor initiating cell markers such as CD133, Oct4, and others (Fig. 2b; [14]). The expression of such stem cell markers is enhanced under tumor sphere conditions, whereas addition of serum downregulates the stem cell genes and upregulates differentiation markers such as the astrocytic protein GFAP (Fig. 2b).

14. In addition to the isolation of primary glioblastoma cells from human tumor specimens, our method could be used to re-isolate glioblastoma cells from orthotopic or subcutaneous tumor xenografts. In this case, the cultures will initially be a mix of murine and human cells, but usually after two passages the culture becomes homogeneous and only consists of human tumor cells, which outgrow the mouse cells.

15. The described protocol has been tested on numerous brain tumor samples and in the majority of cases has led to the successful isolation of primary glioblastoma cells. However, not all cultures will start growing, or the cells could stop proliferating after one or two passages. This usually depends on the quality

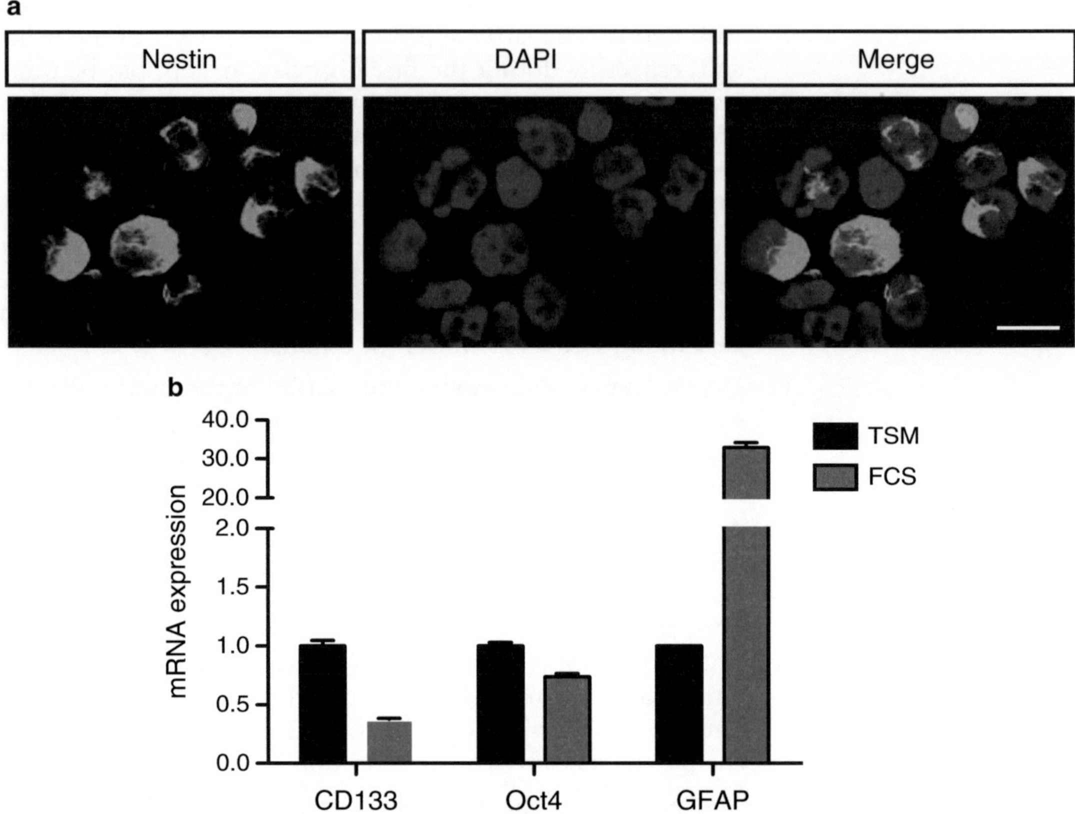

Fig. 2 Primary glioblastoma cells express cancer stem cell markers and retain their differentiation capacity. (**a**) Tumor spheres of primary glioblastoma cells were spun on a microscope slide using a Cytospin centrifuge and stained for the neural stem cell marker nestin (*green*) and DAPI to visualize the nucleus (*blue*). Scale bar 20 μm. (**b**) Quantitative real time RT-PCR of primary glioblastoma cells grown under sphere conditions (in tumor sphere culture medium, TSM) or as adherent cultures supplemented with 10 % fetal calf serum (FCS). Under sphere conditions the cancer stem cell marker CD133 and the embryonic stem cell marker Oct4 are highly expressed, whereas their expression decreases when cells undergo differentiation in the presence of FCS. By contrast, expression of the glial lineage differentiation marker GFAP is highly upregulated by FCS addition

of the tumor sample, as well as the fraction of brain tumor initiating cells in the tumor specimen and their self-renewal capacity. We therefore encourage the researcher to process several new tumor samples if the method does not work at first try.

Acknowledgements

This work was supported by grants from the DFG (AC 110/4 (KFO210); AC110/3-1, 3-2 (SPP1190); the ECCPS (EXC 147)), the Deutsche Krebshilfe (107231), the von Behring-Röntgen

Foundation, LOEWE (OSF), and the German Ministry of Education and Research (BMBF) within the National Genome Network (NGFNplus) and Brain Tumor Network.

References

1. Stupp R, Hegi ME, Mason WP et al (2009) Effects of radiotherapy with concomitant and adjuvant temozolomide versus radiotherapy alone on survival in glioblastoma in a randomised phase III study: 5-year analysis of the EORTC-NCIC trial. Lancet Oncol 10:459–466
2. Singh SK, Clarke ID, Terasaki M et al (2003) Identification of a cancer stem cell in human brain tumors. Cancer Res 63:5821–5828
3. Singh SK, Hawkins C, Clarke ID et al (2004) Identification of human brain tumour initiating cells. Nature 432:396–401
4. Filatova A, Acker T, Garvalov BK (2013) The cancer stem cell niche(s): the crosstalk between glioma stem cells and their microenvironment. Biochim Biophys Acta 1830:2496–2508
5. Garvalov BK, Acker T (2011) Cancer stem cells: a new framework for the design of tumor therapies. J Mol Med (Berl) 89:95–107
6. Visvader JE, Lindeman GJ (2012) Cancer stem cells: current status and evolving complexities. Cell Stem Cell 10:717–728
7. Lee J, Kotliarova S, Kotliarov Y et al (2006) Tumor stem cells derived from glioblastomas cultured in bFGF and EGF more closely mirror the phenotype and genotype of primary tumors than do serum-cultured cell lines. Cancer Cell 9:391–403
8. Li A, Walling J, Kotliarov Y et al (2008) Genomic changes and gene expression profiles reveal that established glioma cell lines are poorly representative of primary human gliomas. Mol Cancer Res 6:21–30
9. Galli R, Binda E, Orfanelli U et al (2004) Isolation and characterization of tumorigenic, stem-like neural precursors from human glioblastoma. Cancer Res 64:7011–7021
10. Bao S, Wu Q, Sathornsumetee S et al (2006) Stem cell-like glioma cells promote tumor angiogenesis through vascular endothelial growth factor. Cancer Res 66:7843–7848
11. Reynolds BA, Weiss S (1992) Generation of neurons and astrocytes from isolated cells of the adult mammalian central nervous system. Science 255:1707–1710
12. Heddleston JM, Li Z, McLendon RE (2009) The hypoxic microenvironment maintains glioblastoma stem cells and promotes reprogramming towards a cancer stem cell phenotype. Cell Cycle 8:3274–3284
13. Li Z, Bao S, Wu Q et al (2009) Hypoxia-inducible factors regulate tumorigenic capacity of glioma stem cells. Cancer Cell 15:501–513
14. Seidel S, Garvalov BK, Wirta V et al (2010) A hypoxic niche regulates glioblastoma stem cells through hypoxia inducible factor 2α. Brain 133:983–995
15. Soeda A, Park M, Lee D et al (2009) Hypoxia promotes expansion of the CD133-positive glioma stem cells through activation of HIF-1alpha. Oncogene 28:3949–3959
16. Calabrese C, Poppleton H, Kocak M et al (2007) A perivascular niche for brain tumor stem cells. Cancer Cell 11:69–82
17. Bao S, Wu Q, McLendon RE et al (2006) Glioma stem cells promote radioresistance by preferential activation of the DNA damage response. Nature 444:756–760
18. Eramo A, Ricci-Vitiani L, Zeuner A et al (2006) Chemotherapy resistance of glioblastoma stem cells. Cell Death Differ 13:1238–1241
19. Cheng L, Huang Z, Zhou W et al (2013) Glioblastoma stem cells generate vascular pericytes to support vessel function and tumor growth. Cell 153:139–152
20. Soda Y, Marumoto T, Friedmann-Morvinski D et al (2011) Transdifferentiation of glioblastoma cells into vascular endothelial cells. Proc Natl Acad Sci U S A 108:4274–4280
21. Wang R, Chadalavada K, Wilshire J et al (2010) Glioblastoma stem-like cells give rise to tumour endothelium. Nature 468:829–833
22. Tafani M, Di Vito M, Frati A et al (2011) Pro-inflammatory gene expression in solid glioblastoma microenvironment and in hypoxic stem cells from human glioblastoma. J Neuroinflammation 8:32
23. Wu A, Wei J, Kong LY et al (2010) Glioma cancer stem cells induce immunosuppressive macrophages/microglia. Neuro Oncol 12:1113–1125
24. Yi L, Xiao H, Xu M et al (2011) Glioma-initiating cells: a predominant role in microglia/macrophages tropism to glioma. J Neuroimmunol 232:75–82

Index

A

B

C

Ivan N. Rich (ed.), *Stem Cell Protocols*, Methods in Molecular Biology, vol. 1235,
DOI 10.1007/978-1-4939-1785-3, © Springer Science+Business Media New York 2015

D

E

F

T

U

V

W

X

MIX
Papier aus verantwortungsvollen Quellen
Paper from responsible sources
FSC® C105338

If you have any concerns about our products,
you can contact us on
ProductSafety@springernature.com

In case Publisher is established outside the EU,
the EU authorized representative is:
Springer Nature Customer Service Center GmbH
Europaplatz 3, 69115 Heidelberg, Germany

Printed by Libri Plureos GmbH
in Hamburg, Germany